Walter George McMillan

A Treatise on Electro-Metallurgy

Walter George McMillan

A Treatise on Electro-Metallurgy

ISBN/EAN: 9783337249601

Printed in Europe, USA, Canada, Australia, Japan

Cover: Foto ©berggeist007 / pixelio.de

More available books at **www.hansebooks.com**

A TREATISE

ON

ELECTRO-METALLURGY:

EMBRACING

*THE APPLICATION OF ELECTROLYSIS TO THE PLATING, DEPOSITING,
SMELTING, AND REFINING OF VARIOUS METALS, AND TO THE
REPRODUCTION OF PRINTING SURFACES AND
ART-WORK, ETC.*

BY

WALTER G. M^CMILLAN, F.I.C., F.C.S.,

CHEMIST AND METALLURGIST TO THE COSSIPORE FOUNDRY AND SHELL-FACTORY; LATE
DEMONSTRATOR OF METALLURGY IN KING'S COLLEGE, LONDON.

With Numerous Illustrations.

LONDON:

CHARLES GRIFFIN & COMPANY,

EXETER STREET, STRAND.

1890.

PREFACE.

In the following pages I have endeavoured to systematise and explain the various processes of Electro-Metallurgy as far as possible. Believing fully that in teaching and writing upon such subjects a *technological* rather than a *technical* treatment is required, I have tried so to set the matter before the reader that, even if he be a novice, he may be led to take an intelligent interest in any practical work upon which he may be engaged; but I have avoided the accumulation of a mass of unnecessary descriptive detail, which would only tend towards confusion, and which would be dictated by common sense to any who have grasped the principles involved. In many cases, however, success is in a large measure dependent upon strict attention to mechanical detail; and here I have not hesitated to introduce such instructions as I believed needful to guide the worker in the special operations in hand, while indicating the reasons which should enable him to apply them to processes of kindred character. In short, I have aimed at a combination of theory and practice.

The necessity for at least a fair knowledge of Chemistry and Electricity has led to the introduction of a chapter dealing in an elementary fashion with such laws as are required for an understanding of our subject; but it is not pretended that this chapter shall in any degree supersede the text-books upon these sciences; it is rather intended to lead up to them. In treating of the sources of current, especially the dynamo-electric machine, I have dwelt longer upon the general theory of construction and use as applicable to all, than upon the special modifications adopted by different inventors and manufacturers.

In addition to the journals, the following works among others have been consulted, and my general indebtedness to these authors must here be thankfully recorded :— Fontaine's *Electrolyse*, Gore's *Art of Electro-Metallurgy*, Japing's *Elektrolyse Galvano-plastik und Reinmetallgewin-nung*, Napier's *Manual of Electro-Metallurgy*, Roseleur's *Manipulations Hydroplastiques*, Schaschl's *Galvanostegie*, Thompson's *Dynamo Electric Machinery*, Urquhart's *Electro-typing* and *Electro-plating*, Volkmer's *Betrieb der Galvano-plastik mit Dynamo-Elektrischen Maschinen zu Zwecken der Graphischen Künste*, Wahl's *Practical Guide for the Gold and Silver Electro-plater and the Galvano-plastic Operator*, Watt's *Electro-deposition*, Weiss's *Galvano-plastik*, and Wilson's *Stereotyping and Electrotyping*. My thanks are also due to the Brush Electric Light Corporation, Messrs. Hoe & Company, Messrs. Siemens Brothers, and Messrs. Townson & Mercer for diagrams of apparatus. I must further acknowledge my obligations to Professor Sylvanus Thompson for his kind permission to describe and figure a special form of switch which he has in use.

WALTER G. McMILLAN.

COSSIPORE, CALCUTTA,
September, 1890.

CONTENTS.

CHAPTER I.

INTRODUCTORY AND HISTORICAL.

CHAPTER II.

THEORETICAL AND GENERAL.

CHAPTER III.

Sources of Current.

CHAPTER IV.

General Conditions to be Observed in Electro-Plating.

CHAPTER V.

PLATING ADJUNCTS AND DISPOSITION OF PLANT.

CHAPTER VI.

THE CLEANSING AND PREPARATION OF WORK FOR THE DEPOSITING-VAT, AND SUBSEQUENT POLISHING OF PLATED GOODS.

CHAPTER VII.

THE ELECTRO-DEPOSITION OF COPPER.

CHAPTER VIII.

ELECTROTYPING.

CHAPTER IX.

THE ELECTRO-DEPOSITION OF SILVER.

CHAPTER XVI.

THE RECOVERY OF CERTAIN METALS FROM THEIR SOLUTIONS OR WASTE SUBSTANCES.

CHAPTER XVII.

THE DETERMINATION OF THE PROPORTION OF METAL IN CERTAIN DEPOSITING SOLUTIONS.

CHAPTER XVIII.

A GLOSSARY OF SUBSTANCES COMMONLY EMPLOYED IN ELECTRO-METALLURGY.

ADDENDA.

LIST OF TABLES IN THIS TREATISE.

A MANUAL

OF

ELECTRO-METALLURGY.

CHAPTER I.

INTRODUCTORY AND HISTORICAL.

THE word *metallurgy* is understood to mean the art of working metals—extracting them from their ores and preparing them for application to the varied uses of daily life. By analogy the term *electro-metallurgy*, originally suggested by Smee, might reasonably be expected to imply such extraction and preparation effected with the aid of electricity. This, however, is, strictly speaking, but one section of the subject, and, indeed, regarded from the standpoint of commercial practicability, it is one of the most recent developments of the art; for the economical application of electricity to the recovery of metals from their ores by the separate-current process was scarcely possible until the invention of the dynamo-electric machine had placed a cheap source of electric energy at the disposal of the metallurgist. Just as the science of metallurgy also is but a branch of that of chemistry, and becomes elevated from an art to a science, in proportion as the laws of chemistry are made to regulate its processes, so the science of electro-metallurgy is dependent on the laws of chemistry and electricity, and will make more rapid progress as the accurate study and application of these laws are made to take the place of the "rule of thumb" methods, which are the inevitable outcome of the tentative experiments made in the early dawn of an art.

Accepting, then, the broader use of the term, electro-metallurgy may be defined as the science which treats of the application of electrical methods to the separation or to the solution of metals from substances containing them, and (perhaps we may add) to the treatment of metals for certain specific purposes in the arts.

1

Scope.—Thus, the electro-metallurgist may be called upon to deposit metals with any of the following objects:—(1) To obtain a coherent and removable deposit on a mould, the form of which it is desired to reproduce with accuracy; this process is termed *electrotyping.* (2) To obtain a thin, but perfect and adhesive, film upon a metal of different character, in order to impart to it acid- or air-resisting, or æsthetic properties, in which it was naturally deficient. This is known as *electro-plating.* (3) To obtain the whole of a given metal from a substance containing it, either as a substitute for extraction by smelting, or for analytical or refining purposes. It will be evident that in the first two of these, the interest centres in securing the exact condition of deposit which is best suited to the work in hand; whilst in the third, it is of paramount importance that the metal shall be completely separated, leaving no residue in the material from which it was to be extracted. Finally (4), he may be required to dissolve metals, either to remove an existing coat of one metal from the surface of another, or to effect the complete or partial solution of a homogeneous body superficially, as in the case of electro-etching.

Early History.—The history of the art is interesting, but perhaps too much involved to render anything more than the following brief sketch of value to the probable readers of this volume.

The fact that certain metals become superficially coated with other metals when plunged into suitable solutions was known to the ancients, and such a covering of iron swords and shields with copper by immersion in copper solutions was described by the Greek historian Zosimus in the fifth century. Paracelsus, who lived in the beginning of the sixteenth century (1493–1541), ascribed the apparent change of iron into copper, when dipped into the blue waters of Schmöllnitz in Hungary, to an actual transmutation of metals, a view which found favour even at a much later period. But although these may be considered as the beginnings of electro-metallurgy on the chemical side, it was not until the lapse of two centuries and a half from the latter date that the application of electricity to the deposition of metals became possible. Let us then glance at the gradual growth of electrical knowledge and its adaptation to the requirements of "electro-deposition." Such a retrospect cannot embrace any long term of years; for, although the attractive force of rubbed amber was known to the ancients, it awoke only a wondering interest until 1647, when Otto von Guericke first constructed a machine which exhibited the phenomenon in an intensified degree; the unknown force received the name electricity (from *elektron* = amber), electrical machines were gradually improved, and in 1752 Franklin demonstrated the identity of the electric spark with the lightning flash. But, in spite of the marvellous disruptive effect of these "frictional" machines, the actual quantity of electricity which could thus be generated was very minute, and could not avail for the deposition of metals from solutions. Its destructive power was derived from the enormous *potential* or "pressure" at which it acted, and no electrolytic effect could possibly have been observed except by a most careful experimenter actually searching for such a manifestation.

In the year 1759 Galvani made his celebrated discovery that a metal wire at one end touching the lumbar nerves of a recently-killed frog, and at the other the muscles of its leg or thigh, caused a rapid muscular contraction. Finding the same phenomena producible with the aid of a frictional machine, he was led to connect the two incidents, and to ascribe the former to the action of electricity resident in the animal itself. Volta, on the contrary, finding—as, indeed, Galvani had done before him—that if two wires of different metals were used, the contractions became more vigorous, concluded that the electrical energy was due rather to the action of the wires than to any property inherent in the animal tissue. Led on by this assumed production of electricity by contact of dissimilar metals, he constructed the series of zinc and copper discs, separated by moist cloth, which bears the name of the Voltaic pile, and with which, for the first time, comparatively large currents, though of very low potential, were obtainable, such as might be applied to the purposes of electro-metallurgy. Meanwhile, Fabroni in Italy, and Wollaston, Davy, and others in England, showed that oxidation, or rusting of the zinc, invariably attended the production of electricity in this way, and ascribed the latter to chemical action.

First Electrical Battery.—In 1800 Volta replaced the pile by his "crown of cups," in which each pair of copper and zinc plates was separated not by damp cloth, but by acidulated water placed in a series of vessels, the copper of each intermediate vessel being connected by a wire with the zinc of the next, leaving a free or unattached copper plate at one end of the series and a corresponding zinc plate at the other, these terminal plates being, of course, equivalent to those of the "pile." This, then, was the original electric battery, the discovery of which has led to the invention of the art of electro-metallurgy.

Separation of Metals.—In the same year Nicholson and Carlisle succeeded in decomposing water, or, in other words, *depositing hydrogen*, by means of the source of electricity thus placed at their disposal ; and in 1803 Cruickshanks, of Woolwich, constructed a large battery of considerable power, with which he deposited, or "revived," as he termed it, many metals from their solutions, and even proposed the use of an electrolysing current in quantitative chemical analysis.   Meanwhile, in 1801, Wollaston had obtained a coating of copper on silver, sufficiently adherent to allow of burnishing, by introducing the latter metal, in contact with one more oxidisable, into a solution of copper, thus forming a small electric battery in the depositing liquid itself.

In the *Philosophical Magazine*, Brugnatelli, in 1805, described the gilding of two large silver medals by means of the Voltaic pile and a solution of "ammoniuret of gold," and also the silvering of platinum surfaces, at the same time directing attention to the gradual solution of the plate through which the electric current entered the liquids.   Then Davy in 1807 made his grand discovery of the alkali-metals, potassium and sodium, by electrolytic isolation.

Magneto- and Dynamo-Electric Machines.—The knowledge of the relation between electricity and magnetism gained in 1820, both by Oersted's researches on the action of the electric current upon a compass-needle, and by the success of Arago in magnetising a steel needle by means of the current, may perhaps be regarded as the primary step towards the invention of magneto-electric machines, the first of which was constructed by Faraday in 1831; it consisted of a copper disc rotated between the poles of a horse-shoe magnet, with the necessary fittings for taking off the current thus generated.   In the same year Faraday observed the mutual action of electric currents, and the conditions governing the formation of induced currents, and thus, as it were, paved the way for the subsequent invention of the dynamo-electric machine.   Faraday's magneto-electric machine was not sufficiently powerful to have any practical value, but in the following year Pixii produced a machine of this character; and this may perhaps be regarded as the prototype from which the subsequent generators of this class have been evolved.

Thermopile.—The thermopile, another source of electrical energy which has been more or less largely used in electro-metallurgical work, especially abroad, owes its origin to Seebeck's observation, in 1822, that a current is produced by heating a compound bar of bismuth and copper at the junction of the metals, provided that the free ends are connected by a metallic wire.

Ohm's Law.—In 1827, Ohm enunciated his great fundamental law, which governs all electrical work, formulating, as it does, the relation between strength or volume of current, electrical "pressure," and the resistance of bodies to the passage of the current. Seven years later, Faraday demonstrated the relation between the strength of the current, and the amount of any metal electrolytically deposited by it, and proved that the quantity of electricity flowing in a given circuit could be measured by the amount of metal which it could deposit in a known period of time. It is by the systematic and intelligent application of these laws that the electro-metallurgist of to-day is able to arrange his plant with scientific accuracy, instead of by mere rule of thumb.

Copper-coating by Bessemer.—In 1831, Bessemer had coated articles composed of an alloy of lead, tin, iron, and antimony with a film of copper by simple immersion in a solution of a copper salt; but finding, as he describes in a letter published by Watt (*Electro-deposition of Gold, Silver, &c.*, p. 60), that the metal was not adherent, he tried for and obtained better results by placing the objects on a copper, iron, or, better still, zinc tray, and then sinking them in the liquid. In this way he formed a small battery *in situ*, as we have seen Wollaston had done in 1801.

Becquerel's Electrolysis Works.—Becquerel, in 1836, was the first actually to apply the principles of electrolysis to the treatment of natural products for the recovery of the metal contained in them. He even planned out works upon a commercial scale for the treatment of complex minerals containing copper and silver, but they were never erected, owing to the prohibitive expenditure of battery zinc involved in the process.

Daniell-Battery.—In the same year a new era was started by Daniell's introduction of his two-fluid battery, which placed a very constant current at the disposal of the electro-metallurgist, and almost immediately produced a ripe harvest of results. De la Rue at once took the first unconscious step in the direction of electrotyping, when he observed that the copper, which is deposited in the cells of the Daniell battery whilst in use, exactly reproduces upon its surface every line or scratch upon the copper plate on which it forms. Intent, however, on other objects, he failed to follow up the line of research thus indicated.

Elkington's Process.—In 1838, the Patent Office Records show that Elkington, who had two years previously protected a process of gilding for copper or brass objects by simple immersion in a solution containing gold, produced a method of zinc plating, analogous in principle to that of Wollaston's, by which the copper, brass, or iron to be coated was immersed in contact with a more oxidisable metal in a solution of that which it was desired to deposit, thus forming a galvanic cell in the depositing bath itself; and so for the first time the deposition of one metal upon another, through the galvanic action produced by the solution of a third, became the subject of a patent specification.

Earliest Electrotypers—Their Rival Claims.—De la Rue, as we have just seen, had already in 1836 indicated the possibility of copying uneven surfaces by electro-deposition, but had missed the practical application of his discovery. But three years later three individuals almost simultaneously, and it would seem quite independently, publicly described processes of electrotyping. These three were Jacobi of St. Petersburg, and Spencer and Jordan in England. The tale of these rival inventors has often been told; it is briefly as follows :—Professor Jacobi published a method of converting into relief, by galvanic means, even the finest lines engraved upon a copper plate, thus producing a printing surface suitable to the requirements of the printer. An account of this process found its way into the pages of the *Athenæum* on May 4, 1839. On the 8th of May following, Spencer gave notice to the Liverpool Polytechnic Institution of his intention to read a paper before that body on the "electrotype process;" but this paper was not read until September of the same year. Meanwhile, however, the account of Jacobi's discovery had been copied into the *London Mechanic's Magazine*, and had called forth a letter from Jordan, dated May 22nd, but not published until June 8th, in which he described his experiments in the same field, which were begun in the summer of 1838. He clearly set forth here the method which has since been known as the single-cell process of electrotyping, and claimed the possibility of multiplying engraved plates, typographical matter or medals, by forming galvano-plastic matrices on the object, and using the "negative" copy thus obtained to reproduce the original form; and he even suggested making tubes by depositing copper around a wire or metallic core which could subsequently be removed.

Strangely enough, neither of these accounts received public attention, and the matter remained unnoticed until the end of September, when Spencer's paper was read. This paper is especially interesting, because it shows how the process of electrotyping was gradually developed in his hands, mainly by an attentive and patient examination into the causes of a series of apparently minor phenomena observed in September, 1837, while experimenting with a single voltaic cell, consisting of a copper plate in copper sulphate solution connected by copper wire to a zinc plate immersed in a solution of common salt. The starting-point on the road to the new discovery was the observation that certain spots on the copper plate, which had accidentally been overlaid with molten sealing-wax, received no metallic deposit when placed in the cell; and he was thence led to attempt the formation of designs in relief, for use in the printing press, by coating a copper plate with an insulating varnish and then tracing the desired pattern by scratching the varnish completely away at the required points, and finally building up a deposit of copper upon the portions of the metallic plate thus exposed. While experimenting in this direction, he made the important observation that the nature of the deposited copper was dependent on the degree of "intensity of the electro-chemical action," or, as we should say, on the strength of the current, strong currents giving rapidly deposited but highly crystalline and friable metal. He now met with difficulties, which, however, were not insurmountable, and even led to further triumphs; he found that the deposited copper would adhere perfectly only to an absolutely clean surface of the same metal. Thus, when he required to obtain perfect adhesion between the metals, he first cleaned the copper surface with nitric acid, whereas if he wished afterwards to separate the deposited metal from the original plate he coated the latter with the thinnest possible film of bees' wax, previous to exposing it in the battery-cell. The subsequent application of a gentle heat enabled the two plates to be separated with the greatest facility. Next, when using a penny-piece instead of a copper plate, he observed that the inner surface of the copper sheath with which it had been coated bore a perfectly sharp copy, in *intaglio*, of the image and letters which were in *relief* on the coin itself; in this way he obtained matrices from which the original could be faithfully reproduced. Now, ascertaining that copper would deposit as readily upon lead as upon itself, he secured exact copies of coins, of set-up type, or even of wood-blocks, by pressing them upon sheets of lead and depositing copper upon the indented lead matrix so prepared. And finally he found that clay, plaster of Paris, wood, or other non-conducting materials could be covered electrolytically with copper, if they were first coated with a conductive film of bronze-powder or gold-leaf.

It would appear hopeless to determine the question of real priority between the three inventors. Jacobi seems to have been the first to publish an account of his researches, and so far his claim is good; Spencer next declared his intention to describe his experiments, but was forestalled by Jordan; on the other hand, Spencer claims to be the earliest experimenter in the field, and his investigations appear to have been deeper and more fully developed than those of either Jacobi or Jordan. It is doubtless one of those frequently recurring instances, wherein the progress of knowledge has led several men to a simultaneous but independent development of the same line of thought; and in such cases credit must be ascribed to each, but the palm awarded to the most thorough and painstaking.

Murray's Blackleading Process.—The immediate result of Spencer's paper was the creation of a sudden mania for electrotyping, the simple and inexpensive character of the necessary apparatus enabling amateurs of all grades to vie with the fresh race of operatives which sprang up at the birth of a new industry. Evidence of this is to be seen in the rapidly increasing number of patents which were now applied for in this branch of the Arts. With so many workers in the same field, it would indeed be strange if the scope of the work were not quickly and widely enlarged, and existing processes much improved; the year 1840 was accordingly destined to see many improvements effected. The application of the art to the requirements of the printer was in this year made practicable by Murray's discovery, that moulds of non-conducting material could be made to take the deposit of copper by brushing them over with plumbago, so that metallic moulds were no longer essential. In the same year the first published newspaper-print from an electrotype block, is believed by Smee to have appeared in the *London Journal*. Nevertheless, Savage's *Dictionary of Printing*, which appeared in the following year, although it contained many good engravings from electrotypes, exhibited a page of "diamond" type, also printed from an electro-deposited block, but this was so imperfect that, as Wilson has suggested, we may infer that the art of electrotyping formes of small type had not yet attained sufficient excellence to warrant its general application to this purpose.

In 1840, Mason endeavoured to utilise the current generated by the single-cell electrotyping arrangement in a second depositing cell, and thus to carry on two operations simultaneously; and although this method was not practically adopted, it, nevertheless, pointed to the possibility of applying a separate current from sources other than Daniell's battery.

Cyanide Baths.—In this year Wright, after experimenting with many solutions, discovered the use of the cyanide bath for the production of thick deposits of gold and silver, in place of the thin films obtainable by simple immersion. The invention was patented and at once put in operation by the Messrs. Elkington, who were foremost in this field at the time. At the end of the same year, de Ruolz patented in France the use of similar solutions, not only for gold and silver, but for platinum, copper, lead, tin, cobalt, nickel, and zinc. The following year witnessed the publication of a very complete work on electro-metallurgy by Smee, and this was followed by several others in rapid succession.

Elastic Moulds.—Leeson, in 1842, greatly advanced the appli-cation of electrotyping to the reproduction of works of art by the use of elastic moulds made of glue and gum, which thus enabled objects of intricate or undercut design to be faithfully copied and indefinitely multiplied; and by the insertion of leading wires in the mould to distribute the current more uniformly, and hence, to facilitate a higher degree of simultaneity in the deposition. In the following year Montgomery proposed the application of gutta-percha as a moulding medium for slightly undercut objects.

Bright Silver.—From that time the inventions for many years, although numerous enough, had not sufficient novelty to render them worth recording in detail, excepting, perhaps, the important discovery by Milward, in 1847, that the addition of a small quantity of carbon bisulphide to the silver plating baths, caused the deposited silver to show no longer a dead or frosted surface, but to exhibit greater lustre and brilliancy.

Cheap Sources of Electricity.—In 1842 and 1843 respectively, Woolrich patented the use of magneto-electric machines, and Poole, that of thermo-electric piles, for depositing metals; but neither of these seem to have been at the time successfully applied in practice. But the introduction of Pacinotti's dynamo-electric machine, first described by him in *Il Nuovo Cimento* in 1864, of Wilde's magneto-electric machines in the following year, and of Siemens' and Wheatstone's more perfect dynamos, simultaneously invented in 1867, profoundly modified the scope of the art by affording a far cheaper source of electricity than had hitherto been possible. From this time, with the more careful study of the theory of the dynamo, and the consequent improvements in its mechanical and electrical efficiency, there have arisen a host of new machines, constructed especially to satisfy certain specific objects, and approaching much nearer to perfection than did their original progenitors. Dynamo-electric machines are now made to suit the needs of the electro-metallurgist, and thus new fields of labour have been opened, more particularly in the domain of metal refining and smelting; and the readiness with which mechanical energy may now be converted into electrical, renders the utilisation of natural waste water-power thoroughly applicable to these purposes.

Ore-Treatment by Electricity.—The later history of our subject is, in its more important branches, intimately associated with the application of dynamo-electric machinery, and of powerful currents to the treatment of ores, furnace-products, or solutions. Becquerel, as we have seen, failed practically to apply his process for the treatment of ores on account of the expense of the zinc; it therefore remained dormant until 1868, when it was tried in San Francisco by Wolf and Pioche, who seem, however, to have effected but little. Elkington's method of treating copper mattes (impure fused copper sulphide) by using them as anodes, started a new epoch in 1871, and many modifications of this and analogous processes have since been carried into practice. How far this type of ore-treatment may be able to compete with the older smelting methods, can only be considered in connection with the particular circumstances of each individual case, and will be more fully dealt with later in the work.

In another direction, electricity has been applied to the extraction of metals by the passage of a current through a fused salt; it was in this way, in 1854, that Bunsen and Deville reduced aluminium from its combination with chlorine, and recently many arrangements purporting to effect a similar result have appeared in the records of the Patent Office. It is interesting to note that a proposal by Pichon, in 1854, to reduce an ore admixed with a small percentage of charcoal in the electric arc passing between two large electrodes, has found a later development in the electric furnace of the brothers Cowles, patented in 1885, for the treatment of ores containing aluminium and other metals.

Latest Advances.—Of still more recent origin are the applications of electricity to bleaching, and to the treatment of organic bodies for the production of more valuable substances (both rather electro-chemical than metallurgical processes), as well as to melting metals, as in the Siemens' electric furnace, to welding by the heat of the electric arc, to annealing wire, to the production of seamless copper tubes and the like. The last named process, originally suggested by Jordan, has in 1888 been modified and rendered more intrinsically valuable, on account of the superiority of the metal thus deposited, by

Elmore's process of continually burnishing the metal during the whole course of the depositing operation.

The growth of the art on the whole has been rapidly progressive ; a few processes may have been superseded by furnace-methods, which have proved to be less costly, as in the so-called galvanising of iron, for this was at one time conducted by electrolytic methods, as the name itself suggests, but is now effected by dipping into a bath of fused zinc. But the various branches have for the most part steadily gained ground as the work became more reliable and more economically conducted, until the electrotyper and the electroplater in gold, silver, and nickel, occupy a quite important position among the manufacturers of the world.

CHAPTER II.

THEORETICAL AND GENERAL.

It has already been hinted that a right understanding of all the problems involved in the science of electro-metallurgy demands an aquaintance, not only with the manner in which certain forces act upon matter, but with the constitution of matter itself. A brief review, therefore, of a few of the fundamental laws and theories of electrical and chemical science naturally finds a place at this point; although for a full explanation of these subjects reference must, of course, be made to the text-books devoted specially to them.

Matter—Force.—It must be clearly understood, then, that *matter* (that is, anything which possesses weight) is variously constituted. It is within our common experience that different kinds of matter exhibit different properties and characteristics, or as we say are made of different materials; and it is the object of chemical science to teach us what the materials are and how they behave when brought into contact with one another. *Force* has no material constitution, and therefore no weight, and is only made known to us by its action on matter. *Physical* (or *mechanical*) *forces*, as they are termed, may affect the relative position or the outward shape and appearance of material substances, but *chemical force* affects the very ingredients of which the substance is composed, and governs the more intimate mutual relationship between different bodies.

Conditions of Matter.—We are conscious that various kinds of matter may exist at different times in certain distinct forms—solid, liquid, and gaseous—but these are solely physical, not chemical, differences (the constitution or component parts of the substance remaining unchanged throughout), and are brought about by physical means, such as alteration of heat or pressure, so that a return to the original conditions is accompanied by a reproduction of the body in its first form. For example—water at 15° C. is a liquid, but cooled to 0° C. it becomes solid ice, or heated to 100° C. it is converted into gaseous steam; nevertheless if the ice and the steam be respectively brought back to 15° C. they will again form a liquid quite undistinguishable from that originally experimented with. We may imagine, then, that water is made up of a vast number of almost infinitesimal particles, all of which are alike and are rapidly vibrating to and fro; and that in ice they are so packed together with shorter paths of vibration that they will not readily separate, thus causing solidity; but that when heat is applied to them, each particle vibrates through a longer distance, and the different units are farther apart and more free to move among themselves, so that they present the characteristics of a liquid; while above the boiling point the freedom is so great that they are actually carried away as a gas.

Constitution of Matter.—If now we imagine these minute similar particles to be so small that further sub-division by physical means is impossible, we are figuring to ourselves those penultimate particles which, in the language of the atomic theory, are termed *molecules.* It is with the molecule that the physicist has to deal, and it is on the molecule that physical forces act; but the chemist is able to break up each molecule into a certain limited number of smaller particles, which are supposed by this theory to be indivisible, and are hence called *atoms* (*a* = not, *temno* = I cut). Any molecule may contain two or a larger number of these atoms, and the atoms themselves may be similar or dissimilar; there are, indeed, two classes of bodies, the first of which includes those molecules in which all the atoms are alike, the number of substances in this class being, of course, equal to the number of different kinds of atoms in existence, while the other group includes those whose molecules are made up of unlike atoms, and the number of these bodies is almost unlimited, because of the endless combinations possible between different varieties of atoms. In the first class both molecules and atoms are all alike, so that such a substance can contain but one kind of matter, and is hence termed an *element.* In the other group, although the molecules of any substance are similar, the atoms are not; but each atom being indivisible, and consisting of one body only, is an element, and, therefore, the substance is said to be a *compound* of such and such elements. Thus, if a molecule of water could be made to yield its atoms, it would be found to contain two of a gaseous element, hydrogen, and one of another gas, oxygen, each quite unlike water, and unlike the other; and if two molecules could be so treated at the same time, the two liberated oxygen atoms would, if kept apart from the hydrogen, unite to form a molecule of oxygen, while the four hydrogen atoms would form two molecules of hydrogen.

Atomic Weight.—Now, if it were possible to isolate and weigh a number of atoms, it would be found that all those of the same nature would possess equal weight, but that diverse atoms would have unlike weights; and for all chemical and electrolytical calculations this must be thoroughly understood. But although the actual weighing of an atom is still an impossible feat, yet by studying the mutual relations of the elements the chemist is able to estimate the *relative*, though not the *absolute*, weights of different atoms.

Chemical Symbols and Formulæ.—Hydrogen, which is the lightest substance known, being regarded as unity, the *atomic weight* of each element is expressed as a multiple of that of hydrogen. Thus, if an atom of hydrogen be regarded as weighing 1, it is found that an atom of oxygen will weigh 16. Now, as each molecule of water has been shown to consist of two atoms of hydrogen combined with one of oxygen, it is evident that it must contain 2 parts by weight of the former with 16 parts of the latter; and as all molecules of water are alike in composition, it follows that in every 18 parts by weight of pure water there are 2 of hydrogen and 16 of oxygen.

It should be clearly remembered that every true compound, no matter how it is produced, not only contains always the same elements, but contains them in the same proportion. Provided, then, that we are able to determine the nature of the different atoms present in a molecule of any substance, and the proportion in which they are there, we can calculate with precision the percentage of each element contained in it; and, conversely, if we know the proportionate weights of the constituents of a given compound, we can at once determine the number of each kind of atom in the molecule. Hence it is possible to assign a definite chemical formula to every compound; that of water might be written "2 hydrogen : 1 oxygen," implying that there are two atoms of the former to one of the latter; but in chemical work to write down the names of the elements in full would be both tiresome and clumsy, so that chemists are in the habit of using a system of shorthand notation, which is both simpler and more scientific. It consists in selecting a symbol, usually the first letter, or the first with some specially suggestive subsequent letter, of either the English or Latin name of the substance, to represent each element; and this is understood to stand for, not an indefinite amount, but for 1 atom; and, therefore, so many known parts by weight of the element. Thus "H" implies 1 atom or 1 part by weight of hydrogen; "O," 1 atom or 16 parts by weight of oxygen; "Fe" (*Ferrum*), 1 atom or 56 parts of iron;

"Sn" (*Stannum*), for 1 atom or 118 parts of tin; and so on. By such a system of notation, then, we are able to express both the nature and the proportionate composition of any substance, and so, for example, to ascribe the formula "H_2O" to water.

It should now be evident that elements can combine with others only in proportions which are multiples of their respective atomic weights, because no fraction of an atom can enter into the constitution of a molecule. But further than this, the proportionate numbers of each kind of atom in any molecule are no mere arbitrary or accidental figures, but are regulated by definite laws; and it is the fixedness of these laws which enables us to predict the exact weight of any metal which should be deposited by a given current in a known time.

Valency.—We have seen that water contains 1 atom of oxygen united with 2 of hydrogen; but in hydrochloric acid (HCl) there is 1 of chlorine, not with 2, but with only 1 of hydrogen, while in ammonia (NH_3) we find 1 of nitrogen with 3 of hydrogen, and in marsh gas (CH_4) 1 of carbon requires 4 of hydrogen. Here we observe chlorine to be typical of a class of elements which combine with hydrogen in equal atomic proportion, oxygen typical of a class where the ratio is 1 : 2, nitrogen of one where it is 1 : 3, and carbon of one where it is 1 : 4. The terms *monovalent, divalent, trivalent*, and *tetravalent* are applied to the elements included in these classes respectively. The words monatomic, diatomic, &c., are sometimes used, but those indicating *valency* are preferable. All elements, however, are not capable of combining with hydrogen; and to determine the valency of these, it is simply necessary to ascertain how many atoms of hydrogen are replaced in any compound by the element in question. For example, common salt is a compound of the metal sodium (Na) with the gas chlorine in the atomic proportion of 1 : 1, the formula being NaCl. But 1 atom of chlorine combines with 1 atom of hydrogen in hydrochloric acid, and, therefore, 1 atom of sodium is equivalent to 1 atom of hydrogen, inasmuch as each requires 1 atom of chlorine to combine with it—thus, sodium, like hydrogen and chlorine, is monovalent. This equivalency is clearly seen by the following reaction :—

1 hydrogen ⎫ added to (1 sodium) yield ⎰ 1 sodium ⎱ and (1 hydrogen).
1 chlorine ⎬   ⎱ 1 chlorine ⎰
Hydrochloric acid „   „   „   *common salt*   „   „

or putting it in the form of an *equation*,

$$HCl + Na = NaCl + H.$$

Again, in oxide of sodium, $Na_2 O$, 1 atom of oxygen combines with 2 of the metal, just as it does with 2 of hydrogen; and by throwing metallic sodium into water, the latter is decomposed and part of the hydrogen is liberated as a gas, while an equivalent of sodium takes its place to form caustic soda. This exchange is thus represented * :—

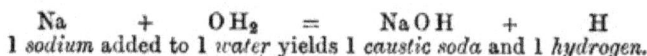

$$Na \quad + \quad O H_2 \quad = \quad Na O H \quad + \quad H$$
1 *sodium* added to 1 *water* yields 1 *caustic soda* and 1 *hydrogen.*

In both these instances also, therefore, we find evidence that sodium replaces an equal number of hydrogen atoms, and is monovalent.

The atom of the metal zinc (Zn) replaces 2 hydrogen atoms, as is seen in the following equations representing the action of hydrochloric acid and of water respectively upon metallic zinc:—

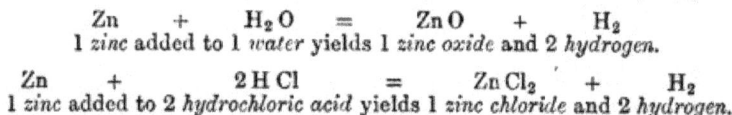

$$Zn \quad + \quad H_2 O \quad = \quad Zn O \quad + \quad H_2$$
1 *zinc* added to 1 *water* yields 1 *zinc oxide* and 2 *hydrogen.*

$$Zn \quad + \quad 2 H Cl \quad = \quad Zn Cl_2 \quad + \quad H_2$$
1 *zinc* added to 2 *hydrochloric acid* yields 1 *zinc chloride* and 2 *hydrogen.*

Zinc is therefore a divalent element.

Similarly, aluminium may be shown to be trivalent, 1 of the metal replacing 3 of hydrogen; or 2 of the metal replacing 6 of hydrogen as indicated—

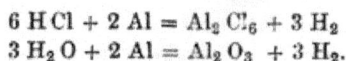

$$6 H Cl + 2 Al = Al_2 Cl_6 + 3 H_2$$
$$3 H_2 O + 2 Al = Al_2 O_3 + 3 H_2.$$

Other elements are tetravalent, pentavalent, or hexavalent, according as they replace 4, 5, or 6 atoms of hydrogen in compounds. To save the repetition of the full term monovalent or divalent element, &c., it is often more convenient to classify the elements as *monads, dyads, triads, tetrads, pentads,* or *hexads.*

Now it sometimes happens that a single element may form two classes of compounds, in which the valency of the element is different; but these classes are quite distinct from one another, for although they are interconvertible, yet they do not

---

* In writing equations, a figure written before a group of symbols means that all the elements symbolised up to the first stop are to be multiplied by the number represented by that figure; while a small figure below the line, to the right of a symbol, describes the number of those atoms to be taken into consideration. 2 H Cl means 2 molecules of hydrochloric acid (2 atoms of hydrogen and 2 of chlorine). $H_2 O$ means 2 atoms of H with 1 of O; while $3 H_2 O$ means 3 molecules of water containing 6 atoms of H and 3 of O.

arbitrarily pass over from the one to the other, but tend to remain separate, even, perhaps, through a long cycle of chemical changes; but one or other of these classes is generally more stable than the other—that is, less liable to change—and is, therefore, the one more commonly met with. Thus, copper (Cu = *cuprum*) forms the more usual group of compounds in which it replaces 2 of hydrogen (Cu O = cupric* oxide, Cu Cl$_2$ = cupric chloride) and is generally recognised as a dyad, but it may, under certain circumstances, behave as a monad and form the group of compounds of which cuprous* oxide = Cu$_2$O, and cuprous chloride = Cu$_2$Cl$_2$, are typical members. To express the valency of an element, it is therefore necessary to know with which class of compounds we are dealing.

Elements.—In the following table will be found the names of the elements at present known, together with their symbols and valencies, their atomic weights, and their *equivalent* weights. By the latter term is meant the number of parts by weight of an element which are required to take the place of 1 part of hydrogen, or of 1 equivalent of any other element; and these numbers are evidently found by dividing the atomic weight (*i.e.*, the weight of an atom) of the element by the number of hydrogen atoms which are equivalent to it. Thus, when any element interchanges with another in a compound, it is always in the proportion of a multiple of their equivalent weights.

In this table, the names of the more common metallic elements and of those metals which, as such or in compounds, are most largely used in electro-metallurgy, are printed in small capitals, while the non-metallic elements are distinguished by italics.

Heat-evolution of Combinations.—We have used the term *chemical force* in the earlier part of this chapter, and we must now devote a short time to the observation of the comparative effects produced by various chemical changes and combinations.

It must first be understood that when any chemical combination occurs, heat is always evolved, owing to the conversion of chemical energy into its equivalent of heat-energy; and that just as given substances always combine in perfectly definite proportions, so each of these combinations is attended by the

* When an element forms two classes of compounds, that class in which the greatest proportion of oxygen or other equivalent substance is combined with it, is distinguished by the suffix -*ic* attached to the name of the element, while the other class takes the termination -*ous*. When the name alone is used, the more common group is usually understood—for example, *copper oxide* would imply *cupric oxide*, Cu O.

2

TABLE I.—LIST OF THE ELEMENTS.

Name of Element	Symbol	Valency	Atomic Weight	Usual Equivalent Weight
ALUMINIUM,	Al	3	27	9·0
ANTIMONY,	Sb	3 and 5	122	40·6
Arsenic,	As	3 and 5	75	25·0
Barium,	Ba	2	137	68·5
BISMUTH,	Bi	3 and 5	210	70·0
Boron,	B	3	10·9	3·6
Bromine,	Br	1	80	80·0
CADMIUM,	Cd	2	112	56·0
Caesium,	Cs	1	133	133·0
Calcium,	Ca	2	40	20·0
Carbon,	C	4	12	3·0
Cerium,	Ce	2	92	46·0
Chlorine,	Cl	1	35·5	35·5
CHROMIUM,	Cr	3	52·5	17·5
COBALT,	Co	2	59	29·5
COPPER,	Cu	2 and 1	63·5	31·75
Didymium,	D	3	96	48·0
Erbium,	E	3	169	56·3
Fluorine,	F	1	19	19·0
Gallium,	Ga	3	69	23·0
Glucinum,	Gl	2	9·3	4·7
GOLD,	Au	3	196·6	65·5
HYDROGEN,	H	1	1	1·0
Indium,	In	3	113·4	37·8
Iodine,	I	1	127	127·0
Iridium,	Ir	4	197	49·2
IRON,	Fe	2 and 3	56	28·0
Lanthanum,	La	2	92	46·0
LEAD,	Pb	2	207	103·5
Lithium,	L	1	7	7·0
MAGNESIUM,	Mg	2	24·3	12·1
MANGANESE,	Mn	2	55	27·5

Name of Element	Symbol	Valency	Atomic Weight	Usual Equivalent Weight
MERCURY,	Hg	2 and 1	200	100·0
Molybdenum,	Mo	6	96	16·0
NICKEL,	Ni	2	59	29·5
Niobium,	Nb	5	97·5	19·5
Nitrogen,	N	3 and 5	14	4·6
Osmium,	Os	6	199	33·3
Oxygen,	O	2	16	8·0
PALLADIUM,	Pd	2 and 4	106·5	26·6
Phosphorus,	P	3 and 5	31	10·3
PLATINUM,	Pt	4 and 2	197	49·2
POTASSIUM,	K	1	39·1	39·1
Rhodium,	Ro	3	104·3	34·8
Rubidium,	Rb	1	85	85·0
Ruthenium,	Ru	4	104·2	26·0
Selenium,	Se	2	79·5	39·7
Silicon,	Si	4	28	7·0
SILVER,	Ag	1	108	108·0
SODIUM,	Na	1	23	23·0
Strontium,	Sr	2	87·5	43·7
Sulphur,	S	2	32	16·0
Tantalum,	Ta	5	182	36·4
Tellurium,	Te	2	129	64·5
Thallium,	Tl	1	204	204·0
Thorium,	Th	1	119	119·0
TIN,	Sn	2 and 4	118	59·0
Titanium,	Ti	4	50	12·5
Tungsten,	W	6	184	30·7
Uranium,	U	4	240	60·0
Vanadium,	V	3	51	17
Yttrium,	Y	3	89·6	39·0
ZINC,	Zn	2	65	32·5
Zirconium,	Zr	4	89·5	22·4

evolution of a perfectly constant amount of heat-energy. The amounts of heat evolved by different combinations are extremely varied, but that given out during one particular combination is quite constant, no matter in what way the union is effected. Then by comparing all the elements except one as to their behaviour in combining with that one other element, it is possible to arrange them in a series in which the first member evolves the greatest amount of heat during its combination, and the last gives out the least quantity. Further, if different individual group elements be taken, and the remaining elements be combined with them, one with one, several of these lists may be compiled, each of which represents the heat-value of the various combinations of the single elements with one of the others. On comparing these lists, it will be found that the main order of the different substances will be approximately the same in all, but that one or two elements may alter their relative position in certain lists, which represent the effect of their combination with special elements for which they have "great affinity." Thus, certain elements may be said to be more energetic than others in all combinations, while others may be regarded as, so to speak, more sluggish—evolving far less heat in any reaction in which they may play a part.

By way of example, the following table illustrates the number of heat-units emitted during the combination of certain of the more common elements with oxygen and chlorine respectively. From this table it will be seen, for example, that 65 pounds of zinc combining with 16 pounds of oxygen evolve sufficient heat to raise the temperature of 86,400 pounds of water 1 degree centigrade.

Law of Selection in Chemical Union.—This table shows that the strongest combination of chlorine with any metal (i.e., the one which evolves most heat in its production) is potassium, next to that is ranked sodium, while zinc stands lower, and silver lower still, on the list. Now, speaking generally, if there be a limited amount only of an element such as oxygen in a condition to combine with an excess of various metals, the strongest combination will always form first; then, if there be any of the deficient element left, it will combine with the element which affords the next strongest combination, and so on. In other words, if a choice of combinations be presented, that which produces the greatest evolution of heat will usually take precedence of all others. By way of example, let us suppose that a limited amount of chlorine is brought into contact in a closed space with an excess of potassium, sodium,

TABLE II.—SHOWING THE NUMBER OF HEAT-UNITS EVOLVED
IN CERTAIN COMBINATIONS.

NAME OF ELEMENT, AND NUMBER OF ATOMS COMBINING.		HEAT-UNITS EVOLVED IN COMBINATION WITH	
		1 Atom of Oxygen, O.	2 Atoms of Chlorine, Cl_2.
Potassium, ...	K_2,	97,200	211,220
Sodium, ...	Na_2,	100,200	195,380
Calcium, ...	Ca,	131,360	170,230
Strontium, ...	Sr,	130,980	184,550
Barium, ...	Ba,	130,380	194,250
Manganese, . . .	Mn,	*94,770	111,990
Zinc, ...	Zn,	86,400	97,210
Hydrogen, ...	H_2,	68,360	†44,000
Iron, ...	Fe,	*68,280	82,050
Tin, ...	Sn,	*68,090	80,790
Cadmium, ...	Cd,	*65,680	93,240
Cobalt, ...	Co,	*63,400	76,480
Nickel, ...	Ni,	*60,840	74,530
Lead, ...	Pb,	50,300	82,770
Copper, ...	Cu,	37,160	51,630
Mercury, ...	Hg,	30,660	63,160
Silver, ...	Ag_2	5,900	58,760
Gold, ...	$\frac{2}{3} \times$ Au,	* − 8,790	15,210

* These oxides are hydrated; the remaining compounds are anhydrous.
† This number refers to gaseous hydrogen chloride.

zinc, and copper. The potassium will have the highest tendency
to combine with chlorine, because, in doing so, the greatest
amount of heat will be evolved, and potassium chloride will be
the chief result of the reaction. When all the potassium is used
up, and not until then, if all the metals may be supposed finely
divided and well mixed together, sodium chloride will be formed,
then zinc chloride, and finally, if there be any chlorine left,
copper chloride. Further than this, if any of the pure elements
be placed in a fused compound of any element occupying a lower
position in the above list, it will tend to liberate the latter

element in the metallic state, and itself to take its place in the compound, because in this exchange heat will be evolved; thus, metallic zinc dipped into fused silver chloride forms zinc chloride (because the heat of combination of $Zn + Cl_2$ is greater than that of $Ag_2 + Cl_2$) and sets free metallic silver, while on the contrary, metallic silver is, of course, without action on fused zinc chloride.

Electro-Chemical Series.—A list of elements arranged in such order that any one will thus displace from its compounds any other lower than itself on the list, but will be without effect on any compound of an element occupying a higher place, is termed an *electro-chemical series*. For every group of compounds (oxides, chlorides, sulphates, &c.) such a list may be formed; and the elements at the head of the list are termed *electro-positive*, expressed by the algebraic sign for *plus*, " + ," while those at the lower end are *electro-negative* and are represented by the *minus* sign, " − "; but these terms are purely relative. For example, in the following list, which expresses the electro-chemical order of the elements in relation to sulphuric acid, zinc is less readily acted upon by the acid than potassium, but more so than copper; it, therefore, occupies an intermediate position, and while it is electro-positive in regard to copper, it is electro-negative as compared with potassium. This fact can be readily demonstrated by dipping a piece of zinc into separate solutions of potassium, zinc and copper sulphates; in the two former no action will be observed, while in the latter zinc will slowly be dissolved, and a corresponding deposit of copper will be found on the remainder of the zinc. And it should be noted that for every equivalent (32·5 parts by weight) of zinc dissolved, an equivalent (31·75) of copper will be separated, as we have already indicated. The metals as a class are electro-positive, the non-metals electro-negative.

TABLE III.—Arrangement of the Common Metals in Electro-Chemical Series.

+

Potassium.	Cobalt.	Silver.
Sodium.	Nickel.	Platinum.
Magnesium.	Lead.	Gold.
Manganese.	Tin.	Hydrogen.
Zinc.	Bismuth.	Antimony.
Iron.	Copper.	The metalloids or
Cadmium.	Mercury.	non-metals.

−

It should be observed that the order of the metals in this list is nearly in accordance with that in the table of heat-evolutions on p. 20, as indeed would be expected from our knowledge of the laws alluded to at that point. The relation of the electro-chemical arrangement of the metals to their heats of combination readily explains also the fact, that in different solutions the order of sequence is not always absolutely identical. It is of great importance to remember this in electro-metallurgical work, especially of an experimental nature, as the numbers representing these "heats of combination" have in many cases been accurately determined and are published in works on thermo-chemistry; and since, as we shall presently see, electrolytic deposits on metals, which are themselves capable of decomposing the plating liquid, are generally of inferior character, the cause of many a failure might be explained by a reference to thermo-chemical data.

Correlation and Interconversion of Forces.—We have seen that during chemical combination, chemical force is converted into and rendered evident as heat-energy; and the amount of heat thus evolved may be regarded as an exact measure of the chemical energy of the combining elements. All forces are interconvertible, but it is no more possible to *create* force or energy than it is to produce or to destroy matter; for the manifestation of any force, such as heat or electricity, is merely a conversion into it of some other form of energy. In the dynamo-electric machine, electricity is "generated," but it is at the expense of an exact equivalent of mechanical force supplied by the steam engine, and the current of electricity produced may in turn be converted into heat-energy in over-coming the resistance of the incandescent lamps, or into chemical energy by decomposing chemical compounds and precipitating the constituents in the electrolytic bath in circuit. We are rarely able to convert the whole of the amount of one kind of energy into any other, where and when we require it, as some portion is always lost to us in a form which we cannot utilise, as, for example, that which is converted into heat by the friction of the bearing or moving parts of the machinery; nevertheless no portion of the energy is destroyed, it is simply converted into another form, which happens at the time to be useless to us. In this way, for example, a small fraction of the electrical energy supplied to the plating-vats is lost as heat in overcoming the resistance of the leading wires and circuit to the passage of the current.

Interconversion of Chemical and Electrical Energy.—Let us suppose that a zinc plate is placed in dilute sulphuric acid; as it slowly dissolves away, a rise of temperature is observable, due to the evolution of heat, which may be shown by careful experiment to be precisely proportional to the weight of zinc dissolved. If now a plate of copper be immersed in the same liquid, but not touching the zinc strip, as in fig. 1, it will neither be attacked by the acid itself, nor will it in any way affect the relations between the zinc and the liquid. But if the two metals be connected by a long coil of thin wire, as shown in fig. 2, the heat-measuring apparatus at once indicates a great diminution in the quantity of heat evolved in the liquid in a given time, although the solution of the zinc is proceeding even more

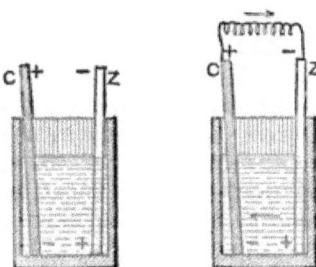

Fig. 1.—Copper-zinc cell, un-connected.

Fig. 2.—Copper-zinc cell, con-nected.

rapidly than before. Either then the heat is not being emitted, or it is being dissipated in some way. But now the long coil of thin wire is found to possess new properties; it attracts light particles of iron brought near to its ends, while it influences the set of a compass-needle, and further it becomes distinctly warmer than the surrounding air. Here then is the solution of the problem : directly the two plates are connected by the wire a current of electricity is set up, which is regarded as flowing from the zinc through the solution to the copper, and thence from the copper through the wire coil to the zinc again ; and this current, meeting with resistance in all parts of the circuit, causes heat to be evolved in all parts by the conversion of electrical energy, but chiefly in the portion which opposes the greatest resistance, namely, in the thin wire. The proof of the existence of this electric current is seen in the magnetisation of the coil. Two similar pieces of zinc connected in this way would produce no current, but any two unlike metals would give results analogous to those from the copper and zinc. Hence it may be generally stated that *when two different metals are metallically connected together in a solution which is capable of acting chemically upon at least one of them, an electric current is set up which will pass in the liquid from the more electro-positive metal to the other, and complete the circuit by returning along the metallic connection, in the inverse direction, to the positive plate.* It is further noticeable that the greater the electro-chemical difference between the two metals used, the more vigorous will be the action, the more rapid the solution of the electro-positive element, and consequently the greater will be the heat and electrical manifestation in the connecting wire. This is due to the greater difference of potential between any pair of metals widely separated in the electro-chemical series, than that between any pair more nearly adjacent.

Electrical Potential.—*Potential* is a term equivalent to *head* or *pressure* as applied in speaking of water-power, and denotes the tendency to start the flow of a current. When a piece of dry glass rod is rubbed with warm silk, the former receives a charge of positive, the latter one of negative, electricity; and each may be made to yield its charge to an insulated brass ball. The ball which now carries the positive charge is said to be at high potential—as though the electricity had been pumped up to a great height — while the other is at low potential; then when the two spheres are caused to approach, the vigour with which the positive electricity leaps forward to neutralise the negative electricity (or rather perhaps to restore equilibrium), depends upon the difference between the two potentials. The action is quite analogous to the flow of water from a high cistern to one at a low level, the greater the difference between the heights (or potentials), the more vigorous and irresistible will be the current of water in the connecting pipe. And just as the sea level is taken as that to which all heights are referred in hydrostatic calculation, so the earth is regarded as the zero of electrical potential; and the current flows from a point at high positive potential to one of lower potential (but still positive), or from one at negative potential to another still lower, in the same way that water flows equally readily from a high mountain to a lower one, or from an upper to a lower drift in a mine. If the ends of the copper and zinc strips in fig. 1 are attached to pieces of wire but not connected together, it may be demonstrated that while they are in the liquid the wire attached to the copper has a slight charge of positive electricity, while that fastened to the zinc is negative, as though the tendency to oxidise the zinc were urging the electricity already to its utmost limit; then immediately on connecting the two, the current is started. In the case of the frictional or static electricity referred to above, the charge passes instantaneously from the one ball to the other, and, equilibrium being restored, no further electrical effect is producible; the upper cistern, as it were, has been emptied. But in the copper and zinc voltaic couple, as soon as the first charge has passed from the copper to the zinc, a fresh charge is urged from behind by the oxidation of a fresh portion of the latter metal; and others follow this in such rapid succession that a continuous current is formed. It is no longer a cistern which may be instantaneously emptied, but is rather comparable with a spring, which will continue to flow, so long as any zinc remains undissolved to pump the electricity to the higher level.

Electro-Motive Force.—The strength or force of the current will depend upon the difference of potential between the two metals employed; and the force which tends to impel the current along its path is termed the *electro-motive force* or, as it is frequently written for convenience, "E.M.F." Thus it should now be clear that the electro-motive force is connected with the difference of potential, and is indeed proportional to it; and that it is ultimately dependent on the difference in oxidisability between the metals in use, or on the amounts of heat evolved by them respectively, in undergoing the particular chemical reaction which is taking place between the metals and the liquid.

Electrolysis.—Following up the investigation of the pair of metals referred to in figs. 1 and 2, it will be remarked that while the metals are isolated from one another, bubbles of hydrogen gas are constantly formed upon the surface of the zinc, and that these rapidly find their way to the surface of the liquid and escape into the air; they are due to the decomposition of the sulphuric acid by the zinc, expressed in the following equation :—

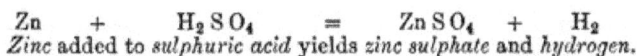

$$Zn \quad + \quad H_2SO_4 \quad = \quad ZnSO_4 \quad + \quad H_2$$
Zinc added to sulphuric acid yields zinc sulphate and hydrogen.

Or, for the sake of simplicity, the water only may be supposed to be decomposed thus—

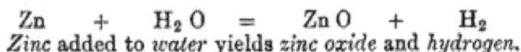

$$Zn \quad + \quad H_2O \quad = \quad ZnO \quad + \quad H_2$$
Zinc added to water yields zinc oxide and hydrogen.

the zinc oxide immediately dissolving in the sulphuric acid to form zinc sulphate.

The exchange of zinc and hydrogen can readily take place because heat is evolved in the process. But on connecting the two strips with wire, the hydrogen-evolution on the zinc plate entirely ceases, and is transferred to the copper. This effect, which clearly marks the introduction of new conditions, is due to the electric current that is set up. It is a simple case of *electrolysis* or electro-deposition.

We have already seen that a current of electricity passed through a coil of wire endues it with new magnetic properties; but when it is passed through a compound liquid, it tends to destroy the compound, and to deposit certain of its constituents on the surfaces through which it enters and leaves the solution. Thus in the cell to which we are referring, while no current was passing, the bubbles of hydrogen, indicating the progress of a chemical change, appeared on the zinc; but when the current began to flow, the elements of each molecule of water that was decomposed were distributed, the hydrogen to the copper plate from which it escaped, and the oxygen to the zinc plate with which it at once combined to form zinc oxide, and this in turn dissolved in the acid liquid as zinc sulphate. It is universally true that when any solution is electrolysed (that is, decomposed by the electric current), the electro-positive element, or metal, is deposited on the plate towards which the current flows in the bath, and by which it leaves the solution; while the electro-negative element is thrown down on that by which the current enters the liquid. Within the copper-zinc cell the current flows from the zinc to the copper; and, as hydrogen is the electro-positive element of water (or sulphuric acid), it is on the copper that this gas is liberated, while the oxygen or electro-negative element is set free upon the surface of the zinc.

Had the zinc plate been immersed in a solution of copper (*e.g.*, copper sulphate), instead of in sulphuric acid, not hydrogen but copper would have been deposited upon its surface, when isolated, or upon that of the copper plate when the two were placed in connection. Any metal will thus substitute itself for any less positive metal in a suitable solution, and will deposit that metal upon its surface, or upon any other less electro-positive metal which may stand in contact with it in the same solution. So lead and iron may be immersed separately in a solution of silver. On reference to the table given above, it will be seen that of these three, iron is the most electro-positive and silver the least so. The result will, therefore, be that, so long as they remain isolated, silver may deposit upon both strips; but as soon as metallic contact is made between them, the iron alone will dissolve, and the silver will now deposit only upon the lead. Had plates of iron and silver been substituted,—while they stood alone silver would have been thrown down only on the iron, but on coupling them together the current set up would flow from the iron within the solution to the silver, depositing oxygen upon the iron with which it would combine, and silver upon the silver. In the same way lead and silver plates would precipitate the silver from the liquid upon the lead when separated, but upon the silver when united.

But the same law of deposition holds good when a current of electricity is passed through a separate solution, independent of the dissolving cell, which thus becomes a *battery.* If the current available have a high electro-motive force, it will effect decomposition in the external liquid as readily as in the original solution, and will deposit the electro-negative element on the surface by which it enters that liquid or *anode,* as it is termed, and the electro-positive body, or metal, at the *cathode* or surface towards which it flows, and through which it leaves the solution. These separated elements are sometimes termed the *ions,* that which forms at the anode being the *anion,* that at the cathode the *cation;* the metals, therefore, as a class would be cations, the non-metals anions. The anode and cathode form the *electrodes.*

When a number of different metals are connected together in the dissolving cell, the most electro-positive metal will be first acted upon to the practical exclusion of all others. When this has been dissolved away, the next electro-positive metal will be attacked, and so on, because the most positive metal evolves most heat during combination, and because that reaction usually occurs which is accompanied by the greatest heat-evolution. Very practical applications of this law have been made, as, for example, when pieces of (electro-positive) zinc have been attached to the (electro-negative) copper bottoms of ships to prevent their corrosion by sea-water, an action which could not then take place until the zinc had been quite dissolved away; or when iron surfaces are protected with a coating of zinc by "galvanising." Conversely a current passing through a mixed solution will generally * deposit first the least electro-positive metal, then the next higher in the series, and finally the most electro-positive; for it should be observed that to decompose a given substance requires an expenditure of energy equivalent to the heat generated in its formation ; and just as with a choice of reactions, that *combination* will occur which *sets free the greatest amount* of heat-energy; so when there is a choice of *decompositions,* that will be effected which requires the *least expenditure* of energy ; and in *electrolysing (i.e.,* decomposing by electricity) a mixed solution, least energy is, of course, needed to break up the compound which evolved the least heat during its formation. Zinc, iron, lead, copper, and silver, bound together by wire and

---

* Under certain conditions, which will be considered hereafter (p. 34), it is possible to deposit alloys or mixtures of metals from a compound solution.

placed in a bath of acid, would, therefore, dissolve in the order
in which they are enumerated, while a current passing through
a solution containing all these metals would deposit first the
silver, then the copper, and finally, the others in inverse order.

In electrolysing a solution, then, by means of a current derived
from chemical action, the chemical force in the battery is converted into electrical energy, and is pumped by the electro-
motive force from the copper strip of the battery, along a wire
to the anode of the solution, and is there reconverted into
chemical force during electrolysis, by the separation of the
elements contained in the liquid. Let us imagine that the
plates by which the current enters and leaves the *electrolyte* (or
liquid which is being electrolysed) are made of platinum or any
other metal, which cannot be attacked by the solution or by the
liberated elements; then, if the chemical force required to dis-
sociate (or drive apart) the elements which are combined in the
electrolyte is greater than the chemical force of the battery-cell,
it is evident that no dissociation can occur. Hence there is a
practical limit to electrolytic decomposition. There is another
point of view from which this phenomenon may be observed,
which may serve to make it more clearly understood. Let us
suppose that a certain pair of
metals, C and Z (fig. 3), are placed
in a battery-cell, B, and joined with
wires to the platinum plates, O
and M, respectively, of an electro-
lysing-cell, E. Here Z is the
electro-positive metal; so that the
current flows in the cell, B, from
Z to C, and thence through the
wire to the plate, O, on which the
oxygen or non-metal of the electrolyte would be separated, and then
through the liquid to M, which
should receive the deposit of metal,
and finally returns to the plate, Z,
of the battery. But it is evident that as soon as the cathode, M,
is completely covered with metal, the electrolyte cell itself will
become practically converted into a battery, of which the deposit
on M may be supposed the positive metal, and the platinum, O,
the negative. The result of this would be to send a current,
produced by the dissolving deposit on M, and pumped by the
electro-motive force due to the different potentials of M and O,
from M through the liquid E to O, thence through the wire

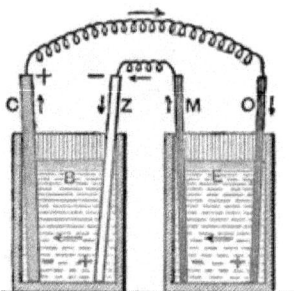

Fig. 3.—Copper-zinc cell connected with electrolysing
cell.

to C, and through the battery solution B to Z, and thence again finally to M ; in other words, the resulting current would be in the opposite direction to that produced by the battery ; and if the *opposing electro-motive force* of the new electrolyte cell is greater than the electro-motive force of the battery, the action of the whole arrangement would be reversed until the deposit on M had re-dissolved again. It is, therefore, easy to understand that if the *counter E.M.F.* (or opposing electro-motive force) of the electrolyte-cell be greater than the original E.M.F. of the battery, no decomposition can occur, for any metal deposited would re-dissolve as rapidly as it separated. Thus the same conclusion is arrived at, that *in order to produce an electrolytic deposit, the decomposing current must have a higher electro-motive force than that which is set up between the deposited metal and the metal of which the anode is made.*

If, instead of insoluble platinum strips, pieces of the same metal that is to be deposited are used, then any current whatever—irrespective of pressure—may suffice to decompose the solution ; because, although metal is being deposited, and therefore a certain amount of energy is absorbed at the cathode, yet the same weight of metal is being constantly dissolved by combination with the non-metal deposited at the anode ; and the work that is absorbed at the cathode by decomposition is thus exactly counterbalanced by that which is produced at the anode by the re-formation of the original substance. To take the other view of the case, the metal which is deposited is of the same nature as the anode itself, and between strips of the same metal no opposing E.M.F. is produced. The function of the current is simply to provide the means for destroying the equilibrium of the electrolyte bath, and to transfer, as it were, the metal by degrees from the anode to the cathode. Finally, to illustrate the matter by a practical example :—a solution of copper sulphate electrolysed between a copper anode and cathode. Copper will be deposited on the copper cathode and will still be opposed to a copper anode, no current can thus be produced by the two plates, no counter E.M.F. is set up, and the feeblest external current suffices to effect the electrolysis. But if platinum plates be used, then copper is deposited on the cathode, and thus practically forms a sheet of copper which is now opposed to the platinum anode ; at once an opposing E.M.F. is produced, equal to the difference of potentials between copper and platinum in a copper sulphate solution. Now, if the battery-current have a higher E.M.F. than this, it will overcome the opposition, and copper will be deposited continuously ; but if it have a lower electro-motive force, no deposition can possibly occur, and the platinum cathode remains uncoppered.

When soluble anodes of a different metal to that which is to be deposited are used the case is less simple, owing to the gradual introduction of the metal dissolved from the anode into the electrolyte. Here two classes of problems may arise.

(1.) *If the metal of the anode be more electro-positive than that undergoing deposition.*—In this case, provided the cathode is made of the depositing metal, no separate battery is needed, because a simple dissolving cell is formed, and the mere connection of the two plates causes the anode to dissolve, and to deposit the metal from the solution upon the cathode, as in the case of lead and iron in a silver solution, which we have instanced on p. 26. But if a separate battery be employed, its current is re-inforced by that generated in the electrolytic cell, and the action proceeds more rapidly than it could otherwise do, until the solution is exhausted of the metal which is being separated. When this point is reached the current of the battery will begin to deposit the metal which has passed into solution from the anode, provided the E.M.F. be sufficiently high; but if, on the other hand, no battery be applied, all action must cease as soon as the whole of the first metal has been deposited.

(2.) *If the anode be less electro-positive than the metal undergoing decomposition.*—Here, provided the E.M.F. of the battery-current suffice to initiate electrolytic action, the required metal will be deposited for a time, but gradually a more electro-negative body diffuses through the liquid as it dissolves from the anode, and will, of course, be ultimately deposited instead of that of which the solution was originally composed.

Electric Conductivity.—We have spoken freely of electricity passing through matter, but have not yet noted that it cannot do so with equal readiness in different media. There are many substances which absolutely check the progress of the electric current, and are hence termed *non-conductors* or *insulators;* they comprise such bodies as sulphur, pitch, shellac, glass, ebonite, sealing-wax, gutta-percha, paraffin-wax, oils, and the like; others are partial conductors, such as wood, and allow a certain amount of electricity—especially, of course, that at high potential—to pass through them; while others transmit electricity with facility, and are termed *conductors.* But there are an infinity of grades even among conductors. The metals as a class are the best, and second, but after a long interval, are acidulated water and aqueous solutions. Of the metals, silver is the best conductor, and perfectly pure copper ranks nearly equal with it; regarding the conductivity of these as represented by 100, the relative conductivity of the principal metals is given in the following Table:—

TABLE IV.—Showing the Relative Electric Conductivity
of certain Metals.

Name of Metal.	Mean Conductivity.	Authorities.	Name of Metal.	Mean Conductivity.	Authorities.
Silver, . . .	100·0	B,D,L,M,P,W.	Antimony, . .	4·2	M,W.
Copper, . .	100·0	B,D,L,M,O,P,W.	Mercury, . . .	2·5	B,M,P.
Gold, . . .	80·6	B,D,L,M,O,P,W.	Bismuth, . . .	1·2	M.
Aluminium, .	55·1	M,W.	Graphite, . . .	0·07	M.
Sodium, . .	37·4	M.	*Alloys, &c.*		
Zinc, . . .	30·2	B,M,O,W.	Cu with 4°/₀ Si,	75·0	W.
Cadmium, .	23·7	M.	Cu ,, 12 ,, Si,	54·7	W.
Potassium, .	20·8	M.	Cu ,, 9 ,, P,	4·9	W.
Platinum, .	16·7	B,D,L,M,O,P,W.	Cu ,, 10 ,, Pb,	30·0	W.
Palladium, .	16·4	D.	Cu ,, 10 ,, Al,	12·6	W.
Iron, . . .	16·4	B,D,L,M,O,P,W.	Cu ,, 10 ,, As,	9·1	W.
Tin, . . .	15·2	B,M,O,W.	Cu ,, 20 ,, Sn,	8·4	W.
Thallium, .	9·2	M.	Cu ,, 35 ,, Zn,	21·1	W.
Lead, . . .	8·8	B,M,O,W.	Cu ,, 50 ,. Ag,	86·6	W.
Nickel, . .	7·9	W.	Au ,, 50 ,, Ag,	16·1	W.
Arsenic, . .	4·8	M.	Sn ,, 12 ,, Na,	46·9	W.

Physicists have obtained very varied results in estimating the conductivities of metals, probably owing to want of care in selecting perfectly pure specimens; for, remembering that a mere trace of impurity is often sufficient to lower the conductivity (or, in other words, to *increase the resistance*) of a metal by an altogether disproportionate amount, it is to be regretted that careful analyses of the samples operated upon were not made, and the nature and quantities of impurities published with the result of the physical observations. In this Table, whenever possible, the mean results obtained by Becquerel,

Davy, Lenz, Matthiessen, Ohm, Pouillet, and Weiller have been taken; in many cases, however, measurements have been made by only one or two of these, and in others a single result obtained by one of them is so divergent from the corresponding numbers quoted by the others as to point to an error of observation, and to render its omission desirable in calculating the mean: for this reason we have indicated the sources from which the figures have been derived by the initial letters of the observers' names.

The conductivity of liquids is much lower, that of dilute sulphuric acid being 0·000133, and copper sulphate solution 0·00000542, that of silver being taken as 100; while pure water itself is a non-conductor. In passing through any substance, electric energy must always suffer loss, owing to partial transformation into heat in overcoming the resistance of the conductor, which is in its effect similar to that of friction on mechanical energy; hence, for conducting wires only that material which offers least resistance (or friction) to the path of the current should be used. Silver is, of course, placed out of the field on account of its costliness; but copper, which is practically as good, is readily obtainable, and should, therefore, have the preference over all metals for this purpose.

Electrolytic Conduction.—Here we have been dealing with *metallic conduction*, which means that the electric energy is merely transmitted through a substance without producing any other effect than the conversion of a small proportion into heat, due to the resistance of the conductor. But there is another, or *electrolytic conduction*, which has only been satisfactorily observed to take place in compound liquids, and in this case the passage of the electric current from one particle to another is accompanied by a *polarisation*, or change in the condition of the liquid, which is only made apparent by the separation of the constituents of the electrolyte at the *poles* (*i.e.*, the anode and cathode). This electrolytic conduction may occur in solutions of solid substances or in fused bodies, but it is essential that they shall be compounds (mercury is a liquid conductor, but being an element it cannot be electrolysed and, therefore, conducts metallically), and that they shall be able to conduct the electricity. In electrolysing fused salts it may sometimes happen that the ions dissolve instantaneously in the liquid and, diffusing through it, reunite as rapidly as they are separated; and thus to all appearances the fluid is conducting metallically rather than electrolytically. Fused alloys of metals appear actually to conduct metallically only, as up to the present no attempt to separate the constituents by electrolysis have proved successful, though it is of course possible that the re-diffusion of the separated metals just described may account for the failure of these attempts.

The conditions, then, for electrolysis by the battery or dynamo are, that the solution should be able to conduct electricity ; and that the electro-motive force of the battery used should be in excess of that set up between the separated ions ; and, further, that the ions, or at least the cations, should exist as such in the electrolyte without bringing about secondary actions or decompositions.

When solutions of a metallic salt are electrolysed, two substances—the salt and its solvent—exist side by side in the liquid, and the question arises as to which of these is the electrolytic conductor ; and this question has an important bearing on the deposition of alloys. For example, in passing a current of moderate strength through a solution of copper sulphate in water, undoubtedly copper alone appears at the cathode and oxygen at the anode. This may be, and at first sight apparently is, due to the decomposition of the copper sulphate, $Cu\ SO_4$, according to the equation.

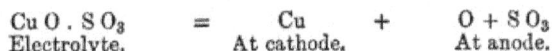

$$Cu\ O\ .\ S\ O_3\ \ \ =\ \ \ Cu\ \ \ +\ \ \ O + S\ O_3$$

Electrolyte.　　　At cathode.　　　At anode.

Thus, copper is set free at the cathode and sulphuric acid is liberated at the anode. But if water were the electrolyte which suffered decomposition, the same final result would still be apparently obtained ; oxygen would in any case separate at the anode, but the hydrogen of the water liberated in the "nascent" condition on the cathode would at once exchange places with the copper in the adjacent solution, re-forming water, and depositing the copper upon the cathode plate. These two reactions would thus be expressed—

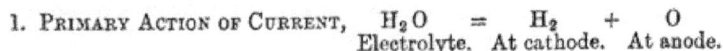

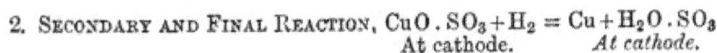

1. PRIMARY ACTION OF CURRENT,　$H_2 O\ \ \ =\ \ \ H_2\ \ \ +\ \ \ O$

Electrolyte.　At cathode.　At anode.

2. SECONDARY AND FINAL REACTION, $Cu O\ .\ SO_3 + H_2 = Cu + H_2 O\ .\ SO_3$

At cathode.　　　　At cathode.

Thus, the final result is oxygen at the anode, and copper and sulphuric acid at the cathode. Thus, if the former alternative be correct and copper sulphate is electrolysed, sulphuric acid would be set free only at the anode, while in the latter case it would appear only at the cathode. Arguing in this manner, Professor Lodge has put the question to the test, and has found that the acid is produced at both electrodes, but in greater quantity at the anode; it is, therefore, evident that while copper

3

and water are simultaneously decomposed, it is the former which conveys electrolytically the major portion of the current. Moreover, it is well known that a very powerful current passed through the solution may cause a simultaneous separation of copper and hydrogen at the cathode—a fact which is quite in conformity with the above view; because when the action is violent, a portion at least of the disengaged hydrogen would be hurried away before it had time to reduce the copper salt.

Deposition of Alloys.—This multiple electrolysis may serve to explain the fact of the deposition of alloys, which, at the first glance, is difficult to understand, in view of the theoretical consecutive separation of metals from mixed solutions in ascending order of electro-positiveness. In the light thrown on the subject by Lodge's experiments, it would thus appear probable, that in a mixed solution of copper and zinc sulphates, both metals may be deposited; but that the zinc instantaneously acts on the copper solution around and redissolves, whilst it precipitates at the same time an equivalent of copper in its place, so that only the latter metal is finally separated. But the deposition of the alloy may be understood in any of the following cases :—(1) If the current be so powerful that the secondary action of the more electro-positive metal on the solution of the other have not time to perfect itself. (2) If the particular solution used be of such a character that the more electro-positive metal, when it is separated, cannot chemically attack the solution of the more electro-negative, because the heats of formation of the two salts are so nearly identical that the same electro-motive force is needed to deposit each. (3) If the proportion of the more electro-negative metal in the liquid be so small, compared with that of the other, that the solution around the cathode is exhausted of the former more rapidly than its replacement is possible through diffusion; so that the electro-positive metal, having no possibility of exchanging places with it, remains undissolved. Thus, from a mixed solution of copper and zinc, brass might be deposited if the current were so strong, and, therefore, the decomposition so rapid, that the separated zinc had not time to precipitate the copper around it, before it was protected by a further electro-deposit of copper; or if such compounds were selected that the copper salt could not be decomposed by metallic zinc; or if the quantity of zinc in the solution were so great as compared with the copper, that there was insufficient copper in the solution around the anode to take the place of the mass of zinc deposited. It should further be remarked that probably a small proportion of heat is generated in the alloying of certain metals, and that this would provide an additional tendency to deposit the alloy, sufficient, perhaps, to determine its production in certain cases.

The comparative electro-positiveness of two metals in a given solution may be ascertained by connecting the metals by wires to corresponding slips of copper immersed in copper sulphate solution, and noting on which strip a further copper deposit is produced; the slip on which the copper is precipitated will, of course, be that which is joined to the more positive metal; or it may be determined by connecting the plate with a wire, and finding, with the aid of a galvanometer, in which direction the current flows, remembering that outside the solution it must always pass from the electro-negative to the positive metal.

Units of Measurement.—Before closing this chapter we will refer briefly to some of the units of measurement employed in practical work.

In dealing with electric currents there are two distinct factors to be taken into account—quantity and potential. The potential, we have already seen, is the "pressure" under which the current commences to flow, the *quantity* or *intensity* refer to the actual amount or "volume" which passes in a given time, irrespective of pressure. And these terms, pressure and volume, perhaps explain better than any others the sense of the two words in question, inasmuch as they effect a comparison with hydraulic power, which is a more tangible force, so to speak, and, therefore, better understood. The power of overcoming resistance increases with the potential in electrical as in hydraulic work : thousands of cubic feet (volume) of water may be flowing in a given time through a large pipe with but a slight fall (potential), and yet but little force is required to check the flow ; while a few cubic inches conveyed along a small pipe, if only from a sufficient height, may sweep away every obstacle placed in the path and will thus cause a current to flow, even through a great resistance. But in electro-depositing, provided that there is just sufficient electro-motive force to overcome the resistance, the amount of work done is exactly proportional to the flow of electricity in a given time ; and this follows naturally from the laws of current-production. A given amount of zinc dissolved in the battery produces a given amount of heat, which is converted into electricity, and can then, in the electrolyte, only yield, at most, its equivalent of chemical energy. In reality the amount deposited is somewhat less than that theoretically predictable by reason of loss of electrical energy as heat due to the resistance of the circuit.

The unit of electro-motive force (pressure), then, is the *volt*, and is approximately equal to the E.M.F. of a single Daniell's cell.

The unit of resistance is the *ohm*, which is equal to that of a column of mercury 106 centimetres long by 1 square millimetre in cross-sectional area.

The unit of current (strength) is the *ampère*, and is that produced by a potential of 1 volt acting through 1 ohm.

In scientific work the metrical system of weights and measures is adopted.

The unit of weight is the *gramme* (= 15·432 grains); ten, hundred, and a thousand grammes being termed dekagramme, hectogramme, and kilogramme; and $\frac{1}{10}$th, $\frac{1}{100}$th, and $\frac{1}{1000}$th being denominated decigramme, centigramme, and milligramme, respectively.

The unit of length is the *metre* (= 39·37 inches), and the multiples and fractions have the same Greek and Latin prefixes respectively as those of the unit of weight.

In cubic measurement a special term *litre* is used to denote the cubic decimetre, or 1000 cubic centimetres (= 1·76 pints).

The unit of heat is the *calorie*, and represents the amount of heat-energy which must be expended in order to raise the temperature of 1 gramme of cold water 1° Centigrade : some authorities use a kilogramme-degree unit, but the gramme-degree system is more general, and is the one adopted in this work, whenever it may be necessary to use it. Other units are used in different countries, as, for example, that of the pound-Fahrenheit-degree, intended to adapt itself to the English systems of measurement.

The "pressure of heat" or temperature used by preference in scientific work is the *centigrade* scale which denominates the freezing point of water zero, and the boiling point 100°, the space between these points being divided into 100 equal degrees. In the Fahrenheit scale, largely used in England, the freezing point is 32°, the boiling point 212°, the space between them having 180 degrees. In the Reaumur system, adopted widely on the Continent, the freezing point is zero, and the boiling point 80°. The first two of these systems alone will be referred to in this book, and on page 354 will be found a Table of Comparison of Temperatures recorded by each.

In addition to these units there are several of practical utility which may well be mentioned.

The electrical unit of quantity is the *coulomb;* it is the volume of current equal to that of 1 ampère passing through a circuit for one second of time.

A current of 1 ampère at the pressure of 1 volt is termed a *watt*; it is a most useful unit for comparing different currents, and is really the product of volume into pressure.

The English *horse-power* (H.P.) is taken at 550 foot-pounds per second, and is thus equivalent to raising 550 pounds through one foot, or one pound through 550 feet in a second. (The French H.P. is 542·48 foot-pounds per second.)

Finally, to connect heat-energy with mechanical work, Joule found that 1 calorie (gramme) is equal to 0·424 kilogramme-metre or 3·066 foot-pounds—*i.e.*, 1 calorie liberated per second is equal to 0·00557 H.P.

CHAPTER III.

SOURCES OF CURRENT.

Relative Efficiency and Cheapness.—From what has been said in the last chapter, it will be apparent that some other form of energy must be converted into electrical energy in order to yield the current required for electro-plating; and we can now understand why the only electrical current originally available, which was generated from mechanical force through friction between unlike bodies, although capable of exhibiting wonderful disruptive effects, owing to its high electro-motive force, was useless for electro-metallurgical work by reason of its deficiency in intensity—that is, volume. Then with the invention of the battery, which aimed at the conversion of chemical into electrical energy, currents of low electro-motive force but great volume were producible, and these soon found an application in plating and electrotyping. But in nearly all batteries zinc is used up, and as a given weight of zinc can only produce a given (in many instances not very different) weight of deposited metal, it is too costly to be used in some branches of electro-metallurgy. The dynamo-electric machine, which is able to convert mechanical force directly into a current of high intensity but low electro-motive force, is now largely used, not only in this direction, but also as a substitute for battery-power generally; because the combustion of the coal in the boiler of the steam engine takes the place of the oxidation of the zinc in the cell of the battery, and is a far cheaper agent. Nevertheless, as the steam engine utilises only a very small fraction of the total heat produced by the combustion of the coal, some means of directly converting heat into electricity is even now a desideratum; at present, the thermo-electric-pile does indeed effect this object, but it is not economical, and is not used for the generation of large currents, nor is it extensively employed in any way.

The manner of producing currents by the battery, the dynamo, and the thermopile, will be shortly discussed in this chapter.

THE VOLTAIC OR GALVANIC BATTERY (*named after Volta and Galvani*).

Principle.—The principle of the galvanic battery has already been explained, namely :—That when any two metals are placed in a fluid which can attack at least one of them, and are connected by any conductor or set of conductors, an electric current is at once set up, which flows within the cell from the more positive metal to the other, and in the opposite direction through the conductors ; and that the greater the distance between the two metals on the electro-chemical scale, the higher will be the electro-motive force of the current produced.   Zinc being a common metal, comparatively inexpensive, and, at the same time, very fairly electro-positive, is almost universally adopted as the positive element.   The electro-negative element and the exciting fluid are, however, frequently varied, such modifications constituting the chief differences between the types of cell commonly employed.

Local Action on Zinc.—Whenever commercial zinc is used as the positive plate in an acid solution, it must be well *amalgamated* with mercury.   Impure zinc dipped alone into acid evolves clouds of hydrogen gas from its surface ; but, if a copper plate connected to the zinc be immersed in the same liquid, all this hydrogen evolution should be transferred to the copper strip (*vide* p. 25) ; yet, as a matter of fact, much still continues to be formed upon the zinc.   This is to be explained by the presence of minute specks on the surface of the zinc of more electro-negative bodies (*e.g.*, lead and iron), which, being impurities in the body of the metal and, therefore, in contact with the mass of the zinc, and also exposed to the acid liquid, act (each for itself) as minute independent batteries, with the zinc as the dissolving element, so that from each of them hydrogen is continuously evolved.

This *local action*, as it is termed, is objectionable, because it entails a loss of zinc without any equivalent of work done outside the battery, all the energy being expended in heating the materials of each of these minute and local circuits within the cell ; this, of course, represents so much waste.   Further, the phenomenon is not merely temporary, for the electro-negative impurities, although perhaps almost invisible, are not attacked by the liquid ; and the gradual removal of the zinc by solution only results in laying bare a greater number of them, thus increasing rather than diminishing the extent of the local action during the progress of the discharge.   The most obvious remedy is to use only the purest zinc, but this is costly; while the protection afforded to the metal of commerce by a thin coat of mercury is found to impart equal efficiency and to be more economical. To amalgamate the zincs, clean them well; then with a piece of flannel or sponge tied to the end of a stick, wash them with dilute sulphuric acid (1 of acid to 15 or 20 of water); then pouring a few drops of mercury upon the centre of each, rub it cautiously over the plate until the whole surface presents a silvery brightness. Mercury will adhere only to perfectly clean metallic surfaces, hence the necessity for the preliminary washing with acid; if, however, the zinc be very dirty or at all greasy, this treatment will not suffice, and it should be first immersed in warm caustic soda or potash solution. This will remove all grease; then, after washing thoroughly in water, the plate should be flooded with the acid and rubbed with mercury as before. An excess of mercury is to be avoided, as it soon penetrates the zinc and renders it brittle. The plates should be often examined, and re-amalgamated as soon as dark spots are seen upon the surface, or when gas is evolved from them during the working of the battery. The acid should not be allowed to touch the hands more than necessary; and if it fall upon the clothes it should be at once neutralised by a drop of strong ammonia, or it will in course of time produce a red stain, and finally destroy the cloth.

Dry Pile.—Originally copper and zinc were used with dilute sulphuric acid, the earliest form of battery being the *dry pile*, in which copper and zinc discs, separated by moistened circular pieces of flannel (in the order—copper, flannel, zinc, copper, flannel, zinc, &c.), were clamped together in a wide glass tube until the whole was filled; wires were then soldered to the last disc at either end, and caps were fastened on to maintain the whole arrangement in position; on connecting the end wires a current could at once be produced. But the volume of current thus obtainable was still very small, although the electro-motive force might be comparatively great owing to the arrangement of the discs in continuous series, as will be explained later in the chapter.

Wollaston's Copper-zinc Cell.—The simple copper-zinc cell was a distinct advance on the pile. It was made by placing a strip of copper in dilute sulphuric acid opposite to one of zinc, and connecting the two strips with the wire which was to complete the circuit. It was, in fact, the cell shown in fig. 3. But this battery, too, is very weak, owing to the low electro-motive force produced by these two metals immersed in dilute sulphuric acid; it is true that by placing several cells in series the potential is raised, but this is neither convenient nor economical, and the original *Wollaston* cell is rarely, if ever, seen now.

Polarisation.—One of the chief objections to cells of this class is that the hydrogen gas evolved at the copper plate clings to it and cannot readily escape. This formation of hydrogen films on the surface of the negative strip is termed *polarisation*, and presents a double disadvantage: firstly, the mere formation of a film of any gas upon the metal prevents the contact of its whole surface with the liquid, and, by interposing a layer of an insulating medium, reduces the surface area of the plate as much as if a portion of it were coated with an acid-resisting varnish; and in the second place, a far more serious difficulty is found in the tendency of the separating (positive) hydrogen to set up a cell on its own account, in co-operation with the zinc which now forms the negative plate; and this secondary current will, of course, start from the hydrogen to the zinc in the liquid, and from the zinc to the copper (and thence to the hydrogen in contact with it) outside—that is, it will be in the reverse direction to the original current, and by partially opposing and neutralising it, must necessarily reduce the resultant current from the battery. Thus a copper-zinc cell may for the first few seconds of use give a fairly strong current; but as soon as the action has proceeded sufficiently far to give rise to a production of hydrogen on the copper plate, it becomes polarised, and the opposing electro-motive force which is set up greatly impairs the efficiency of the cell.

Parts of Battery.—It must here be explained that in describing a battery, the more oxidisable or electro-positive metal (that, in fact, which by dissolving produces the current) is termed the *positive plate;* but as the current outside the solution starts from the other plate this second one is called the *positive pole.* In the copper-zinc cell, the zinc is the positive plate, but the wire attached to it is the negative pole; and the copper, while it is termed the negative plate, becomes the positive pole outside the liquid. This is not likely to be forgotten if the facts be clearly grasped:—(1) That the plate from which the current starts, whether within the solution or without, is invariably designated positive; (2) that the current always starts from the electro-positive element to pass through the intervening liquid of the cell, and always, therefore, starts from the electro-negative element to pass through the wires and outside-connections of the battery; and (3) that the starting and finishing points within the liquid are termed the positive or negative plates, metals, or elements, and the starting and finishing points outside the solution are called the poles. The whole path of the current is called the *circuit*,—that in the cell being the *internal*, and that outside the *external* portion; and any conductive connection between the two plates is said to *complete the circuit*. When the two plates themselves come into contact within the cell, or they become united, either in the interior of the cell or close to the poles, by any thick metal substance which presents practically no resistance to the current, the battery is said to be *short-circuited*. When this short-circuiting takes place in any way while the current is flowing through the ordinary circuit, which probably presents a much higher resistance, nearly the whole of the current is diverted from the desired path, and passes through the easier passage, thereby causing a cessation of work and a great waste of battery-zinc; for when there are two possible paths open to a current, it will divide itself between the two in inverse proportion to their resistances. Thus, let us suppose a current of 200 ampères to arrive at a point in the circuit at which a junction is reached, so that it may continue to flow to another point in the circuit, either by one channel with a resistance of 1 ohm, or by a second with a resistance of 99 ohms; the result will be that the path having 1 ohm resistance will carry $\frac{99}{99+1}$th of the current, or 198 ampères, while the other takes the remaining $\frac{1}{100}$th, or 2 ampères.

Reduction of Polarisation.—A large number of voltaic cells have been invented to meet the different requirements of practice, but for the most part with the object of producing either a constant current (that is, a current of uniform strength for a long time), or a current of high potential. In any case, polarisation must be minimised, and this is effected either by mechanical or by chemical means. The mechanical method consists in so arranging the cell that the hydrogen shall escape from the surface of the negative plate immediately it forms; the chemical method seeks to prevent the separation of the hydrogen altogether, by means of substances which combine with it to form harmless compounds, or which cause the deposition of another suitable metal in its place. The latter or chemical system of overcoming polarisation has the added advantage that it increases the chemical activity in the cell and thus adds to the electro-motive force. The substances employed for converting the hydrogen into a neutral substance are usually

of an oxidising character and thus effect their object by the
formation of water; and this operation may be conducted either
by immersing both plates in a single oxidising solution, or
by dipping the negative metal only in the oxidant, which
must then be contained in a porous vessel placed within that
holding the dilute acid and the positive element. The former
class are termed *single-fluid cells*, the latter *two-fluid cells*.
Other arrangements are possible but are not largely used, and
will not be dealt with here—for example, those with two liquids
(in different compartments) and one metal, or the gas battery
of Grove.

Smee's Cell.—Of single-fluid cells in which polarisation is
mechanically remedied, the most noteworthy is that invented
by *Smee*. In this battery a plate of silver is placed in dilute
sulphuric acid between two plates of zinc, but the silver plate is
coated superficially with a very thin layer of platinum, so that
the hydrogen may escape with comparative readiness from the
slightly roughened surface which is thus produced. But even
with this device a certain amount of hydrogen must exist on the
surface of the silver so long as the cell remains in action. The
separation and escape of the gas, however, occur with fair regularity, so that although the electro-motive force rapidly decreases
during the first few seconds of use, it then remains constant at a
point considerably higher than it would if the zinc were opposed
to a pure silver plate, and yet higher than if it were opposed to
copper, on account of the greater electro-chemi-
cal difference between the two former metals
than between the latter. But the mechanical
de-polariser is palliative only, not remedial.

The annexed sketch of the Smee-cell (fig. 4),
shown in section, illustrates its construction.
The zinc plate, Z, is doubled, so that it may
surround the silver on two sides and thus utilise
both surfaces of the latter. It may be formed of
two zinc plates mounted with the platinised
silver between them in a wooden frame, which
being a very feeble conductor may carry away a
minute fraction of the current, but serves to
hold the metals in position, so that a quite thin
sheet of silver may be employed without fear of
its bending out of shape and making a short
circuit. The zinc alone dissolves, and requires,

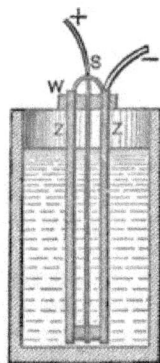

Fig. 4.
Smee-cell.

therefore, to be initially of fair thickness; the silver plate may
theoretically be only sufficiently stout to carry the current

without presenting undue resistance, for the E.M.F. of a cell is
independent of the dimensions of its elements. The loss of
energy by conduction through the wood may be obviated by
steeping the frame in melted paraffin, or by coating it with
shellac varnish. The acid, which is prepared by *carefully* (*vide*
p. 320) adding 1 part of strong sulphuric acid with constant
stirring to 12 parts of water, is contained in a glass or stoneware
vessel of suitable shape. These vessels are generally rectangular,
and may be made to hold one pair or any greater number of the
zinc-silver plates, which are supported by resting the wooden
framework upon the tops of the side walls of the acid trough.
The Smee-cells are very weak, the electro-motive force of one
being only about 0·47 volt; and they evolve much hydrogen,
the escape of which into the air in a constant succession of
minute bubbles, each mechanically carrying with it a trace of
sulphuric acid, causes a slightly unpleasant choking sensation in
breathing the atmosphere of the room in which they are working.
But, on the other hand, they are compact and very simple in
construction, and if kept clean are not liable to become dis-
ordered; while the escape of the hydrogen produces a peculiar
hissing sound, which, to the practised ear, may indicate by its
varying intensity any accidental derangement either in the cells
themselves or in the baths which they are electrolysing. It will
cease altogether if the circuit is broken and the current has ceased
to flow, or it will grow louder if a short circuit has formed at any
point, and so diminished the resistance and increased the activity
in the battery. This cell is still largely used by electrotypers on
account of its regularity and simplicity when properly super-
vised.

Daniell-Cell.—The form of battery invented by Professor
Daniell is one which, by reason of its constancy, is well suited
to electro-depositing work. In it hydrogen is not deposited at
all, but is made to displace metallic copper from its solution, by
immersing the negative plate in a saturated solution of copper
sulphate. If both plates were placed in this liquid, then, as soon
as the circuit was broken and they remained isolated, the zinc
being more electro-positive than copper would at once commence
to exchange with that element contained in the solution, a part
of the zinc forming soluble zinc sulphate, and an equivalent
amount of copper precipitating in a pulverulent or spongy form
on the surface of the remaining zinc. This film, being pervious
to liquid, would still allow the copper solution to attack the zinc
core, and by "local action" would even facilitate the exchange of
the metals. The battery would be rendered wasteful of zinc, and,

indeed, useless. To obviate this, the copper solution with its electro-negative plate of copper is separated from the simple acid liquid containing the zinc by a porous partition of some kind, generally of baked but unglazed earthenware, which will permit diffusion, but not free interchange, of liquids through its pores. This is usually effected by placing the zinc in a porous vessel within the outer cell containing the copper; thus, a "two-fluid cell" is produced. The action which takes place may be expressed as follows—number 1 representing the condition of the cell when it is standing inactive, number 2 that when the circuit is completed and a current is being generated :—

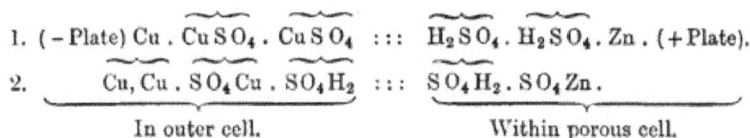

1. $(-\text{Plate})\ Cu\ .\ CuSO_4\ .\ CuSO_4\ :::\ H_2SO_4\ .\ H_2SO_4\ .\ Zn\ .\ (+\text{Plate}).$

2. $\quad Cu, Cu\ .\ SO_4 Cu\ .\ SO_4 H_2\ :::\ SO_4 H_2\ .\ SO_4 Zn.$

| In outer cell. | Within porous cell. |

These changes or *reactions* may be described in the form of equations thus,

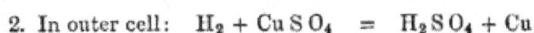

1. In porous cell: $Zn + H_2SO_4 = ZnSO_4 + H_2$

2. In outer cell: $H_2 + CuSO_4 = H_2SO_4 + Cu$

or the ultimate reaction of the battery may be summed up in one equation,

$$Zn + CuSO_4 = ZnSO_4 + Cu.$$

Thus, the hydrogen finds its way, as it were, through the liquid in the pores of the inner cell-wall, exchanges places with the copper, and produces a practical transference of sulphuric acid from the outer to the inner cell; and as the acid on the one side of the porous division becomes neutralised by zinc, the strength of the copper solution on the other side is rapidly diminished by the deposition of its metallic constituent. The complete exhaustion of either side would set a limit to the utility of the cell; but long before this point is reached the power of the battery is much lessened. As it is chiefly essential that copper alone, and no hydrogen, should be separated, care must be taken to keep the liquid in the inner cell saturated with copper sulphate, by suspending crystals of the solid copper salt (*blue vitriol*) just below the surface of the liquid, so that the solution may be replenished as fast as it is impoverished.

The usual form of the *Daniell-cell* is indicated in fig. 5. It consists of a cylindrical copper vessel holding a saturated solution of copper sulphate, and acts both as a containing vessel and as the negative plate of the battery. Around the top is a perforated copper, ring-shaped trough, T, which is below the level of the liquid, and is always kept filled with crystals of the copper salt. Within this vessel is the porous clay-cell, P, closed at the bottom but open above, and in this is the amalgamated zinc rod or plate, Z, immersed in dilute sulphuric acid (1 acid : 10 water), or sometimes in a solution of zinc sulphate. Instead of using a copper external vessel, a thin sheet of rolled copper may be bent into cylindrical shape and placed in a stoneware or glass jar; the spare crystals are then suspended in muslin bags within the outer cell. The prime cost of this latter arrangement is less, but it must be remembered that in the former the copper is the negative element, and the cylinders are, therefore, in no way injured, but on the contrary are thickened by the gradual deposition of metal while the current is passing, and are not appreciably attacked by the dilute acid contained in them.

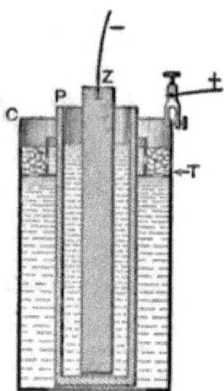

Fig. 5.—Daniell-cell.

Modified Daniell-Cells.—The arrangements for maintaining the supply of copper sulphate crystals are very numerous; for example, Breguet and others have adopted a globular receptacle of glass, which is charged with the salt, filled up with water, and, the neck being closed by a cork with two perforations, is inverted above the inner porous cell, so that the bottom of the cork lies beneath the fluid level within the latter; then as the battery liquid deposits copper it becomes lighter, and floating upwards becomes displaced by the heavy saturated solution in the flask, while the poorer solution, which has found its way into the receiver, again becomes enriched, and is in turn ready to take the place of a further quantity of impoverished liquor. In this cell the copper is placed in the inner compartment, as in fig. 6: the globe, G, is supported by the neck of the porous pot, while between the two is inserted a copper tape, C, which serves for the negative element. Around the porous cell

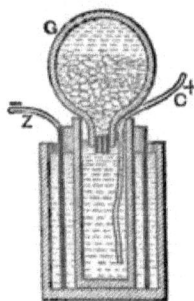

Fig. 6.—Daniell-cell (Breguet's form).

is a bent cylinder of zinc in sulphuric acid; the current is led away by the wires marked + and − attached to the copper and zinc respectively. Somewhat similar to this is the Meidinger-cell, in which a flask containing the crystals dips into a small cup containing the copper plate, resting on the bottom of a glass vessel, the diameter of which is enlarged to contain the zinc cylinder at a point somewhat above the upper portion of the cup. In Kuhlo's cell a copper vessel is used, which carries the perforated trough outside instead of inside, and the zinc is enclosed in parchment paper instead of in a porous cell.

Other modifications in shape may be made. In the pattern adopted by the Post Office Authorities, a long teak-wood trough, coated inside with some water-resisting varnish, such as marine glue, is divided into a number of compartments by alternate divisions of slate and porous earthenware. It contains the zinc and dilute acid in divisions, 1, 3, 5, and the remaining odd numbers; and the copper as copper sulphate in the even divisions, 2, 4, 6, &c. It is simply a conveniently compact and portable method of setting up a battery of several cells. Large Daniell-cells may be made horizontal instead of upright, and without porous divisions, provided they are allowed to rest undisturbed in a position where they will not be subjected even to any considerable amount of vibration, the difference in the specific gravities of the solutions employed sufficing to keep them in their required relative positions. A simple form of such a gravity battery is shown in fig. 7. A wooden trough of convenient size, say 14 inches long by 10 wide and 3 high, is lined with insulating varnish, and provided with a glass partition, G, near to one end, which must reach to the top of the trough, but only to within a quarter of an inch off the bottom. On the bottom rests a large sheet of thin copper

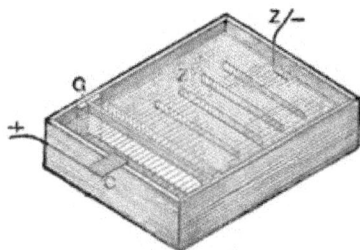

Fig. 7.—Gravity battery.

plate with a narrow strip attached to it, which, passing under the glass partition, is bent up within the smaller compartment to form the positive pole, C. Above this plate, and resting on wooden brackets in the sides of the trough, is a stout grid or plate of cast zinc with perforations to allow of the escape of any gas which might be formed through incorrect working; this should be well insulated from the copper and kept at a minimum

distance of half an inch from it. A zinc strip attached to the
grid serves for the negative pole, Z. A dilute solution of zinc
sulphate, to which a few drops of sulphuric acid have been added,
is poured into the trough until the zinc is covered; the small
compartment is then filled with copper sulphate crystals, and
these, gradually dissolving, form a heavy solution which, finding
its way beneath the glass partition, covers the floor of the vessel
to the height of an eighth or a quarter of an inch. The zinc is,
therefore, now in a dilute zinc sulphate, the copper in a strong
copper sulphate solution, and the battery is ready for use. The
crystals in the smaller division must, of course, be renewed as
they gradually dissolve away. If well attended to, this form of
the Daniell-cell will yield good results, but it is essential to avoid
all agitation which would tend to mix the solutions and bring
copper-bearing liquids into contact with the zinc; a parchment-
paper tray to receive the zinc plate will minimise this risk, but
in any case the grid should be carefully removed from time to
time and cleansed from any copper deposited upon it.

Indeed, in all forms of the Daniell-cell the zinc plates should
be constantly inspected, as not infrequently splashes of copper
liquids find their way into the outer compartment; then, when
the cell is at rest, a portion of the copper is at once deposited on
the zinc by simple electro-chemical exchange, and local action
being set up the amount of copper separated rapidly increases.
As soon, therefore, as the red-brown colour of metallic copper is
observed upon any portion of the zinc, it must be removed, and
the surrounding liquid examined; even the merest shade of blue
colour in the solution indicates the presence of copper, and points
to the necessity for its renewal. When copper is thus found
in the outer vessel, the porous cell should be most carefully
examined, as it may have become cracked, and would thus
permit a comparatively free interchange of liquids.

The Daniell-cell has an electro-motive force of about 1 volt,
varying with the strength of the acid, the nature of the solvent
and the like, from 0·98 to 1·08 volts. The E.M.F. usually accepted
is 1·079 volts, and this is so near to unity that it is regarded as
a standard-cell with which other batteries may be compared. It
is well suited for electrolytic work, especially on account of the
extreme constancy of the current, but requires greater care and
attention than the Smee-cell, to which, however, it is in many
respects preferable.

Grove-Cell.—In the *Grove-cell* a different depolarising system
is adopted. Instead of substituting a more negative metal for
the gaseous hydrogen, Grove uses a medium containing an excess

of oxygen which oxidises the hydrogen into the innocuous compound water, as rapidly as it is deposited upon the negative plate. This medium is nitric acid, the chemical action being represented as follows :—

1. In outer cell: $Zn + H_2 SO_4 = Zn SO_4 + H_2.$
2. In inner cell: $H_2 + 2 H N O_3 = N_2 O_4 + 2 H_2 O.$

Or in one equation—

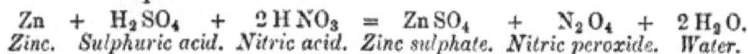

$$Zn + H_2 SO_4 + 2 H N O_3 = Zn SO_4 + N_2 O_4 + 2 H_2 O.$$
Zinc. Sulphuric acid. Nitric acid. Zinc sulphate. Nitric peroxide. Water.

The nitrogen peroxide ($N_2 O_4$) is a gas which at once dissolves in the nitric acid, imparting to it a red tint at first, and after-wards a deep green colour. It is this gas which, after the acid is saturated with it by the long-continued action of the battery, is seen to come off in the form of dense red-brown fumes possessing a suffocating odour.

Fig. 8.—Mode of connecting a pair of Grove-cells.

The negative plate is no longer of copper, which would be vigorously attacked by the nitric acid, but of platinum in the form of foil, as this metal entirely resists the action of the acid. Fig. 8 shows (in section) two Grove-cells set up in series to illustrate the method of connecting them. Within a glass or glazed earthenware outer vessel is the dilute sulphuric acid containing the zinc plate bent into the shape indicated, in order to give a larger surface by surrounding the negative plate, and for convenience of connecting with other cells. Enfolded by the zinc is the porous compartment containing the strip of platinum foil, P, in strong nitric acid. The platinum is not attacked when in use, and is only deteriorated by repeated mechanical manipulation in setting up or taking down the battery; but its high value adds greatly to the prime cost of the cell, and for this reason the Bunsen-cell, with its block of carbon in lieu of platinum, is generally preferred for commercial purposes.

Fig. 9 gives a general view of the external appearance of the Grove-cell, showing one element complete, with the attachments of the next platinum plate on the one side, and of the zinc with its porous compartment on the other.

Bunsen-Cell.—The *Bunsen-cell* has the same reaction and, therefore, practically the same electro-motive force as Grove's,

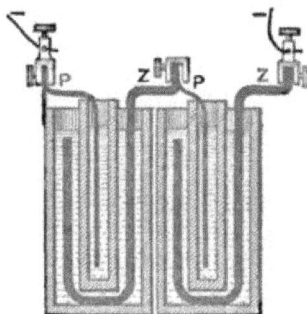

4

viz.—nearly twice that of Daniell's (1·8 to 1·9 volts). In a circular stoneware jar (fig. 10) is contained the porous cell with its rod of gas-carbon and strong nitric acid, while around the

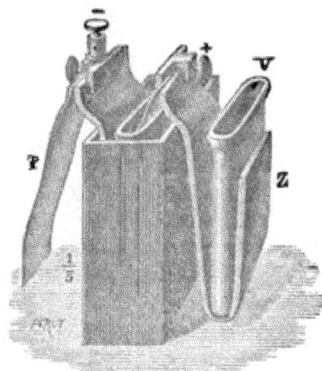

Fig. 9.—Grove-cell (external view).   Fig. 10.—Bunsen-cell.

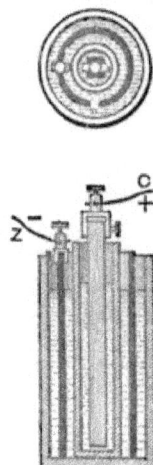

porous pot is a bent cylinder of zinc immersed in dilute sulphuric acid. The carbons are usually cut from the blocks of the deposit which gradually forms on the interior of gas-retorts, and is known as gas-carbon or retort-carbon. It is extremely dense, strong and durable; but is sufficiently porous to allow the acid to pass upward by capillary attraction, and so to attack the brass *binding screws*, which form the connection between the battery plates and the wires through which the current is conveyed to its place of application. To prevent this action, D'Arsonval recommends steeping the extreme upper end of the carbon for a short time in melted paraffin, then depositing a thin film of metallic copper on its surface by electrolysis, and plunging the whole of this portion beneath melted type-metal; the type-metal will thus form a covering which is in contact with the carbon at the edges of the joint, and must effectually prevent the corrosion of the binding screw, while maintaining perfect electrical connection between the parts.

In both Grove's and Bunsen's cells the exciting fluid may be prepared by adding 1 part of sulphuric acid to 19 of water. The nitric acid must initially be highly concentrated, the specific gravity being not less than 1·3082 (= 34° Baumé).

The tendency of the reaction is to produce nitric peroxide,

which finally escapes as a gas, and water, as we have already seen; thus, not only is the nitric acid constantly being weakened by decomposition, but the remainder is at the same time diluted by the water which is formed. In this manner the strength of the acid rapidly falls off, and its oxidising efficiency is lessened. By the time it has been reduced to a density of 30° Baumé (sp. gr. = 1·2624) the electro-motive force of the cell becomes quickly diminished, and when it has reached a strength of 28° Baumé (sp. gr. = 1·2407) it is no longer serviceable, and should be at once renewed. These cells present many advantages; they have a high electro-motive force, and are fairly constant for a few hours, but they require to be filled with fresh acids daily. The nitric acid, however, is a constant source of danger in inexperienced hands on account of its extreme corrosiveness; and the red nitric peroxide gas, which is constantly evolved, is not only disagreeable, but if inhaled for any time or in any quantity is positively injurious. Whenever either of these forms of cell is used it should be placed in a well-ventilated space outside the work-room, or in a cupboard provided with the means of discharging the gases into a flue or directly into the outer air. In spite of these drawbacks they are largely used for nickel-plating and for some other classes of work, especially, perhaps, for those of an experimental nature.

Bichromate-Cells.—Some operators have substituted chromic for nitric acid, because the products of decomposition are solid and remain dissolved in the liquid instead of passing into and contaminating the atmosphere; a suitable solution, proposed by Higgins, consists of 1 part of potassium bichromate dissolved in a mixture of 3 parts of sulphuric acid with 9 parts of water. The disposition of such a cell is similar to those which we have just described.

But there are single-fluid *bichromate-cells* in which no porous division is needed, and these have the advantage of a lower internal resistance. The best proportions for the exciting fluid are, perhaps, those which correspond to the equation representing its action as follows :—

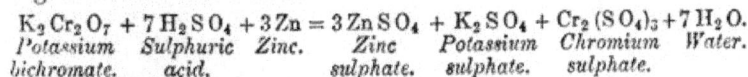

$$K_2 Cr_2 O_7 + 7 H_2 S O_4 + 3 Zn = 3 Zn S O_4 + K_2 S O_4 + Cr_2 (S O_4)_3 + 7 H_2 O.$$
Potassium   Sulphuric   Zinc.   Zinc   Potassium   Chromium   Water.
bichromate.   acid.   sulphate.   sulphate.   sulphate.

This would require about 3½ ounces of potassium bichromate, and 8½ ounces of sulphuric acid to be dissolved in a quart of water. As the zinc is here actually in the strongly-oxidising solution during the action of the battery, and as a very notable amount of solvent action would take place even when the battery

is at rest (though by no means comparable with that which
would be set up in a single-fluid nitric acid cell),
means must be devised for removing the zinc
from the liquid directly the current is broken.
This is generally accomplished by drawing or
winding them up out of the acid. In the bottle-
form of this cell (fig. 11) two long strips of carbon,
united by a metallic connection above, are fastened
(parallel to one another) to a vulcanite stopper,
and are there connected with the binding screw
+ ; these form the negative element and pass to
the bottom of the bottle; between them is a short
thick strip of zinc attached to a brass rod passing
stiffly through the centre of the ebonite cork
and connected with the binding screw —. The
zinc is entirely insulated from the carbon by the
ebonite, and may be drawn out of the solution by means of the
brass rod as soon as its services are no longer required. In the
trough-form of cell (fig. 12) a series of carbon-and-zinc couples

Fig. 11.—Bichro-
mate of potash
cell (bottle-form).

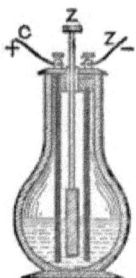

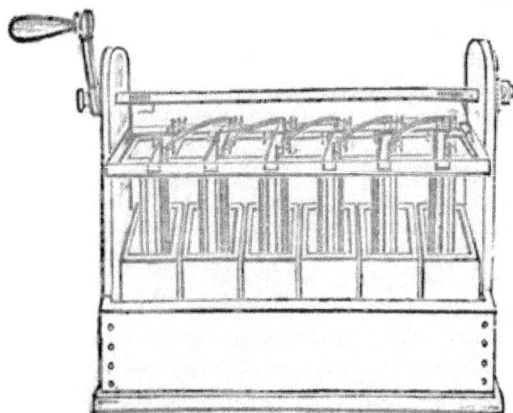

Fig. 12.—Bichromate of potash battery (trough-form).

are immersed in the acid mixture contained in a long rectangular
trough, from which they can be withdrawn when the battery is
not in use. These batteries are simple, give a high electro-motive
force (1·9 to 2 volts), and have a very low internal resistance; but
the chromic solution, although emitting no fumes, is highly corrosive, and, moreover, must be constantly watched, for as it
becomes weakened by use the "current pressure" rapidly declines.

Leclanché-Cell.—The number of other batteries in use for

various purposes is immensely large and is yearly increasing, but only a few are of practical use to the electro-metallurgist. Many are too weak, others are strong at first but speedily lose their power, while others again which, by their strength and constancy are well adapted to our purpose, are too expensive or too trouble-some for daily use under the conditions of the workshop. Thus, to take a single example, the Leclanché-cell, economical enough as to prime cost and in use, gives a current which, at first powerful, rapidly declines owing to polarisation, but almost as rapidly recovers itself when standing at rest; it is thus very suitable for intermittent or irregular work, such as the ringing of electric bells, but is useless for electro-plating.

Practical Hints on the Use of Batteries.—In all dealings with batteries scrupulous cleanliness in every detail is necessary. The cells and the metals must be thoroughly cleansed before commencing work; and it is of the highest importance that all metallic connections in the external circuit, such as the junctions of the plates with the binding screws or with one another (when placing them in series), be kept perfectly bright. A film of oxide or tarnish between two surfaces through which a current must pass increases the resistance of the circuit to the passage of the current, and so, while retarding the action of the battery, adds to the amount of electrical energy converted into useless heat, and thus practically wasted. Porous cells must be kept clean; they are apt to become choked with insoluble deposits which add greatly to their resistance; and in the Daniell-battery nodules of copper frequently form upon their surfaces,—these must be watched for and removed. When taking a battery to pieces after use, in order to replenish and re-fit it, it is well to soak the porous cells for some time in water, in order to extract the salts contained in them, and never to allow them to become dry until they are thus purified.

A fairly pure zinc should always be used; it may be in the condition of rolled sheet, but is more usually cast into the desired shape; in any case the whole surface must be thoroughly, but not excessively, amalgamated. As soon as bubbles of gas are observed to form upon the zinc, it should be carefully examined, as this is a certain sign that some more electro-negative substance is present, which may either be an impurity in the metal itself, when re-amalgamation will remedy the defect; or it may be due, as in Daniell's cell, to a metal deposited from the battery solutions, in which case it must be removed by cutting, scraping, or filing. Old or worn-out zincs should be saved, as they contain much mercury which is recoverable (see p. 308).

No metallic or semi-conductive connection between the two plates of a battery is permissible, because such a connection forms a short circuit, which diminishes or destroys the current flowing in the external circuit. Absolute *insulation* (or electrical separation) between the two poles, and between the wires at all intermediate points in the circuit, must be preserved; and the higher the electro-motive force of the current, the greater is its power of overcoming resistance, and, therefore, the more carefully must the insulation be ensured. Dry wood, which is a partial conductor for electricity of very high tension, is practically an insulator for ordinary battery currents; but when moistened with acid or any of the battery solutions, it becomes capable of causing very considerable current leakage.

The battery solutions require careful watching. Any tendency to crystallise or deposit solid matter upon the sides or bottom of the cell must be checked; it may be due to the use of too concentrated solutions at the outset, which is curable by the addition of water; or it may be that they have become saturated and worn out, when they, of course, require to be renewed. The quantities to be used in making up solutions should be weighed or measured; this occupies but little more time, and in the result is more reliable than rule-of-thumb work or mere guess measurements. If the solutions made up from a crystalline salt be turbid, they should be filtered through a blotting-paper cone inserted in a glass funnel (metallic funnels should not be used, as they are liable to be corroded by the fluids passing through them). A circular filter-paper is readily made to fit the funnel by folding it first across one diameter, as shown at A B in division 1 of fig. 13; then on folding it again at right angles, as at C D in No. 2, it has the form of No. 3; now on inserting the finger between the folds of the paper it may be opened out to the conical shape depicted in No. 4, and is thus ready to place in

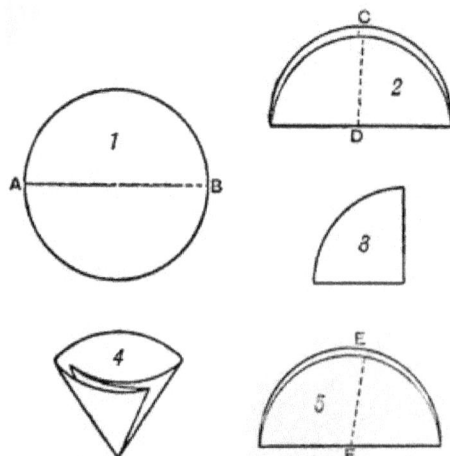

Fig. 13.—Mode of folding filter-paper.

the funnel. If, however, the paper should not fit well into the cone of the latter, it may be refolded along the line, E F, as in No. 5, or along any other suitable line, and may thus be adapted to suit a funnel constructed with any angle at its apex. Strongly-acid solutions, such as those used in the bichromate battery, cannot be thus filtered, as they destroy the paper ; but the solution of the potassium bichromate may be passed through a filter before adding the acid to it. If it be necessary to clear any solution which attacks paper, a plug of spun glass or of asbestos may be lightly rammed into the apex of the funnel, and will form an efficient filtering-medium in lieu of paper.

It is a good plan to place the battery in a chamber apart from, but adjoining, the depositing room, remembering that the greater the length of copper wire to be traversed by the current between the battery and the vats, the greater will be the loss of energy through the resistance of the circuit. By keeping it outside, however, the operator is not annoyed by the fumes or acid spray produced by certain forms of cell; and there is less danger of mistake or of accident in manipulating the battery and depositing solutions. If circumstances compel the double use of the same room, the battery should on no account be placed above or too near the electrolytic vats, but should be fitted up in a place easy of access, and preferably in a ventilated cupboard, as already explained. It will be found convenient to set up the battery in proximity to a sink provided with a good water supply, so that every facility may be afforded for cleansing the different parts when the work of the day is over. In charging two-fluid cells, no drops of the depolarising fluid must be permitted to enter the zinc compartment.

On the Fittings and Connections of a Battery.—It will be remembered that the electro-motive force of a voltaic cell is dependent on the heat-energy produced by the combinations taking place within it; hence the size and shape of a cell may greatly modify the volume of current generated by it, but are without influence on the pressure or E.M.F. Other things being equal, the volume of any current may be increased by raising the electro-motive force or by lowering the total resistance of the circuit, as will be more fully explained in the next chapter (*infra* p. 10); now, the internal resistance of a battery is increased, either by expanding the distance between the two plates, because the current has to traverse a longer distance through an inferior conductor; or by using smaller battery plates, because the area of the liquid conductor is thus reduced, and the current has to flow through a narrower channel. By increasing the resistance by either of these methods, the total current must be diminished while the electro-motive force is constant. Conversely, the use of larger plates (that is, of larger cells) or the nearer approach of the metals in the cell, while leaving the pressure unchanged, increases the volume of current by diminishing the resistance to its path ; in other words, the output in ampères is increased but the E.M.F. in volts is constant. With any given type of cell, then, it is possible to alter the current strength but not the electro-motive force ; to effect the latter, changes in the chemical conditions are necessary.

Modes of Arranging Cells.—By the use of more than one cell the electro-motive force as well as the current-intensity may be altered at will. If two similar cells be joined, as shown in fig. 14, copper to copper and zinc to zinc, the electro-motive force is no more altered than would be the total fluid pressure produced by placing two pails of water side by side upon a level floor in place of one ; for both the cells are yielding the same pressure of electricity, and the mere coupling them *in parallel arc*, as it is termed, is only equal in effect to increasing the size of a single cell, which as we have seen is without influence on the E.M.F. But if the cells be disposed as

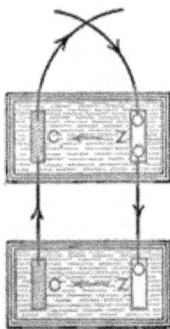

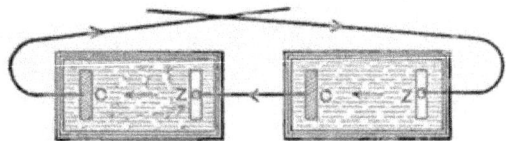

Fig. 14.—Two cells with like metals con- nected.

Fig. 15.—Two cells with unlike metals connected.

in fig. 15, with the copper of one joined to the zinc of the next, and the free elements connected to the main circuit, the current generated in the first cell has to flow through the second, and that of the second through the first in order to complete the whole circuit, with the result that the total electro-motive force is doubled. This arrangement, which is termed coupling *in series*, is exactly analogous to lifting the one pail of water above referred to and placing it upon the other, when the pressure is of course doubled. In parallel arc the chemical energy of each cell is placed upon the same level, and can, as it were, pump the electricity only to the same height ; but in series the energy of one cell pumps the electricity to one level, and then that of the second cell starting from this point raises it as high again. And so in setting up any number of cells, if placed all parallel, the

E.M.F. is only that of one cell, but the internal resistance is reduced, as it would be in one large cell of the same type; while if all are arranged in series the E.M.F. will be raised in direct proportion to the number of cells in use. Intermediate dispositions to suit special requirements may be made by grouping several cells together in parallel arc, and then uniting this group as a whole in series with other similar groups.

Let us with the aid of the diagram, fig. 16, consider the possible groupings of the 6 cells, A, B, C, D, E, F, each of which has, let us say, an electro-motive force of 1 volt and an internal resistance of 1 ohm. One such cell, short-circuited by a piece of metal having practically no resistance, would give a current of 1 ampère; for Ohm has ennunciated the law that the volume of current in a given circuit is proportional to its electro-motive force, and inversely proportional to the total resistance, which is expressed by saying that the current volume is equal to the electromotive force divided by the resistance, or more shortly by formulating

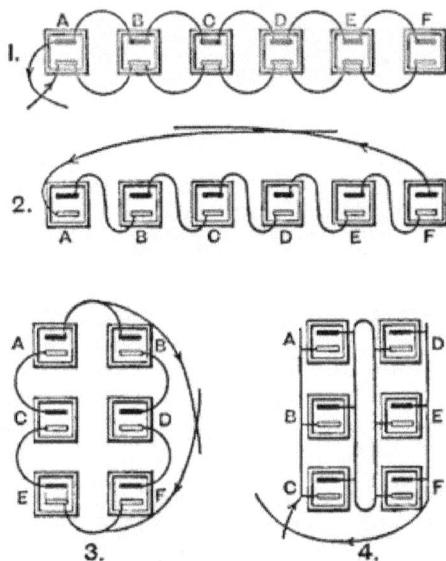

Fig. 16.—Modes of grouping six cells.

$C = \dfrac{E}{R}$. In other words, *Ohm's law* may be written for any

circuit as ampères $= \dfrac{volts}{ohms}$. When the 6 cells in fig. 16 are

placed parallel, as shown in position 1, all the coppers are united together, and as a group are opposed to all the zincs similarly united; the E.M.F. is unaltered, and still stands for the whole group at only 1 volt; but the resistance is divided by 6, because there are now 6 equal internal passages for the current instead of 1, and it is therefore ⅙ ohm instead of 1 ohm; so that now the volume of the current flowing through a short circuit is

$\dfrac{1 \text{ (volt)}}{\frac{1}{6} \text{ (ohm)}} = 6$ ampères. But if the 6 cells are joined in series, as

in position 2, the E.M.F. is multiplied by 6 and is $6 \times 1 = 6$ volts, while the resistance is also multiplied by 6, and is now $6 \times 1 = 6$ ohms, because now the current has to flow through all the cells in series, and is, therefore, checked by the resistance of each; the cur-

rent yielded on short-circuiting is, therefore, $\dfrac{6 \text{ (volts)}}{6 \text{ (ohms)}} = 1$ ampère,

or the same as that of a single cell! (Note, however, that this is only on short-circuiting—*vide infra*.) Now, by joining the pairs A and B, C and D, E and F, each respectively in parallel, and connecting the groups A B, C D, and E F in series, as in position 3, the electro-motive force of each parallel pair is only 1 volt; but there are three pairs in series, so that the total electro-motive force is 3 volts; while the resistance of each pair is of course half that of one cell, or $\frac{1}{2}$ ohm, and the sum of the three resistances is $\frac{3}{2}$ ohm, so that the ultimate current on short

circuit is $\dfrac{3}{\frac{3}{2}} = 2$ ampères. And, finally, by arranging the cells, as

in position 4, with the three cells A B C parallel, and united to form a series with the group D E F also placed parallel, the electro-motive force of each group is 1 volt and the total pressure is 2 volts, while the resistance of each is $\frac{1}{3}$ ohm and the total

resistance $\frac{2}{3}$ ohm, so that the resultant current is $\dfrac{2}{\frac{2}{3}} = 3$ ampères.

The volume of current quoted in each case is, as we have stated, that produced when the external resistance is *nil*; but this condition never obtains in practice, and the anomaly of six cells giving only the same current as would be afforded by one is not met with when the external resistance is high, for then the value of the increased electro-motive force is felt. Thus the current from one of these cells acting through an external resistance of 100 ohms (making with the internal resistance of the cell a total resistance of $100 + 1 = 101$ ohms) gives an ampèreage equal to

$\dfrac{1 \text{ (volt)}}{101 \text{ (ohms)}} = \dfrac{1}{101}$ ampère, whereas the six cells in series would

give $\dfrac{6 \text{ (volt)}}{106 \text{ (ohms)}} = \dfrac{6}{106}$ ampère; thus the current ratio in the two

cases is $\dfrac{1}{101} : \dfrac{6}{106} = 106 : 606$; that is to say, there is now 5·72 times as much current produced from the six cells in series as there is from one alone. On the other hand, the arrangement of the six cells in parallel, so advantageous on short circuit, is little superior to a single cell when the current is to be passed through so high a resistance as 100 ohms. In this case the total resistance is $100 + \frac{1}{6}$ ohms; the total E.M.F. is 1 volt; the total current

$$\frac{1}{100 + \frac{1}{6}} = \frac{6}{601}$$ amperes; and the ratio of efficiency of 1 cell to 6

in parallel is only $\frac{1}{101} : \frac{6}{601} = 601 : 606$, or $1 : 1.01$ instead of

$1 : 5.72$ as when they are set up in series.

From this it will be seen that to obtain the highest effect from a number of cells, they must be united in series when the external resistance is very high, or in parallel when it is very low. The highest efficiency of all is to be obtained when the cells are so grouped that the internal resistance of the battery is equal to the external resistance of the circuit.

But although the coupling-up of cells in series may be desirable by affording the greatest possible volume of current, such a disposition is by no means the most economical. A certain weight of zinc, in dissolving, produces a certain E.M.F.; and in one cell yields a certain volume of current capable of doing a certain quantum of work. And where one cell only is employed, a given weight of zinc, in dissolving, deposits electrolytically a definite equivalent weight of any metal in the plating-baths. But when cells are multiplied in series, an equal amount of zinc dissolves in each, but only the one equivalent of deposited metal is yielded; the energy developed in the oxidation of the remaining zincs being used in raising the electromotive force,—in pumping the current to a higher level. For example, in the case of the six cells in series in short circuit, the same ultimate current is produced as would be yielded by one cell; but in the former case six times the quantity of zinc is dissolved to produce the same effect. It is clear that, economically, the best result is obtained when the smallest number of cells are placed in series which is compatible with the production of the minimum electro-motive force required for the electrolysis; and the disposition of the cells must be governed by the balance of the two considerations—time and economy.

Connecting Screws.—In making electrical connections between the different parts of the circuit, the use of brass binding-screws is advisable ; these may be procured of any apparatus-maker and of any shape that may be required. Figs. 17, 18, 19, 20, 21, and 22 illustrate six of the more useful forms.

Figs. 17 to 19 represent the commoner methods of uniting plates or bars to wires. Fig. 17 is employed in the Bunsen battery to join the carbon-block to the terminal wire; but it is equally applicable to the uniting of any thick substance to a wire, the former being held by the screw at the side of the large clamp, the latter by that in the spherical portion above. Fig. 18 may be screwed into a metal block, and is commonly used to join the negative wire to the zinc of the Daniell-cell. Fig. 19 is employed as the terminal binding-screw in a Grove-cell, and will

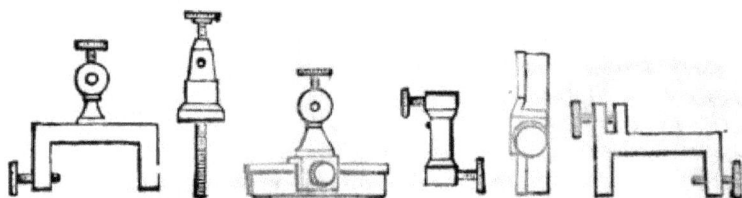

Fig. 17.   Fig. 18.   Fig. 19.   Fig. 20.   Fig. 21.   Fig. 22.

Forms of binding-screws for batteries.

serve to connect any thin sheet to a wire. Fig. 20 illustrates the arrangement for uniting two wires end to end. Figs. 21 and 22 represent systems of coupling plates or blocks, the former as applied to two thin sheets, the latter to a thick block with a thin plate. Other forms may, of course, be devised to suit special requirements, but these will answer most of the purposes to which they may be put.

Switch-Boards.—Often for practical purposes, but for experimental work more especially, it is desirable to have at hand a rapid method for altering the arrangement of a group of battery cells, so that any desired combination in series and parallel are may be obtained. The constant alteration of battery connections is tedious and clumsy ; while a switch-board used by Professor Sylvanus Thompson is at once simple and efficacious. This instrument is made with a series of binding-screws, and of accurately-ground brass plugs fitting into cavities between adjacent brass blocks, such as those employed in the manufacture of resistance-coils for electrical measurement. Fig. 23 represents a modification of Thompson's switch-board, which we have in constant use; it is precisely the same as the other in principle,

but is of cheaper construction, and may be made by any carpenter.
A is a plan; and B, a sectional elevation of the board. Two
parallel strips of brass plate, p and n, are let into the surface of
a varnished board, and are connected each with a terminal binding-screw at the end (+ and −). At even distances along the
surface of each brass rod
is a series of upright split
brass tubes, as at s', s' in
the section B; the tubes
on p being spaced midway
between those on n, as
shown. Along the lines,
P and N, are similar rows
of upright split tubes, s, s,
each of which is connected
by metal wire, with a corresponding horizontal split
brass tube, s'', s'', as at a,
b, c, a', b', c', &c. The connection is made as shown
in the figure, B; every

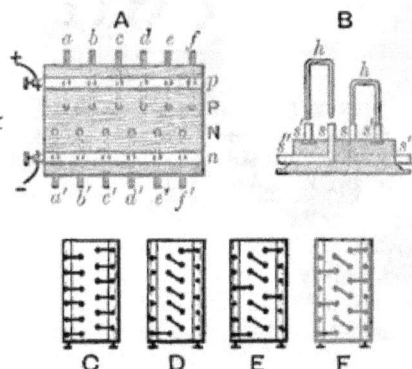

Fig. 23.—Thompson's switch coil.

pair of split tubes, s' and s'', must be perfectly insulated from
every other pair, and from the series, p and n; there is, therefore, no brass rod running along the lines, P and N. The
upright tubes on the row P correspond exactly to those in the
row p; those in N to the others in n; and the space between
any corresponding tubes, s and s', and also the diagonal space
between s of row P and s of row N, must be equal in every
case to those between the tubes on rows p and n. Twelve
U-shaped pieces of brass rod, h, h, are prepared of such size
that they fit with each limb into one of any adjacent pair
of vertical split tubes (the object of the splitting is to give
sufficient spring to ensure a tight fit, and therefore good
contact); a like number of short straight brass rods are
prepared for the horizontal tubes, s'', s''. The necessity for
even spacing of the tubes is evident, if the handles, h, h, are
to be interchangeable.

Each cell of the battery is now connected by wires to the
apparatus, the positive (copper) pole of the first to the s'' tube,
marked a, the negative (zinc) pole to that marked a'; the second
is similarly attached to b and b', the third to c and c', and so on;
taking care that the positive poles are all connected to one side,
the negative to the other, and that the two poles of every cell
are attached to corresponding tubes (a and a'; b and b'; &c.).

The battery wires may be soldered, each to one of the above-mentioned short straight brass rods, to be inserted in horizontal split tubes. Supposing 6 cells to be thus connected-up, no current can flow even if the terminal wires marked " + " and " − " are joined, because each of the tubes, s, s, is isolated; but by inserting the handles, h, h, in a suitable manner, the various connections are made in any required fashion, and the current passes through the circuit from the terminal + to that denoted by − .

By connecting every s in the P line with the corresponding s' in the p line ; and similarly those in the N row with the others in n, the positive pole of every cell is placed in direct contact with the brass strip, p, the negative with the strip, n; so that all the cells are in parallel arc, and discharge thus through the circuit. This arrangement is shown diagrammatically in plan C. But if only the positive wire of cell a is placed in connection with p, and only the negative wire of f with n, and the inner lines of tubes are joined-up diagonally (a of line P with b' of N ; b of P with c' of N ; and so on), the whole 6 cells are in series, as represented in diagram D; because the copper pole of the first is connected to the outer circuit, while the zinc pole is joined to the copper pole of the next, and so on until the zinc pole of the last is free and is united to the wire of the outer circuit. Diagram E shows the cells, a, b, and c, in series connected in parallel arc with d, e, f, also in series. Diagram F shows each pair, a and b, c and d, e and f, in series, but the three couples in parallel arc. Thus diagrams C, D, E, and F illustrate the four possible methods of combining 6 cells, all parallel; all in series ; two parallel groups of 3 cells in series ; and three parallel groups of 2 cells in series; and these connections may be altered at will in a few seconds of time.

THERMO-ELECTRIC BATTERIES.

The fact, already alluded to in the introductory chapter, that a compound bar, made up by soldering two unlike metal strips end to end, is capable of producing an electric current when the junction is alone heated, renders possible the direct conversion of heat into electrical energy. Instruments used to effect this change are termed *thermopiles* or *thermo-electric batteries*.

Thermo-Electrical Series.—Any two metal bars joined as in

fig. 24 will give a current always in the same direction when the point of union is heated, or in the opposite direction when it is cooled, more than the rest of the bars. In respect of their behaviour when thus treated the various metals behave very differently, and may be arranged in a thermo-electrical series similar to that descriptive of their electro-chemical relations. Any single pair of metals gives a constant electro-motive force so long as the difference between the temperature of the junction and that of the remainder of the circuit is constant; and, moreover, in many instances the E.M.F. produced is nearly proportional to this difference in temperature. The actual electro-

Fig. 24.—Simple Thermopile.

motive force of any pair of metals is extremely minute compared with that of a galvanic battery, so that a large number of couples must be used to produce an equal effect.

In the following table are grouped some of the principal metals with the electro-motive force produced by heating a single couple—made by heating them respectively at their juncture with metallic lead—to a temperature 1° Centigrade higher than the rest of the circuit. The E.M.F. is given not in volts but in *micro-volts* (1 micro-volt is the one-millionth part of 1 volt). In harmony with the electro-chemical system of nomenclature, which designates that metal electro-positive from which the current starts to pass through the liquid in the voltaic cell, a substance is said to be thermo-electro-positive to another when the current passes from it to the second through the heated juncture. In the table lead is taken as the basis for comparing the others, so that all the metals standing above lead are marked with the + sign, which indicates that the current would flow from them to the lead through the joint, while with the negative elements the current starts from the lead. The numbers cannot be accepted as absolutely accurate, or universally applicable, because the behaviour of metals in respect to these currents is greatly affected by the presence of impurities, and because the relations between them are disturbed, and in some cases reversed, at different temperatures. The letters M, B, refer to Matthiessen and Becquerel, from whose results these numbers are taken.

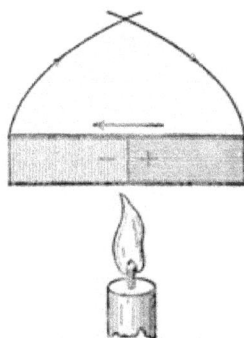

TABLE V.—Showing the Thermo-electro-motive Force of Various Elements in Relation to Lead, Expressed in Micro-volts per Degree Centigrade.

METAL.	Observer.	Micro-Volts.
Bismuth, pressed wire, . . .	M	+ 89 to 97
Bismuth, crystals, ...	M	+ 65 to 45
Bismuth, ordinary, ...	B	+ 40
Bismuth-Antimony Alloy (10 : 1), .	B	+ 64·5
Cobalt, ...	B & M	+ 22
Nickel, ...	B	+ 15·5
German Silver, ...	B M	+ 11·6
Palladium, ...	B	+ 6·8
Mercury, ...	B	+ 3·2
Lead, ...	...	Zero
Tin, pure pressed wire, . . .	M	− 0·1
Tin, ordinary, ...	B	− 0·4
Copper, commercial, ...	M	− 0·1
Platinum, ...	B M	− 0·9
Gold, ...	M	− 1·2
Cadmium, ...	B	− 2·4
Antimony, pure pressed wire, . .	M	− 2·8
Antimony, commercial pressed wire, .	M	− 6·0
Antimony, ordinary, ...	B	− 17·1
Antimony, crystals, ...	M	− 22·6 to 26·4
Silver, ...	M	− 3·0
Zinc, ...	M	− 3·7
Copper, electrolytic, ...	M	− 3·8
Arsenic, ...	M	− 13·6
Iron, pianoforte wire, . . .	M	− 22·6
Red phosphorus, ...	M	− 29·7
Antimony-Zinc Alloy (2 : 1), . .	B	− 99·0
Copper sulphide, ...	B	− 196·7
Antimony-Cadmium Alloy (1 : 1), .	B	− 231·9
Tellurium, ...	M	− 502
Selenion, ...	B	− 807

Thus, for example, a cobalt-lead junction might be expected to give a current of 22 micro-volts for each degree through which the junction was superheated; and similarly a lead-iron junction would give almost the same E.M.F., but the current would flow in the opposite direction. To determine from these figures the difference of potential between any other pair of metals, it is only necessary to deduct the stated E.M.F. of the more negative metal from that of the other, for example:—

(1) When both metals are positive : a nickel-palladium pair should give a current of $15\cdot5 - 6\cdot8 = 8\cdot7$ micro-volts per degree passing from nickel to palladium through the junction. (2) When one is positive and the other negative : a nickel-zinc junction should give $15\cdot5 - (-3\cdot7) = 15\cdot5 + 3\cdot7 = 19\cdot2$ micro-volts, flowing from nickel to zinc. And (3) When both are negative: a platinum-silver couple should show $-0\cdot9 - (-3\cdot0) = 3\cdot0 - 0\cdot9 = 2\cdot1$ micro-volts, the current starting from the platinum. If the junction be superheated 100° instead of 1°, these numbers must of course be multiplied by 100; thus the nickel-zinc pair should yield a current of 1,920 micro-volts or 0·001920 volt, so that it would require 521 of these couples to give the E.M.F. equal to that of a single Daniell-cell.

Neutral Point.—A peculiarity of these couples has already

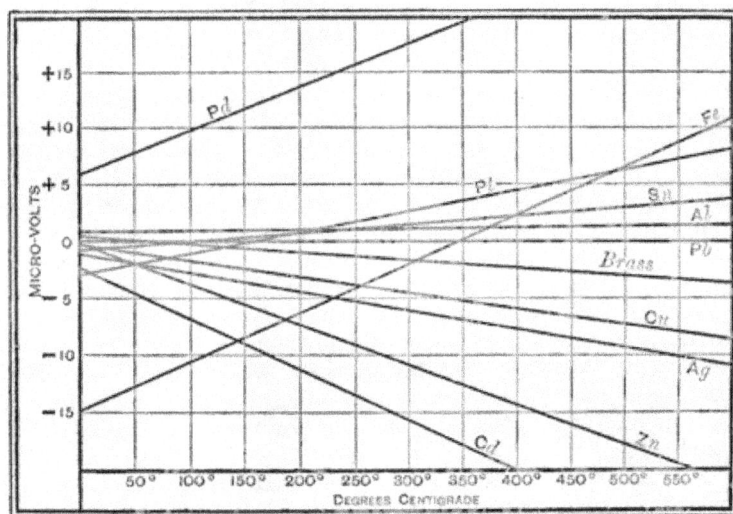

Fig. 25.—Thermal E.M.F. of metals at different temperatures.

been mentioned, namely, that the potentials often vary disproportionately to the rise of temperature. It may so happen that a pair of metals giving a very low E.M.F. per degree at ordinary temperatures, will give a comparatively high E.M.F. at high temperatures; while another couple which started well may gradually fall off as heat is applied, until at a certain temperature (varying with the metals, and known as the *neutral point*) there

5

is no E.M.F., and, therefore, no current; yet, on continuing the application of heat, an E.M.F. is again set up, but it now tends to drive the current in the reverse direction. The behaviour of metals in this respect is best seen by reference to fig. 25. Lead is here adopted as the standard of comparison because, unlike most other metals, hot lead in contact with cold lead produces no difference of potential. To ascertain from this table the effect upon the electro-motive force of a rise in temperature of 1° C. at any given temperature, it is only necessary to find the vertical line corresponding to the required temperature, and then, noting the points at which the sloping lines representing the two metals cross it, to read off the difference in micro-volts between them with the aid of the scale on the left-hand margin. Thus, for example, an increase in temperature from 0° to 1° in the warmer junction of an iron-copper couple, increases the electro-motive force by $-1 - (-15) = 14$ micro-volts; but a similar rise from 100° to 101° only gives $-2 - (-11) = 12$ micro-volts; and from 200° to 201° only $-3 - (-6) = 3$ micro-volts; in all these cases the current flowing from copper to iron through the juncture. But a rise of temperature from 260° to 261° produces no current, this being the neutral point of these two metals (their respective lines cross here), while if the two junctions of the copper-iron pair were at the temperatures of 350° and 351°, there would again be an E.M.F. of $0 - (-5) = 5$ micro-volts, but the current would now be flowing from iron to copper. With some pairs of metals there is practically no neutral point; palladium and iron give a constant rise in potential for each increment of temperature, while with others the neutral point is below the zero on the Centigrade scale; so that the higher the temperature to which the one joint is heated, the greater the efficiency of the pair—e.g., palladium and cadmium.

Further, the total E.M.F. of a circuit may be readily calculated from this table. Of course if the rise in potential were strictly proportional to the increase in temperature, it would be only necessary to discover the difference in temperature between the two junctions of the metal, and multiply it by the number of micro-volts stated per degree; but such a procedure would evidently give very misleading results. To estimate the total E.M.F. from the table, find the diagonal lines representing the two metals, ascertain the *mean* temperature of the two junctions of the couple, and mark the points at which the two diagonal lines respectively cross the vertical line representing the mean temperature, then multiply the number of micro-volts between these two points, by the number of degrees difference between the two temperatures, and the result is the required E.M.F. Thus, it may be required to know the actual E.M.F. of an iron-copper couple of which one junction is at 200° C., the other at 100° C. The mean temperature is 150° C., and the difference of potential between the metals at this point is (per degree) – 2·5 – (– 8·5) = 6 micro-volts. The required E.M.F. is, therefore, 6 × (200 – 100) = 600 micro-volts or 0·0006 volt.

Another most important point, clearly brought out by this table, is that when the neutral point of any pair of metals is above zero Centigrade, not only does an increase of temperature yield no proportionate return in the value of the E.M.F., but as soon as the temperature of the hot end has exceeded that of the neutral point there is an actual decrease of current, because the E.M.F. at the two junctions *are* tending in opposite directions. A single instance—that of lead and iron—will serve both to illustrate and explain this phenomenon. The neutral point of these two bodies is approximately 350° C.

TABLE VI.—Illustrating the Total E.M.F. Produced by an Iron-and-Lead Couple at Different Temperatures.

Temperature.		Total Micro-Volts produced under these Conditions.	Increase of Total E.M.F per 100°.
At Cool Junction.	At Hot Junction.		
Centigrade. 0°	Centigrade. 100°	12·8 × 100 = 1280	Micro-Volts. 1280
0°	200°	10·7 × 200 = 2140	860
0°	300°	8·6 × 300 = 2580	440
0°	350°	7·5 × 350 = 2625	45 (per 50°)
0°	400°	6·4 × 400 = 2560	– 65 (per 50°)
0°	500°	4·3 × 500 = 2150	– 410
0°	600°	2·1 × 600 = 1260	– 890
0°	700°	0 × 700 = 0	– 1260

The maximum current is thus produced at 350°, thence it decreases to zero at 700°, and would then begin to increase again, but in an inverse direction. The important bearing of this inversion upon the choice of couples for the production of thermo-

electric currents, and on the degree of heat to which they should
be submitted is at once evident.

Fig. 26.—Simple thermo-
electric couples.

Mode of Arranging Thermic Couples.—
A variety of thermopiles have been
proposed, differing in the metals employed, in the construction, and in the
manner of applying the heat. The essentials are that the two metals shall be as
far apart in the thermo-electric series as
possible; that there shall be no chemical
or physical reason (too great fusibility
or oxidisability) against the use of either;
and that they shall be conveniently arranged so that the one point of juncture
may be superheated as compared with
the other. The simplest method of arranging a series of couples, so necessary where
each pair produces but an infinitesimal electro-motive force, is
that shown in fig. 26. The rules which apply to the fitting up
of galvanic batteries in parallel or in series hold good equally
with thermo-electric batteries.

Clamond's Thermopile.—The thermopile most commonly used
is that of *Clamond*, in which metallic iron is united to an alloy of

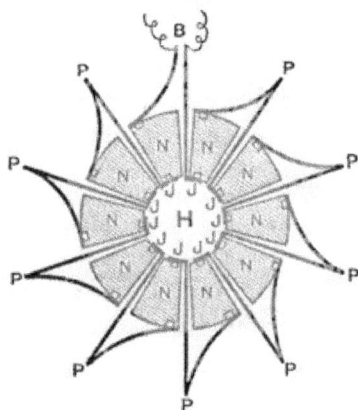

Fig. 27.—Ten thermo-electric couples
in circular form.

antimony and zinc (combined
preferably in the ratio of their
atomic weights, viz.:—122 : 65
or nearly 2 : 1). Fig. 27 shows
the arrangement of ten couples
in circular form: strips of tin-
plate, P, are generally employed for the iron element,
and good contact between the
two metals is ensured by bending the strip into a narrow loop
at one end, placing this portion
in a mould, and pouring the
melted alloy around it, so that
it is actually embedded in the
casting; thin plates of mica are
inserted externally between the
metals to insulate them at all
except the desired points of contact. Each block, N, may be about
$2\frac{1}{2}$ inches long by $\frac{3}{4}$ inch thick, but the dimensions may, of course,
be greatly varied; they are arranged radially with a central space

H, between them through which the heat is applied to the junctions, J. The different pairs are arranged in series by bending the free end of each tin-plate strip and soldering it to the external edge of the adjoining block next in regular succession, a break, B, being left at one point by which connection with the external circuit is made. By this arrangement the points to be heated are brought to a central focus, while the cooled junctures are placed at the outside, where they are most free to radiate away all excess of heat conducted through the metal blocks. Generally there are 10 pairs in the circle, and several tiers of similar circles are piled one upon another, but insulated from each other by a layer of cement composed of powdered asbestos moistened with a solution of potassium silicate. All the pairs in each row are in series, and each tier is placed in series with the preceding one, so that the resulting electro-motive force of a pile of ten tiers of ten blocks each would be 100 times that of a single pair. The source of heat is usually coal-gas, which is burned at jets from a perforated earthenware tube placed upright in the central cavity. In some forms of this pile, the top of the cavity is closed, and a thin sheet-iron tube is inserted between the centre pipe and the metal couples, leaving a space above, so that the products of combustion are led down through the annular space outside the iron lining, and are passed away through a flue at the bottom, the object being the economy of a greater proportion of the heat from the burning gas. Charcoal or coke may be substituted for coal-gas when necessary. A battery of 6,000 pairs, arranged in series and heated by coke, has been found to give an electro-motive force of 109 volts (equal to 100 Daniell-cells) with a total internal resistance of 15½ ohms.

Noé's Pile.—In the *Noé pile* a cylindrical bar of the zinc-antimony alloy is soldered to stout German-silver wires; the end to be heated is fitted with a brass cap, through which passes a stout copper-rod. The juncture of the metals is heated, not by direct contact with the flame or with heated air, but by the conduction of heat along the copper rod, the free end of which is thrust into the flame. The arrangement of a single Noé pair is shown in fig. 28, together with a sketch of the method of fitting several couples in series

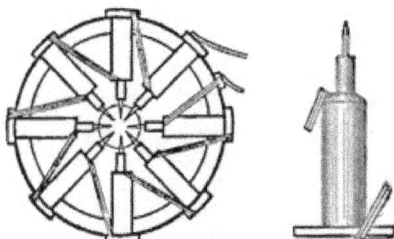
Fig. 28.—Noé pile.

around a central flame. The German-silver combination is greatly preferable to that with iron, and the electro-motive force of a single pair of the Noé pile may in consequence amount, it is said, to even $\frac{1}{15}$ of a volt (by the application of sufficient heat), with an internal resistance of $\frac{1}{10}$ of an ohm.

But even at best the thermopile is very wasteful, and it is questionable whether more than 1 per cent. of the heat-energy expended is recovered in the form of electricity; indeed, Fischer found by experiments, conducted in 1882, that only 0·3 to 0·5 per cent. was utilised. Nevertheless, the study of the subject is very fascinating because there is a direct conversion of heat into electricity; and although at present an almost prohibitive loss is involved, increase of knowledge may be attended by the production of a higher efficiency: the simplicity and convenience of the arrangement would probably then ensure to it a wide application as a source of electricity for electro-metallurgical purposes.

THE DYNAMO-ELECTRIC MACHINE.

In the generation of electric energy the choice at present practically lies between the oxidation of zinc in a battery, and the oxidation of coal, with a conversion of chemical force first into mechanical energy and thence by dynamo-electric machinery into electricity; but here also a large percentage of loss is incurred in the first stage of the conversion, owing to the waste which is inevitable even in the best constructed steam- or gas-engine. Where, however, a steady and constant water-supply is available for actuating water-wheels or turbines, the dynamo is an economical machine, which may convert a very large proportion of an otherwise waste mechanical power into useful electricity.

The prime phenomenon on which the dynamo depends, is the production of electric currents in closed circuits of metal which are caused to move in the vicinity, of either permanent or electro-magnets, or of wires through which another current of electricity is flowing.

Lines of Magnetic Force.—In considering the theory of the dynamo, one must not lose sight of the close connection between electrical and magnetic phenomena. By coiling a piece of wire spirally around a bar of iron, and then transmitting a current of electricity through the wire, the iron is converted into a magnet, and will continue to evince magnetic properties so long as the current passes, but no longer; and a coil of wire through which a current is flowing will itself be magnetic, attracting light particles of iron to itself, and tending to set north and south, even though it have no iron core; such a coil is termed a *solenoid*. If a series of light compass-needles be suspended, as in fig. 29, above a bar-magnet or a solenoid, they will be found to place

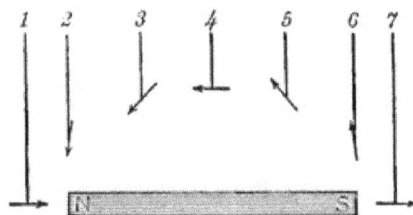

Fig. 29.—Compass-needles and magnet.

themselves at various angles as indicated; the needle marked 4, being central, would have each pole equally magnetised by the equidistant poles of the magnetic bar, and would, therefore, occupy a position parallel to it; but numbers 3, 2, and 1, being each nearer than the last to the north pole, would fall more and more within its influence, and would be less and less affected by the increasingly distant south pole, and would thus tend to set itself more nearly vertical, while the needles 5, 6, and 7 over the other half of the bar would behave similarly. Now, the position of each needle clearly indicates the direction in which the magnet is exerting its forces at that point; and if a map were made, showing the attitude of the needle towards the bar in every conceivable position, the various *lines of force* would be fully indicated. Such a map may be made by sprinkling iron filings on to a sheet of paper spread evenly above a magnet, and then very gently shaking the paper; each filing itself becomes a magnet by virtue of its proximity to the bar, and is indeed a miniature compass-needle. The lines of force thus indicated around a single bar-magnet are sketched in fig. 30. The existence of these lines of force may be thus readily demonstrated, as well as the fact that they are most numerous in the region of the poles. They are always considered to run from the north pole to the south.

Conversion of Alternating into Constant-Direction Currents.— Whenever a coil of wire is caused to cut these lines of force, so that the number passing through the coil is altered by the movement, an electric current is produced in the coil; and the greater the number of lines cut in this way the higher is the electromotive force of the current; the direction of current depends upon that of the lines of force, and upon that of rotation of the coil. A coil cutting an increasing number of lines of force running through it from left to right produces a current in the same direction as it does when cutting a diminishing number running

from right to left; but opposite to that taken by the current
produced when an increasing number passing from right to left,
or a diminishing number from left to right, are entering through
it.   Supposing, for example, the rectangle of wire, A B C D
(fig. 31) to be rotated on its axis, E F, between the poles of the
magnets, N, S, from left to right (so that A B approaches N);

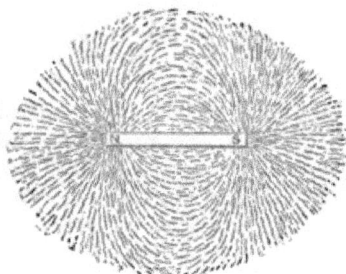

Fig. 30.—Bar-magnet lines
of force.

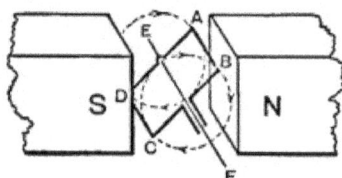

Fig. 31.—Production of currents
in magnetic field.

then as rotation proceeds the loop cuts a diminishing number of
the lines of force running across the space from N to S until it
is in a horizontal position, and the current produced during this
period will pass along the coil in the direction D C B A; now on
continuing the rotation, an increasing number of lines will be
cut, but they will be entering the loop from the opposite face
(from the face which was at first opposite S, but is now gradually
becoming opposite to N); and the result of such increase,
accompanied as it is by an inversion of the direction of the lines
in regard to the loop (the lines still run from N to S, it is only
the loop which has changed position), is to give a current still in
the same direction, D C B A.   But as soon as the coil has passed
through an angle of 180 degrees and is again vertical, but now
with A B underneath, a decreased cutting of the lines of force
begins; and as they are still threading the loop in the same
direction, a sudden reversal in the direction of current occurs
as the vertical point is passed.   At the next quarter turn, from
the horizontal to the original vertical position, fewer lines of
force pass through the loop, but once again their direction
relative to the coil is reversed, so that the current in the latter
maintains its second direction, until when the vertical point is
again reached, it is reversed again and the same cycle of
changes is repeated.   Thus the rotation of such a loop of wire

between the poles of the magnet is accompanied by the production in it of a current of electricity, the direction of which is reversed each time the coil passes through the position parallel to the faces of the magnets. Thus if the loop were extended so as to include an external circuit, the current passing through the latter must change its direction twice as often as the loop rotates; in order to avoid this constant *alternation* of the current, which would be fatal to electro-plating work, a *commutator* is added, which converts the alternate into a *constant* current of uniform direction. The commutator for a single coil,

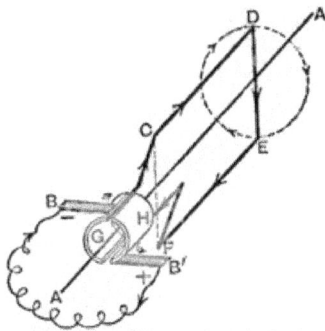

Fig. 32.—Commutator for a single coil.

C D E F, is arranged as shown in fig. 32, and consists of a metal tube split into two equal parts lengthways, the two halves being fastened around a small cylinder, but well insulated from one another; one half is attached to one end of the loop wire, the second half to the other. Copper strips or *brushes*, B B′, are fixed to the frame of the dynamo, so that they press lightly upon the split cylinder at points diametrically opposite to one another; as the tube is divided equally, and as the brushes are parallel, it follows that whatever the position of the cylinder, both brushes are never in contact with the same section simultaneously. To the brushes are fastened the wires connecting with the external circuit, so that any current generated in the coil flows first to one section of the split tube, thence to the brush in contact with it, and so around the outer circuit to the other brush, then through the second section of the tube to the other end of the coil of wire, and so completes the circuit. Now, provided the rotation of the coil around its axis, A A, be always in the same direction, and both the brushes and the breaks in the commutator tube be at the highest and lowest points in the circle, at the moment when the reversal of current takes place, the current in the main circuit flows continually in the same direction; because the brushes come in contact with different sections of the commutator directly the current is reversed, and the same brush is always connected with the descending side of the coil, the other brush always with the ascending portion, no matter whether it be C D or E F. Supposing, for example, the coil is rotating from left to right as indicated, C D is now descending and brush B is touching

section G, and the direction of current may be supposed to be
(C-D-E-F-H) - B' - circuit - B - (G-C), &c.; this direction continues
until a half-revolution is completed and the side E F is uppermost, now as E F begins to descend, the current tends to flow in
the coil in the reverse direction, thus, F E D C; but at this
instant the commutator also shifts, and section G is now in
contact with the brush B' (instead of with B) and H with B;
now, then, the current flows through the circuit in the direction
(F-E-D-C-G) - B' - circuit - B - (H-F), &c., and so on. It is very
clear, then, that in spite of alterations in direction of the current
in the coil, the current invariably flows from the brush B' to
the outer circuit, and from the outer circuit through the brush
B, to which ever section of the commutator happens to be touching it; B' is, therefore, as much the positive pole of the system
we have been describing, as the binding-screw attached to the
platinum plate is the positive pole of the Grove-battery. As a
matter of practice, the brushes are usually given a "lead," that
is, they are slightly in advance of the true vertical diameter of
the coil, for reasons, connected with the mutual reaction of the
currents in the magnets and in the coils, into which it is unnecessary to enter here.

Parts of Dynamo.—In the actual dynamo the number of
these coils is multiplied, and the methods of arranging them
various, the whole system being known as the *armature*. Thus,
in electric generators of this type, there are two main parts,
the *field-magnet* and the *armature* (including the commutator);
and the different types of machine are made by varying the
construction of one or both of these. It should be noted here
that the electro-motive force of a dynamo may be raised either
by increasing the number of lines of force cut by the rotating
rings (*i.e.*, by using a stronger magnet), or by increasing the
number of coils in the armature, or by increasing the speed of
its rotation.

Field-Magnets.—First, as to the *field-magnets*: originally a
coil of wires was rotated in front of the poles of a steel permanent magnet of horse-shoe form; the lineal descendant of
this machine is the small magneto-electric apparatus used for
medical purposes. But the difficulty in obtaining any large
store of magnetism in permanent magnets soon caused their
abandonment in favour of the far stronger magnets made by
circulating a current of electricity around bars of soft iron,
which being more powerful have a greater number of lines of
force, or, as it is termed, a stronger *magnetic field*. All the
dynamo-electric machines belong to this second class, but the

manner in which the magnetisation, or *exciting*, of the soft iron core is conducted is variable. There are four principal methods of construction—

1. *By Separately Exciting the Field-Magnets.*—A current from a battery or from another dynamo circulates around the soft iron core and thus magnetises it; and the whole current generated in the armature is used in the outside or *main circuit*. Such a machine is equivalent to an impossibly strong magneto-machine (with permanent magnets), and has the additional advantage that the extent of magnetisation, and hence the strength of the magnetic field, is not only quite constant, but is under perfect control by altering the volume of the exciting current.

2. *The Series-Wound-Dynamo.*—The whole current generated in the armature passes, not only through the main circuit, but also through the wire surrounding the field-magnets. By this arrangement, if the resistance of the external circuit be increased, the volume of current is, of course, diminished, and this reacts on the field-magnets by exciting them less, and thus producing a weaker magnetic field. The primary action of the machine is due to the small amount of residual magnetism, which remains even in the softest iron long after the exciting current has ceased. As soon as the armature acquires sufficient speed, the immensely weak magnetic field, resulting from the residual magnetism of the iron, causes a minute current to be set up in the coils; and this current in passing around the exciting wires of the field-magnet adds slightly to the magnetism, and thus increases the magnetic field, so that a stronger current is produced in the armature, which again intensifies the magnetic field; thus a constant action and reaction takes place, resulting in a continuous increase of current-strength, until the maximum capacity of the machine is reached. The whole action is, however, complete within a few seconds of starting the machine. A great disadvantage of this class of machine for use in electro-metallurgy is that, if by any cause the dynamo is stopped without first breaking the circuit, the opposing electro-motive force from the plating-vats (*vide* p. 29) causes a current to pass through the wires surrounding the field-magnets of the dynamo, and produces a current in them which overpowers the residual magnetism of the iron-core and reverses the poles of the machine: the consequence of this is that when the dynamo is again started, the current passes through the vats in the wrong direction. Such a result is, of course, fatal to the work of plating, so that the greatest care is needed to guard against this catastrophe when series-dynamos

are employed. It is usual to place in the circuit a cut-out, which automatically breaks the current as soon as it has diminished to a given volume.

3. *The Shunt-Wound-Dynamo.*—In this a part of the current generated passes to the main circuit, while a smaller portion is passed or *shunted* around the field-magnets, which is accomplished by making the wires around the magnets to have a high resistance as compared with the external circuit, and branching off the one circuit from the other, so that the current divides between the two in inverse proportion to the resistances (see p. 42). Hence any addition of extra resistance in the main circuit diverts a greater proportion of the current through the shunt wires of the magnet, intensifying the magnetic field and causing a larger amount of current to be generated, in spite of the higher resistance in the working portions, which is an advantage in dealing with irregular external work, as the machine becomes self-compensating. The starting of the machine is due to residual magnetism in the same way as in the series-dynamo.

4. *Combination- or Compound-Wound-Dynamos.*—By combining any two of these methods in one machine, as, for example, making the magnetisation of the iron dependent partly on series-winding, partly on shunt-, the adaptability of the dynamo to divers purposes is practically unlimited.

Armature.—The *armature* is also capable of modification; it may be a *drum-armature*, made up of a number of coils like that shown in figs. 31 and 32, wound on a rotating cylinder or drum;

Fig. 33.—Ring-armature.   Fig. 34.—Pole-armature.

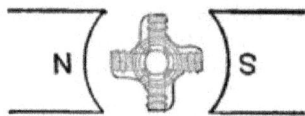

it may be a *ring-armature* (as in fig. 33), in which the wire is wound in smaller coils around a rotating ring; or it may be a *pole-armature*, in which it is coiled around projecting pieces of iron, as shown in fig. 34; or fourthly, it may be a *disc-armature*, with the coils placed close against the faces of a rotating disc. The armature-coils are usually in series in order to increase the total electro-motive force; the different loops might be attached separately, each to the ends of a copper plate, but the arrangement would then be like a number of voltaic cells in parallel arc, the total electro-motive of which is equal only to that of one cell. But by connecting all the coils in a continuous series, as shown for the ring-armature of fig. 35, all the different electromotive forces generated in each coil at any moment, according to the number of lines of force cut by it, are added together and make up the total electro-motive force between the brushes, B, B'. That the number of magnetic lines of force cut by a coil

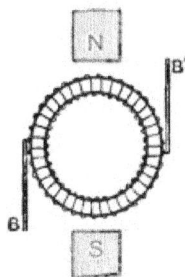

Fig. 35.—Ring-
armature.

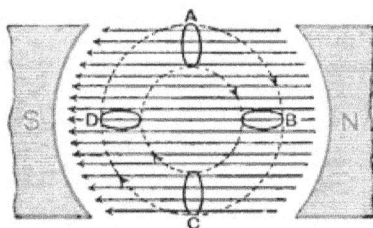

Fig. 36.—Variation in lines of force
cut during rotation.

varies at different periods of its rotation, and hence that the E.M.F. produced by the passage of the same coil through equal angular distances in different parts of the field also fluctuates, will be clearly seen by the diagram (fig. 36). Here the coils A and C, when moved through (say) 5 degrees from their present position, are moving almost parallel to the lines of force, so that in this medial position between the magnets there is almost no current produced ; and it is at these points also that the reversal in the direction of the current is observed. But the coils B and D, when moved through an equal angle from their position on the magnetic axis, are meeting the lines of force nearly at right angles, and thus cut the greatest possible number in a given period. These points, B and D, in the magnetic axis are, therefore, the centres of greatest current-production in a given coil, while A and C, perpendicular to the lines of force, are those of least energy, and all other points between them fall into a regularly graduated series ; so that, starting from the locality of no current-production, A, the electro-motive force in each successive coil increases until the maximum is reached at B, when it diminishes to zero at C, and rises to a second maximum at D, finally falling off to nothing again at A. But supposing the current to be flowing through the coils on the right-hand side of the armature-ring, from a brush placed at A to another at C, it will also flow downwards from A to C on the left-hand side, because, although the direction of current is reversed at C, so also is the position of each coil reversed in relation to the magnetic poles, so that now a reversed current flows practically through a reversed spiral. Thus the current flows in a parallel fashion from one brush to the other through the two halves of the ring, and the electro-motive forces generated in both halves are identical, if the armature is symmetrically wound, and are equal to the sums of those produced in the consecutive coils. Thus the points A and C, where the two parallel currents meet, are those of highest aggregate electro-motive force, so that they are in every way marked out as the positions for the brushes. As, however, it would be mechanically inconvenient to place them at the outside of the ring, it is usual to connect the circuit at intervals with the different sections of the commutator, as shown diagrammatically in fig. 37; the rings are thus divided into a convenient number of groups, and each group (not each ring) is joined to the collector, which consists simply of a series of copper bars insulated from one another, and fastened to the axis of the ring, with one group of coils connected to each ; on this axis the brushes press lightly at the points of least action but highest E.M.F. The brushes only touch two diametrically opposite commutator-bars at the same time, and are thus insulated from all the others, so that the current generated cannot find its way to them except by passing through every coil upon the half-ring. This explanation has referred only to ring-armatures, but the other forms are similarly dealt with.

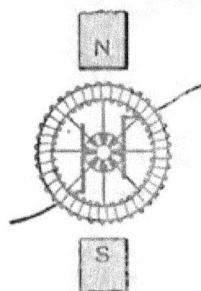

Fig. 37.—Modes of connecting groups of coils to commutator.

For electro-metallurgical purposes the essential recommendation of a dynamo is, that it shall be able to produce a perfectly-steady current of low electro-motive force but large volume. In order to accomplish this, a small number of turns of stout wire or rod (because of its low resistance) is made to take the place of the many windings of wire, which are required to yield the high electro-motive force needed in electric lighting. The electro-motive force is reduced because there are fewer coils, the current volume is increased because the resistance is more than proportionately diminished ; but the product obtained by multiplying the number of ampères by that of the volts (or the number of *watts*, as this product is termed) may be the same in the machine before and after alteration.

A few dynamos used in electro-metallurgical work may be

briefly referred to; after the general principles above set forth, no attempt will be made to give a detailed description of each machine. A few selected drawings are also given, but for the full description of the different instruments reference should be made to Thompson's *Dynamo Electric Machinery*, or other standard works.

Wilde-Dynamo.—The *Wilde-dynamo* (fig. 38), which was one of the first practically used, was made with a small magneto-electric machine (with permanent steel magnets) as an exciter, the current thus produced circulating around the field-magnets of the dynamo beneath, and so generating the stronger current in the armature of the larger machine. Both exciter and generator were united on the same stand, and could be run from the same shafting by using suitable pullies and belts.

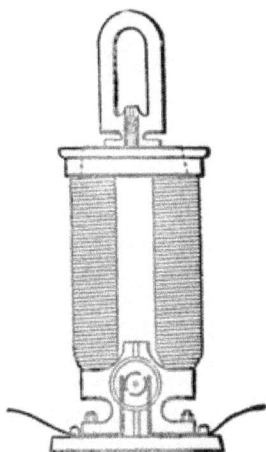

Fig. 38.—Wilde's first dynamo.

Here, the horse-shoe form of magnet was adopted, which has since been retained in a few types of machines, but has been replaced in many by magnets with *consequent poles*.

Such a magnet is made by winding a soft iron coil in one direction (see pp. 92, 93) along one-half of its length, and reversing it in the other, so that the two extremities of the magnet are of like polarity, while the centre has the opposite polarity induced from each end of the bar. Fig. 39 illustrates diagrammatically the principal parts of a shunt-wound-dynamo having field-magnets with consequent poles. FM are the field-magnets wound with wire in reversed directions as described; A is a ring-armature

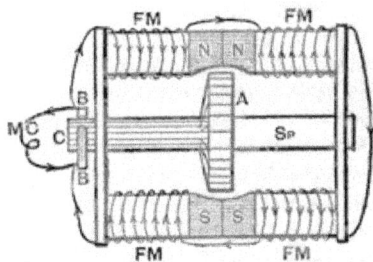

Fig. 39.—Magnet with consequent poles.

with wires leading from each coil to the corresponding section of the collector, C, from which the current is withdrawn by the brushes, B. From the brushes the main circuit, MC, is taken together with a thin shunt-wire required for exciting the electro-magnets. The collector and armature are rotated by the spindle, Sp, which is attached to a pulley outside the machine not shown in the sketch.

Gramme-Dynamo.—A common form of the *Gramme machine* used for electroplating, is illustrated in diagrammatic form in fig. 40. It is series-wound, has consequent poles, and is not

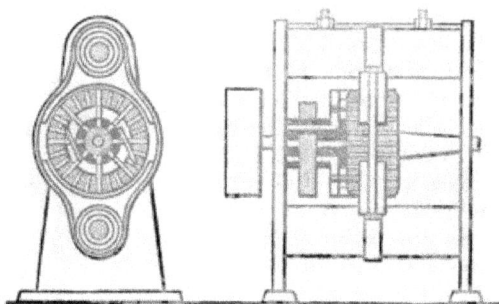

Fig. 40.—Gramme machine.

unlike the machine shown in fig. 39. The left-hand sketch shows a vertical section of the machine through the line of the brushes, the right-hand side gives an outline elevation. The field-magnets are often wound with sheet-copper, instead of wire, to reduce the resistance, and with the same object the armature-coils are of copper rod; they are, of course, more numerous than those in the illustration, in which detail has been sacrificed to clearness; the sections of the commutator must always be so close together that one comes into contact with the brush as soon as the preceding one leaves it, otherwise the current becomes very intermittent, and sparks are produced at the points of contact. In Gramme electro-metallurgical dynamos, the armature is built up of stout copper rods, insulated from one another by bitumenised paper, by laying them lengthwise upon the surface of a drum then winding well-varnished iron wire around the central portion, and enclosing the latter by a second series of copper bars attached by end-pieces to the front of one long bar, and to the back of the next, so that a continuous spiral is formed, of which the front portions are joined up to the commutator.

Siemens-Dynamo.—The *Siemens machine* (fig. 41) is also made up with thick copper armature-plates instead of wire, and in one form the exciting of the field-magnets is effected by a current passing through seven parallel turns of stout copper rod, in order to minimise the resistance. This machine also has consequent poles but uses the drum-form of armature. When wire-wound armatures are employed, the resistance is frequently reduced by arranging the wires, not all in series, but grouped in four parallel circuits.

Fig. 41.—Siemens machine.

Weston-Dynamo.—In the *Weston machine*, a diagram of which is given in fig. 42, six electro-magnets with steel cores are fixed radially in the interior of a cast-iron drum, bolted to a bed-plate also of cast iron. The wires in the alternate electro-magnets are wound in different directions so that adjacent poles shall be of opposite denomination; within the circle formed by the electro-magnets rotates a six-part pole-armature, the wires from each pair of poles diametrically opposite being united, and coupled-up to one section of a three-part commutator. The alternate magnets presenting unlike poles renders

Fig. 42.—Weston machine.

this form of machine different to that which we have been considering, in so far as there were but two reversals in the

6

direction of the current in the twin-magnet-dynamo, while here there are six, one between each pair of magnets, so that a somewhat different form of commutator is needed to rectify this *alternating* current and make it *continuous*.   Weston obviates the danger of reversal of current by the back electro-motive force from the depositing vat on the occasion of sudden stoppages

Fig. 43.—Brush machine.

Fig. 44.—Victoria dynamo.

(*vide* p. 75), by the use of steel cores to the electro-magnets, which have a greater supply of residual magnetism, and by a special form of governor attached to the base-plate, which automatically regulates the current and prevents it falling to a dangerously low electro-motive force.

Shückert-Dynamo.—The *Shückert-dynamo* is in principle similar to that of Gramme, but the armature is contracted in length until it is only a flat ring, and it is almost completely surrounded by broad iron pole-pieces (*i.e.*, continuations of the poles) attached to the magnets.

Brush-Dynamo.—The *Brush machine* (fig. 43) was at a very early date adapted for electro-plating, and in its new form is specially compound-wound, so that it shall give a current of constant potential through very wide ranges of intensity.

Victoria-Dynamo.—Fig. 44 illustrates the "Victoria"-dynamo, which is also adapted by the Brush Electric Light Corporation for electro-metallurgical work upon a large scale, such as for the refining of copper or other metals, or for electro-treatment of ores generally.

Kröttlinger-Dynamo.—The *Kröttlinger-dynamo* is used by many firms abroad ; it is shunt-wound with short wide field-magnets set upright on a base plate, with inwardly curved pole-pieces above, between which the armature rotates.

The number of dynamos in the market at the present moment is very numerous, and most makers are prepared to wind machines upon their system, suitable for electro-metallurgical work ; the above list is, therefore, in no way complete, it has simply included some of the best known and most frequently used generators of this type.

Instructions as to the Management of Dynamos.—In itself the dynamo is a simple machine and should give but little trouble or difficulty, provided that a due amount of attention is paid to its installation and superintendence. It should be firmly set on good level foundations, in a position where it is not likely to be exposed to contact with any of the liquids employed in the shops, nor to acid fumes or dust. It is best located in a dry place, enclosed in a box of wood or glass which may be easily removed for purposes of inspection, and which has openings cut in its side to provide for the passage of the engine belt or for the shaft of the pulley ; it should be in a room adjoining the plating-room, but as near as possible to the vats, in order to minimise the loss of energy due to the resistance of the wires. If erected in the shop it must be in a place removed from splashings, and must be constantly examined. It must be kept scrupulously clean in every part ; the bearings must always be well lubricated with good oil, as the speed of shaft rotation is usually very high (500 to 1000 revolutions per minute), but excess of oil, and especially leakage of oil into the armature or on to the commutator and brushes, must be rigidly guarded against. By the action of internal currents, as well as by friction, the machine is liable to become overheated, and may even be seriously damaged through the burning of the insulating material upon the wires; the temperature of the fixed portions should, therefore, be ascertained while current is being generated by feeling them carefully with the hand. With fair usage the only part needing special attention is the commutator or collector. The brushes must be perfectly flat; when they become ragged or turned up at the tips they should be clamped in a wooden holder and filed carefully; they should rest with a gentle pressure upon the copper strips of the collecting shaft; insufficient pressure combined with an uneven commutator will induce sparking at the brushes, which causes much wear and tear of the parts, whilst excessive friction will wear away the collecting rods unevenly and ruin the brushes, at the same time producing more or less copper dust, which in course of time penetrates into and injures the machine. The brushes should also touch the collector rods at opposite ends of a diameter, and the "lead," already referred to (p. 74) as necessary to the brushes of nearly every dynamo, should be rightly adjusted or else sparking will ensue. Should the dynamo show sparks at the commutator, the lead of the brushes may be altered backwards or forwards until a neutral point is found; they are generally set on a frame by which they are maintained exactly opposite to one another, and by which they may be kept in any desired position. The commutator may be very slightly greased before commencing the day's work, preferably with a little grease or vaseline applied with the finger; cloth or cotton should not be used as fibres may be left behind, and it is essential that no foreign substance find its way to this portion of the machine. Oil must never be applied in any quantity, and the grease only in very minute proportions (they add to the resistance), while black lead or mercury, which some operators have applied to the surfaces, must be avoided altogether, as the latter gradually amalgamates the copper and renders it very brittle, while the former gives a slightly conductive film to the insulating space between the bars of the collecting ring, and so impairs the electrical efficiency of the machines. The brushes must never be lifted out of contact while the dynamo is running and the current passing; the insulation may be, and the commutator certainly will be, damaged by the sparking caused by such treatment. Irregular-ities in the commutator rods, or undue pressure, or sparking at the brushes will gradually wear the shaft to an oval or uneven shape; this fault should be watched for and rectified at once by turning in the lathe to true circular section; such a course should not, however, be necessary for several years.

The dynamo may be driven by any regular source of mechanical energy, either the steam-, gas-, hot air-, or petroleum-engine (or, of course, water-power) may be used, the only requirements being sufficiency of power and uniformity of speed. Frequently spare power is obtainable from an engine used for other purposes, and may be applied with advantage, unless great irregularities in speed are caused by frequent variations in the amount of power absorbed by the machinery to which it was originally adapted.

ACCUMULATORS or *Secondary Batteries*, which are so largely used in other branches of electrical engineering, may be also employed in electro-metallurgical installations, but generally speaking little benefit will be derived from their use.

Planté-Faure Accumulator. — The great advantage of the accumulator is that it converts the electrical energy of the dynamo, at times when it cannot be directly applied, into another form of energy (chemical) which may be, as it were, stored up until a more convenient moment has arrived for its re-conversion into electricity. Several kinds of secondary batteries have been brought before the public, but almost the only one in general requisition is that originated by Planté and afterwards improved by Faure and others. It consists of two lead plates opposed to one another in dilute sulphuric acid. On passing a current through this constantly in the same direction, the anode plate, by which the current enters the solution, becomes superficially converted into lead peroxide (PbO_2) by the oxygen liberated at its surface, and this, being insoluble in sulphuric acid, remains *in situ*, the action penetrating deeper and deeper into the plate the longer the current is continued. The cathode, at which hydrogen is deposited, remains, of course, unaltered. On the cessation of the current, two plates are opposed in the solution, one of metallic lead, the other practically of lead peroxide ; on completing the circuit these act like a *primary* (or ordinary voltaic) battery, the clean lead plate (electro-positive) oxidising and receiving a deposit of the monoxide, which like the peroxide is insoluble, while the peroxide (electro-negative) gives up its excess oxygen and becomes also converted into monoxide. When both peroxide and metal have attained to the same condition the action ceases, but on re-charging by a dynamo, the original arrangement is restored, the lead oxide at the anode taking up oxygen and becoming peroxide, that at the cathode being reduced to the state of metal by the hydrogen liberated upon it. It is thus rehabilitated and is ready for use again, and this cycle of changes may be repeated indefinitely. Now, instead of forming the cell from two clean plates, the positive and negative plates are generally coated with pastes of red lead and sulphuric acid and of litharge and sulphuric acid respectively; cavities are frequently made on the surface of the lead with the object of affording a larger surface, and, more particularly, of maintaining the oxide pastes in position. Plates thus prepared behave at the outset like those which have been in use for some time, and do not require the frequent repetition and reversal of charging which is found necessary to ensure sufficient penetration of the oxide into the substance of the pure metallic sheets. In use, these cells are exactly analogous to ordinary batteries, and obey the same laws as to generation and distribution of current—indeed they are practically nothing but an unusual form of galvanic battery; but they should not be allowed to completely discharge themselves, nor should they be subjected to electrical shocks such as would occur if they became suddenly short-circuited. The electro-motive force of each cell is equal to about 2 volts.

CHAPTER IV.

GENERAL CONDITIONS TO BE OBSERVED IN ELECTRO-PLATING.

Absolute Cleanliness Essential.—In electro-metallurgical work there is one prime necessity which cannot be too often or too strongly insisted upon, and that is absolute cleanliness in every particular. Neglect of cleanliness in the electric generator, and in the connections throughout, causes the current to be deficient and the resistance of the circuit to be increased; neglect of cleanliness in making up the baths introduces impurities into the solutions, and the character of the deposit suffers accordingly; neglect of cleanliness in preparing the articles to be plated for their plunge into the depositing-vat ensures an irregular and non-adhesive coating; neglect of cleanliness after the plating is accomplished is likely to cause rapid tarnishing or rusting of the deposit. In short, want of cleanliness is the cause to which the largest proportion of electrotypers' and platers' troubles may be referred. And in respect of non-adhesiveness of a deposit, it should be noted that the articles to be treated must be *chemically clean;* a trace of grease, producible by mere handling, suffices to ruin the coating, because the precipitated metal will adhere only to perfectly pure metallic surfaces, and the merest film of foreign matter, invisible perhaps to the eye, prevents this adhesion, or weakens it to so great an extent that very slight friction may cause separation to take place.

As also Careful Adjustment of Currents.—It is further essential that the current shall be correctly proportioned to the work required from it and, if for this reason alone, it is to be regretted that old "rule-of-thumb" methods find very general favour. Apparatus for measuring the current is comparatively inexpensive, and the first outlay in instruments will soon be found to repay itself in the greater security insured against accidents due to careless work or imperfect knowledge, and in the immediate and certain indication of a failure in any part of the circuit, which may prove most detrimental to the work if allowed to remain unremedied, but which if attended to in time might exert no evil influence. An *ammeter* (= ampèremeter)

for measuring the volume of current to every bath, or to every
series of baths placed in parallel arc, with one *voltmeter* for
determining the pressure of the current as it leaves the battery,
should be used in all large installations in which it is desired
to produce good work of uniform character, or where varied
work is undertaken. It is true that an experienced workman
may know by inspection whether his baths are in good order;
but the best workman is not infallible, and the use of measuring
apparatus substitutes certainty for uncertainty, besides enabling
the foreman to ascertain at a glance the conditions of work at
any moment. Where, too, a partial failure has occurred, the
remedial measure to be adopted may often be indicated, and the
current be at once restored to its original strength by adjusting
the resistance of the circuit until the ammeter returns to its
first position.

TABLE VII.—SHOWING THE AVERAGE CURRENT VALUES SUITABLE
FOR DEPOSITING CERTAIN METALS.

METAL.	AMPÈRES.		VOLTS.
	Per Sq. Decimetre of Cathode Surface.	Per Sq. Inch of Cathode Surface.	
Antimony,	0·4 – 0·5	0·020 – 0·030	1·0 – 1·2
Brass,	0·5 – 0·8	0·030 – 0·050	3·0 – 4·0
Copper, acid bath,	1·0 – 1·5	0·065 – 0·100	0·5 – 1·5
,, alkaline bath,	0·3 – 0·5	0·020 – 0·030	3·0 – 5·0
Gold,	0·1	0·006	0·5 – 4·0
Iron,	0·5	0·030	1·0
Nickel, at first,	1·4 – 1·5	0·09 – 0·10	5·0
,, after,	0·2 – 0·3	0·015 – 0·02	1·5 – 2·0
,, on zinc,	0·4	0·025	4·0 – 5·0
Silver,	0·2 – 0·5	0·015 – 0·030	0·75 – 1·0
Zinc,	0·3 – 0·6	0·02 – 0·04	2·5 – 3·0

A current which is either too strong or too weak gives unsatisfactory deposits, the most suitable strength in any instance
depending upon the nature of the metal which is being deposited
and that of the bath from which it is separated. A very weak
current causes a slow and in some cases a bad deposit, while
a very intense current renders the metal non-adhesive, and may
even reduce it to a pulverulent form. In the preceding table
are given the quantity and electro-motive force of the currents
which have been found by careful operators to be best suited

for depositing the commoner metals; but such general state-
ments must be regarded only in the light of a guide; every bath
will be found to have its own most suitable current value, which
will probably differ but little from those given above, but which
may be readily determined, once and for all, by the use of the bath.

Electro-Chemical Equivalents.—In dealing with electro-depositing arrangements, Ampère's fundamental law that the current
in ampères is proportional to the electro-motive force in volts
divided by the resistance in ohms ($C = \frac{E}{R}$, see p. 57) must be
borne in mind.   The weight of metal deposited in a given time
is dependent solely on the volume (ampères) of current passing
through the solution.   A *coulomb* of electricity (*i.e.*, 1 ampère
passing for the space of one second of time), according to
measurements made by Lord Rayleigh, deposits invariably
0·000010352 gramme of hydrogen; of any other metal it will
deposit this fraction of a gramme multiplied by the equivalent
weight of the metal, that is, by the atomic weight divided by
the valency of the metal as it exists in the solution.   Thus, a
coulomb of electricity precipitates per second from silver cyanide

$$0\cdot000010352 \times \frac{108}{1} = 0\cdot00118 \text{ gramme of silver, or}$$

$$0\cdot000010352 \times \frac{63\cdot5}{2} = 0\cdot000328675 \text{ gramme of copper from a}$$

solution of copper sulphate; while from a solution of a *cuprous*
salt it would deposit twice the last-named weight of copper,
because in this class of compounds, copper is monovalent, and
the atomic weight is therefore divided by 1 instead of by 2.
The figures obtained by thus multiplying the coulomb-weight-
value of hydrogen by the equivalent weights of the elements,
are termed their *electro-chemical equivalents*, and afford the data
from which the electro-plater may determine the power which
he will require to deposit a known weight of any metal, or to
obtain a given thickness of coating upon a known area of surface.
These numbers are collected together in tabular form on p. 347,
where also will be found the weights of the commoner metals,
expressed both in grammes and in grains, which should be
deposited in one hour by a current of 1 ampère; together with
the thickness of the deposits produced in the same period of time
by a current of 1 ampère per square decimetre, and per square
inch of anode- and cathode-surface.   The number of grammes
precipitated per hour is calculated by multiplying the electro-
chemical equivalent, or grammes per second, by 3600 (the

number of seconds in the hour). The thickness of deposit is found from the weight deposited per hour, taken in connection with its volume;—a cubic centimetre of any metal weighing, in grammes, the number which represents its specific gravity (for example, the specific gravity of electrolytic copper being 8·914, 1 cubic centimetre weighs 8·914 grammes). From this table, then, may be found approximately the time required to produce either a given weight or a given thickness of deposit, provided that in the former case the strength of current, and in the second both current-strength and total area of cathode be known.

Alteration of Resistance.—Since the current depends upon the E.M.F. and the resistance; to multiply the strength of current, and hence, also, the rapidity of deposition, either the electromotive force must be increased or the resistance reduced. Of these alternatives, the latter is more readily modified than the former, and it is usual, for this reason, to introduce into the path of the current an arrangement of wires forming an added resistance, which may be thrown in or out of circuit at pleasure, and will thus produce a commensurate effect on the current volume. Any alteration of the relative positions of the electrodes in the bath produces a change in the electrical resistance, and, therefore, demands thoughtful attention, especially when measuring instruments are not available to indicate the extent of the derangement.

Whenever an anode (or a cathode) is removed from the bath, the conducting surface through which the current enters (or leaves) the solution is diminished, and the resistance is consequently increased; or when the anode is removed to a greater distance from the object being plated, the current has to traverse a more extended length of the solution, which is a very inferior conductor, and the resistance is again increased. In each case the volume of current is correspondingly diminished, and the alteration is at once detected by the retrogression of the pointer upon the scale of the ammeter, or current-measurer. It sometimes happens that the electrodes touch beneath the liquid, or become connected by a fragment of metal, broken off perhaps from a faulty anode or a bad cathode-deposit; the current then ceases to pass through the solution, and finds its way through the short circuit; electrolytic action is of course stopped, but outwardly there may be nothing to indicate the casualty, except in a few instances the behaviour of the battery, which may show signs of unwonted activity; here again the ammeter at once gives the alarm, because the diminution of resistance, owing to the substitution of a metallic for a liquid conductor, causes a great increase of current; it should be observed that this increase of current is accompanied by a greater consumption of battery-zinc, of which none is doing useful work; not only is the whole energy wasted, but the deposit may even be seriously damaged. All this serves to emphasise the necessity for applying measuring-apparatus at any time, but especially when dynamos are used, as they are more liable than batteries to be injured by short-circuiting or unfair treatment.

Again, since increase of resistance is attended by a decrease in current volume, the resistance of the circuit must be minimised in every possible way; the generator must be placed as near to the vats as may be convenient, the copper connecting-rods or leads must be as thick and as pure as possible, and all surfaces of contact, through which the current has to flow, must be perfectly clean and bright, while the electrolyte-solution itself should be chosen with a view to high conductivity. Conversely, all wires must be prevented from mutual contact, except at the desired points of connection, otherwise a short circuit or leak may be set up which permits practically the whole, or at best a large portion, of the current to return by the negative wire to the generator, not only without having done its appointed work, but with introduction of inconvenience and waste by its conversion into heat principally within the battery itself, which now imposes the principal resistance.

Resistance Boards.—The arrangements which are employed to introduce additional resistance into the circuit are usually made of wire, which must not be too thin, or they will become overheated by the passage of the current; they may be constructed of copper, brass, or German-silver wire, of electric-light carbons, or (for strong currents) of moderately thin hoop-iron. Of the three former, German-silver is the least conductive and therefore the best, brass standing second. A length of 20 inches of German-silver wire, or 10 yards of copper wire, either of No. 20 Birming-

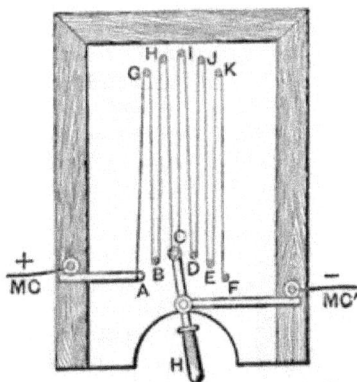

Fig. 45.—Mode of arranging resistance board.

ham wire-gauge, or a foot length of a carbon $\frac{3}{10}$ of an inch in diameter, should each afford a resistance equal to about the quarter of an ohm. Fig. 45 illustrates a convenient method of arranging several coils of wire. The circuit is broken at one point,

and the two ends of the conductor are connected to the system
of resistance wires at MC and MC'; the wire is mounted on a
wooden frame by stretching it backwards and forwards alternately over the metal pins A, G, B, H, C, I, D, J, E, K, and F,
of which A is attached to MC. The pins are so arranged that
the handle H, which alone is attached to MC', may be moved
upon its axis from contact with the MC rod at A, until it touches
the buttons B, C, D, E, or F successively. Each section of the
wire from A to B, from B to C, and so on, should have a resistance of (say) half an ohm. H consists of a flat brass rod with a
wooden holder around the free end : while it rests at A, the
current flows through it directly from MC to MC' without
meeting with any appreciable resistance, but as soon as it is
shifted to B, the current has to pass through the section of wire
passing from A over G to B, and so meets with a resistance
of $\frac{1}{2}$ ohm; by moving the handle until it rests upon C, the
resistance of the length of wire B H C is added to that of A G B,
and thus a total of 1 ohm is now inserted ; similarly each successive move adds an extra $\frac{1}{2}$ ohm until, when H is resting upon F,
the added resistance is $2\frac{1}{2}$ ohms in all ; finally, by shifting the
handle beyond the position F, the current is broken altogether.
This instrument serves, therefore, as a *switch* to start or break
the current, and as a regulator to control its strength.

Galvanoscope.—Of the measuring and detecting instruments
available, the *galvanoscope* or *detector* is one of the simplest : it
is simply a magnetic needle surrounded with a coil of insulated
wire through which the current is made to pass; it cannot well
be employed for actual measurement, but may
sometimes be useful in indicating the direction
of the current, and thus enabling the operator to
determine at a glance and without reference to
the battery or dynamo, which of two wires should
be connected to the anode of the depositing-vat.
This it does in obedience to the law, discovered
by Ampère, that a current flowing in a circuit,
placed around a magnetic needle in a plane perpendicular to that in which the latter is free to
turn, causes the needle to set itself at right
angles to the coil (fig. 46), the south pole being
on that side of the coil from which the current
appears to circulate in the direction of the hands
of a watch. This arrangement will, perhaps,
be better understood by the self-explanatory
diagram (fig. 47). But for ordinary electro-
plating currents, even a detector is unnecessary; a common

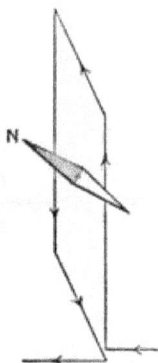

Fig. 46.
Simplest
galvanoscope.

compass-needle held in its case above or below the wire through which the current is passing suffices to indicate its direction by turning on its pivot with the north pole facing to the left or to the right. For this experiment the wire should extend from north to south so as to be parallel to the normal direction of the magnetic needle. The simple rule by which the direction of current may be determined (and this rule applies equally to the last-considered example of a coil of wire) is, supposing a man

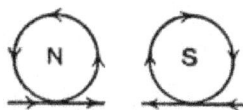

Fig. 47.—Illustrating Ampère's rule.

to be swimming with the electric current, inside the wire, head first, and with his face turned towards the magnetic needle, the north pole of the latter will set so as to be on his left hand. Prof. Jamieson's mnemonic rule is also simple and reliable here; it will be readily understood in reference to figs. 48 and 49. If

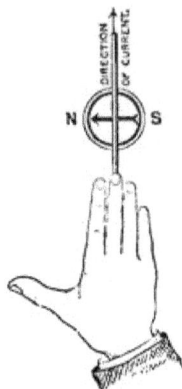

Fig. 48.

Fig. 49.

Direction of magnetic currents.

a compass-needle be placed on one side of a wire through which a current of unknown direction is flowing, and the right hand be placed on the other side of, and with the palm next to, the wire, so that the thumb points in the direction taken by the north pole of the needle; then the current will always be found flowing (from positive to negative) from the wrist towards the fingers. Thus, if the current in a wire run from north to south, the compass-needle will place itself with its north-seeking end pointing east when the compass is held beneath the wire, and pointing west when it is above it.

Galvanometers.—For measurements, elaborations of the compass-needle detector are employed, and are termed *galvanometers*. These are of many kinds, but all depend on the fact that a stronger current always causes a greater deflection of the needle than does a weak current. In its simplest form the galvanometer is like the galvanoscope; it consists of a number of coils of wire surrounding a delicately-poised compass-needle, which is attached to a thin vertical wire passing through a circular card, and carrying above the surface of the latter a light horizontal index-needle or pointer; the card is firmly attached to the coils, and is divided up into degrees, so that the angular motion of the pointer produced by a given current may be measured accurately. The angle traversed by the index is not, however, proportional to the strength of the current, and the instrument must be graduated by ascertaining the varying angles of deflection produced by currents of known strength; for the same volume of current always registers the same number of degrees upon the scale. The extent of the deflection is regulated by the resultant of the two forces—one the directive force of the earth which tends to set the needle north and south, and the other that of the current which tends to place it equatorially. In using this instrument the index must first be allowed to come to rest under the action of the earth's magnetism alone, the coils (and the card with them) are then gradually shifted until the index points to the zero of the scale, then the current is passed around the coils, and the angle through which the pointer has turned

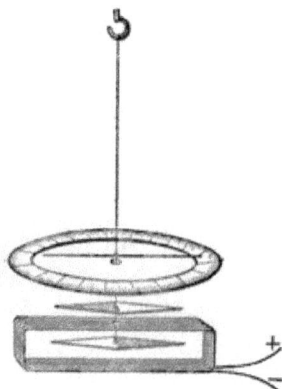

Fig. 50.— Astatic galvanometer.

is measured as soon as equilibrium is restored. It is often inconvenient to be obliged to alter the position of the galvanometer, so that the needle is initially north and south; and an *astatic pair* is with advantage substituted for the single needle; that is to say, two needles equally magnetised are fixed, one at a short distance above the other, but pointing in reverse directions, so that the north pole of the upper one lies above the south pole of the lower. The wire coil is wound around the lower needle only; the object of this arrangement being to eliminate terrestrial magnetism, by causing its opposite directive action on the two needles respectively to produce equilibrium, while it does not interfere with the relation of the needles to the current, because, although they are reversed in position, one lies above and the other beneath the upper portion of the coil, so that the effect of the current is also reversed. The method of fixing the astatic galvanometer is shown in fig. 50 : the two needles are attached to a pointer which rotates above the fixed card-board scale, the whole system of moving parts being suspended by a single filament of raw (natural) silk. This galvanometer must also be graduated.

With a coil of considerable diameter, a magnetic field of uniform intensity is created at the centre, and a very small needle balanced at that point will set, under the influence of different currents, at angles, of which the tangents are proportional to the volume of the directive currents. With such a *tangent galvanometer*, when once the angle of deflection produced by any known current-strength is determined, the value of any other current may be found without further graduation, by observing the angle through which it causes the index to turn ; the strength of the first current is to that of the second, as the tangent of the former angle is to that of the latter. Thus with a table of natural tangents at hand, this galvanometer is practically a direct-reading instrument, and is of more general utility than the astatic form.

Ammeter and Voltmeter.—Modified galvanometers are now made specially for determining the strength of current at a glance ; they are graduated on the scale in ampères, and are known as *ampèremeters*, or better as *ammeters*. By greatly increasing the number of coils in an ammeter, and with this its resistance, it becomes practically a *voltmeter*, and affords a direct reading in volts of the electro-motive force of the current employed. These two instruments are now so simple, compact, and inexpensive that they are within the reach of all users of electricity, and should certainly be included in the plant of every electro-metallurgical work. The ammeter has a very low resistance and is placed in the main circuit, thus the whole strength of the current passes through it, and may be read off at any time ; where several baths are arranged parallel to one another, and the current is divided between them, an ammeter should be included in each section, in order that the operator may be sure that every vat is receiving its due proportion of current. The voltmeter will indicate the difference of potential between any two points of the circuit, but it is usually connected as a shunt, or across from one wire to the other ("parallel" to the vats) in

close proximity to the poles of the generator; the resistance of
the coils is so great, as compared with that of the baths, that
only an infinitesimal portion of the current passes through them,
and the distribution to the baths is unaffected. Nevertheless, it
is customary to attach a small press button on a switch to the
stand of the instrument, so that it is only thrown into circuit
at the moment when the observation is made. A single voltmeter usually suffices for an installation.

Arrangement of Baths in Electro-plating.—In arranging the
baths and selecting solutions the operator must be guided entirely
by the class of work which he proposes to undertake. The
necessity for pure solutions of high conductivity has already been
insisted upon. The chemicals employed should be the purest
obtainable, and the water should, if possible, be distilled; otherwise, rain-water must be used. The solutions must be made
up carefully to the required strength, and watched well, and
occasionally tested while in use to ensure that they do not
sensibly vary in composition, or become excessively alkaline or
acid. Moreover, they must be suited to the nature of the
substance which is receiving the deposit; if this latter metal
should of itself decompose the solution, unaided by any external
current, the resulting deposit will probably be non-adhesive.
Hence, if a very electro-positive metal is to receive a coating of
one which is highly electro-negative, it should first be covered
with an intermediate metal from some solution which it cannot
readily decompose, this metal, in turn, being of such nature that
it will not break up the electrolyte of the ultimate metal to be
precipitated. For example, in coating iron with silver, a preliminary film of copper may be given in the copper cyanide bath,
and the coating of precious metal is readily deposited upon this.
The vats should be thoroughly cleaned out from time to time to
prevent the accumulation of slime, which always tends to collect,
owing to dust and accidental impurities derived from the air,
and to insoluble impurities contained in the anodes, that remain
as a precipitate in the liquid after the rest of the metal has
dissolved. The current should be switched on as soon as the
objects are introduced, or there will be a tendency for the basis
metal to dissolve into and contaminate the bath before it can
be covered with a protecting film; this frequently renders the
introduction of resistances necessary when first placing goods in
the vat, in order to avoid using an excessive volume of current,
as will be explained in dealing with silver (see p. 206).

Arrangement of Baths in Series or Parallel.—When several
baths are used on the same battery-circuit, they may be arranged

either in series, or parallel to one another; but for miscellaneous plating the latter method is superior, although, in a few instances, when the work is quite uniform, as, for example, in electrotyping plates for printers, it may sometimes be advantageous to couple the vats in series, especially if a dynamo be used as a generator. The two systems may, of course, be combined; two or more series, with two or three baths in each, being arranged in parallel arc. When insoluble anodes are employed, it must be remembered that the electro-motive force required for the decomposition will be as many times higher than is required for one couple, as there are baths in series. Two vats in series require twice the pressure needed for one, three vats demand thrice the pressure, and so on. When the anode is soluble and is made of the metal which is being deposited, the absorption of E.M.F. is not great, but the resis-

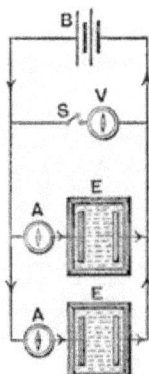

Fig. 51.—Parallel arrangement of vats.

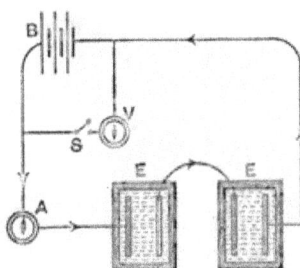

Fig. 52.—Vats arranged in
series.

tance is, of course, multiplied. Placed parallel, the resistance is diminished as the number of vats is increased, while the potential required is constant. Figs. 51 and 52 show diagrammatically the arrangement of vats parallel and in series respectively. B is the battery; V is the voltmeter, which may be thrown into circuit when required for use by means of the switch, S; A is the ammeter, of which there is one in each parallel circuit (with it may conveniently be a set of resistance wires); and E represents the electrolyte or plating-vat.

It was shown on p. 59 that battery-cells coupled in series caused the solution of equal quantities of zinc in each cell, although altogether they could not deposit more than one equi-

7

valent of metal in the plating-bath, so that five times as much zinc was dissolved to precipitate one pound of silver when five cells were placed in series as when one cell alone was made to do the work.   Conversely, in electro-deposition, a given current

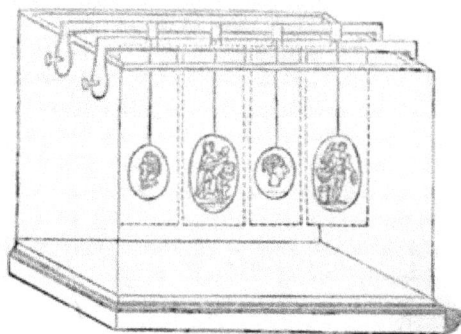

Fig. 53.—Plating-bath.   Parallel arrangement of articles.

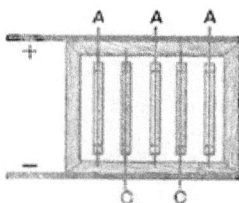

Fig. 54.—Plating-bath. Parallel arrangement of articles.

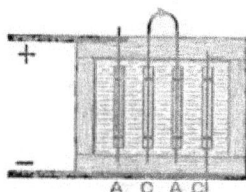

Fig. 55.—Plating-bath. Arrangement of articles in series.

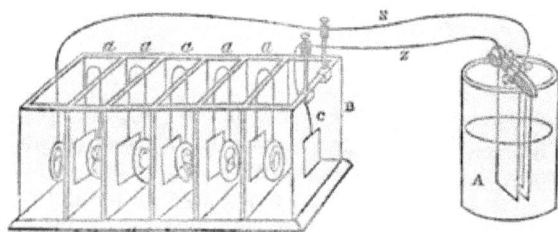

Fig. 56.—Plating-bath.   Arrangement of articles in series.

with sufficient electro-motive force will deposit in each vat as much metal per unit of zinc dissolved in the battery as would be precipitated in a single vat. Just, therefore, as in the battery coupling in series gives an increase of pressure and of power to do work rapidly, so a similar arrangement of plating-baths

involves a greater absorption of pressure, and to a corresponding
extent increases the time required to deposit a given thickness
of metal. The economy of power, apparently promised by the
precipitation of an increased weight of metal per unit of zinc
dissolved, is thus discounted by the inconvenience of a slow
deposit. A further objection to the series system as applied to
the plating-baths is that they become too mutually interdepen-
dent, for any increase in the resistance of one bath, as when a
large article is removed from it, reduces the current-volume in
the whole circuit. It is true that even when the parallel
arrangement is adopted, an alteration of resistance in one vat
also influences the current-strength of the whole system, but the
effect is scarcely appreciable owing to the large aggregate surface
presented in the different electrolytes; the added resistance in one
bath merely causes a greater volume of current to flow through
the others, while on the other hand it tends to diminish the
total current. Where, then, the articles to be plated may be of
every conceivable size, and the superficial areas are difficult to
estimate exactly, the parallel system is preferable; but when the
articles present a fair uniformity of surface, and the current has
sufficient potential, the alternative method may be substituted.
For similar reasons it is best to hang the various articles in each
bath parallel to one another, as shown in figs. 53, 54, and not in
series, so that the current entering the electrolyte by one anode,
passes to its corresponding cathode, and thence by a wire connection to the next anode, and so on, as depicted in figs. 55, 56.

Choice of Anodes.—In the matter of anodes stress has already
been laid upon the necessity for their absolute purity, and complete solubility in the electrolyte; this latter condition is, however, a question to be more particularly regarded in the selection
of the constituents of the bath. Unless under the action of the
current the anodes dissolve freely in the liquid, the latter must
become impoverished and change rapidly in composition; this is
always a source of trouble and annoyance, and where the use of
an insoluble or imperfectly soluble anode is unavoidable, small
quantities of that salt of the metal which is undergoing electrolysis (or of metallic oxide) must be added from time to time to
make good the loss due to deposition. Cast anodes will often
be found more soluble than the rolled metal, as they are more
porous and open in grain, and, therefore, more readily attacked
by the liquid; in some cases they may even be found to become
spongy and friable, a condition which should of course be avoided,
and which indicates the desirability of substituting the rolled
sheet. When there is any difficulty in obtaining pure metal for

anodes, the rolled material may generally be preferred, because the cast plates may contain a large percentage of foreign substances, which would escape detection on merely examining the exterior of the block, whereas any considerable addition of impurities would in many instances cause the metal to break up as it passed through the rolling machinery, so that the mere fact of a metal being in the form of rolled sheet is often a guarantee of at least a fair degree of purity. Cast-iron should on no account be used; it always contains three or four per cent. of the insoluble substances, carbon and silicon, with other bodies, such as manganese, phosphorus, and sulphur; some at least of these are necessary to render it sufficiently fusible to melt in the cupola-furnace. Anodes should usually be of larger size than the cathodes to which they are opposed, so that the greater surface exposed to the solvent action of the electrolyte may compensate for the slower rate of solution as compared with that of deposition.

Spacing of Electrodes—Polished Cathodes.—It has been seen that the resistance of a bath is higher when the electrodes are small, and that it increases as the distance between them is more extended; thus economy is effected by plating many objects simultaneously in parallel arc, and by minimising the distance between them and their corresponding anodes. But there are objections to approximating them too closely—first, they are more liable to come in contact or to be united by a fragment of metal, and thus to produce short-circuiting; and, secondly, if the surface of the object is irregular the deposited metal is liable to be of unequal thickness, because a current passing between points at unequal distances tends to deposit most rapidly upon those portions of the cathode which most nearly approach the anode. By increasing the space separating the two surfaces the irregularities of either have less influence on the deposit, because they are small as compared with the mean distance between them. This difficulty is chiefly experienced in electrotyping, where strong deposits of uniform thickness are required. In depositing copper when the solution is strong and at rest, small projections and striated markings may be observed, which increase in extent as the current is continued. Marks and scratches on the cathode do not become obliterated, but rather accentuated, as the metal is deposited over them; great care must, therefore, be taken that the surface to be coated is not only thoroughly cleansed, but that all tool- or file-marks are completely removed, as there is no remedy when once deposition has begun.

Homogeneous Solutions necessary.—Another difficulty inherent in the process is, that a current long continued with the solution at rest produces a gradual local alteration in the density of the latter. At the anode, metal is constantly dissolving into the surrounding liquid, which thus becomes heavier, bulk for bulk, than the remainder of the bath, and sinks to the bottom ; while at the cathode the liquid is denuded of metal, and from its lower specific gravity rises to the surface. In this manner a very gentle but sure circulation occurs in the vat, producing an undue proportion of metal in solution at the bottom, and of acid at the top. The effect of this is that thicker deposits form on the lower portions of goods immersed in the bath, owing largely to the higher conductivity of the strong solution and the greater proportion of current flowing through it ; moreover, a kind of local action may be set up in the deposited copper-plate, which is resting in two practically different solutions, with the result that the upper portion of the plate in the acid solution tends to dissolve, and to deposit a corresponding proportion of copper upon the lower half. It is probably the steady flow of liquid over the surface of the cathode which gives rise to the striated markings above referred to. The only remedy is to keep the solutions thoroughly mixed by stirring or gently shaking them in any suitable way, and this precaution should never be omitted when thick deposits are required, which necessitate a comparatively long exposure in the bath.

Having seen the causes which operate to produce failure in electro-plating, it remains to be seen how they are avoided in practice.

CHAPTER V.

PLATING ADJUNCTS AND DISPOSITION OF PLANT.

Light and Pure Air.—In arranging an electro-plating establishment, due regard must be had for light and ventilation; with insufficiency of the former bad work is almost sure to result, as it is not easy to judge when the pieces are sufficiently stripped, polished, cleaned, or quicked, and the progress of the deposition cannot be watched with the requisite amount of care; it is often necessary to stop a process immediately upon the appearance of certain signs, indicative of imperfect cleansing or the like, and it is of the highest importance that these characteristics should be noted at once, which cannot be done if the light be deficient. Badly-ventilated rooms are productive of ill-health and disease to the workmen; this maxim, applicable, indeed, to all rooms in which men live or work, must be especially regarded in rooms where batteries or cyanide plating-solutions are in use; the acid fume or spray given off by most batteries is most penetrating and injurious to health, even when considerably diluted with air, as it is likely to be found even in a large room, if it be insufficiently provided with means to carry off vitiated air and supply fresh in its place; moreover, the cyanide solution becomes slowly decomposed by the carbonic-acid-laden air of towns, and evolves the deadly prussic acid gas—in minute proportion, it is true, but amply sufficient to become prejudicial to the well-being and comfort of the operator, when breathed continuously for any considerable period of time.

Arrangement of Rooms.—At least three rooms should be available, if possible, for the purposes of the art. In the first of these the mechanical operations of cleansing and polishing are carried out; these give rise to the production of more or less dust, and should not, therefore, be conducted in the same room with the plating-vats. Again, the engine driving the machinery should be in a separate chamber, apart from the dust of the polishing-room on the one hand, and from the fumes of the vat-room on the other, the power being communicated to the lathes and other machine tools by shafting running between the two rooms. When the dynamo is the source of electric energy, it

should be placed in the engine-room, but as close as possible to
the baths in the adjoining chamber, so that there may be no
great loss of energy owing to the resistance of the copper conducting wires or *leads* to the passage of the electric current.
But if batteries be employed, they should be carefully isolated
from each of the three rooms above mentioned; if a fourth be
not available (a small one will suffice), a corner of the vat-room
should be partitioned off, the chamber or cupboard thus formed
being provided with a separate outlet into the open air, or, better
still, into a chimney, so that all fumes may be at once and completely removed; a sink and a supply of fresh water may be
fitted in this room with advantage. In the third or vat-room
are the potash- and all other baths used in chemically cleansing
the pieces, together with those devoted to plating; an ample supply of water must be available in this department, and a steam-
pipe should convey steam from the boiler—if one be used—to the
potash-tank and steam-heated vessels. Here the various pieces
of furniture should not be crowded together, but an ample
margin of space should be left so that the operator may not be
cramped for want of room. When more than one plating process
is employed, each should be kept to itself, and in large establishments a separate room may be set apart for each.

It is, of course, impossible to lay down any general plan for
the disposition of the plant of an electro-plating shop, because it
must be arranged and modified, not only to suit the work to be
performed, but also the floor-space, shape, and position of the
rooms available. But, to sum up, it may be taken as a general
rule that mechanical work should be isolated from chemical, that
the battery in the one case, or the dynamo and motor in the other,
should be separated from both, that in each room the various
classes of work should be kept distinct, but that where consecutive processes have to be followed, the arrangements for conducting
them should be so placed that, when one stage of the work is
completed, the articles may be conveniently transferred into
position for the next operation.

Drainage of Floors.—The floors should be of stone, asphalt, or
concrete, or they may be covered with lead sheet, so that they
may readily be kept clean, and be non-absorbent of the acids and
chemical reagents splashed upon them; wood, besides being constantly wet, is liable to become rotten by the action of these substances. Trapped gullies or sinks should be placed at suitable
points flush with the floor; they should not communicate with
the house-drains directly, but should discharge into a pipe which
runs outside the house, and delivers *into the open air* above a

second trapped gully communicating with the sewers or drains. In this way the floor may be kept free from accumulations of liquid, and may be readily and perfectly cleansed by flooding it with water, and then sweeping it into the gullies by means of india-rubber "squeegees" or brooms; at the same time, there is no danger of sewer-air contaminating the atmosphere of the room (a fertile source of danger), because there is no direct communication with the drain. Pipes from sinks should discharge into the open air after a like manner.

Ventilation.—To insure the purity of the air, it is not well to trust simply to the ventilation of the room by doors and windows, but a systematic arrangement should be adopted, such as that of Tobin. Several ventilators should be made immediately below the ceiling of the room, by removing a brick, passing a tube through the wall, and bending it upwards in the open air (it may be shielded from rain by a cap placed a few inches above the exit); if possible, one of these ventilating-tubes should pass into a disused chimney-shaft. The vitiated air is thus carried away, and pure air must be admitted at the floor level to take its place; this may be done by carrying two or three pipes through the wall close to the floor, and bending them up in the interior of the room to a height of five or six feet. Fresh air is delivered by them in a manner which does not give rise to draughts. Entering, cold, it flows up these pipes and gradually distributes itself over the chamber, where, becoming heated, it rises to the roof and finds an exit through the upper row of ventilators.

The guiding principles for the sanitary and safe conduct of an otherwise unhealthy occupation are expressed in a few words by saying—ensure abundance of light, water, and fresh air.

ARRANGEMENT OF PLANT FOR ELECTRO-PLATING.

The vats and apparatus used in cleansing are described in Chapter VI.

Vats.—The vats employed to hold the solution for electroplating should be considerably larger than the largest object to be coated in them, and must be made of, or at least lined with, some material which will resist the acid liquid that may be placed in them.

Glass is by far the cleanest and best, but is rarely used, except for very small work, on account of the initial expense and the risk of fracture. For very small objects glass vessels may be had in one piece as circular trays or jars; but for large articles, the bath should be made by joining five plates of sheet glass of the requisite sizes with marine glue, or white lead, or other cement, protected on the inside by a varnish made of asphalt dissolved in benzoline; or of gutta-percha in benzene or in carbon bisulphide; or, in fact, by any water- and acid-proof mixture. The glass should be supported by an outer frame of wood. Slate similarly arranged is a good substitute for glass; but though less brittle, it must still be used with care. For small work stone- or earthen-ware glazed troughs are readily obtainable and are very convenient.

Lead presents the advantage that it is readily formed into any shape. A tank made of this material should be supported beneath and around by a wooden case, so that it will not be subjected to the strain of a mass of liquid within. All the joints in the lead-lining should be made by auto-genous soldering—that is, by melting together the two edges of the lead sheet, in-stead of uniting them with soft solder, which would set up galvanic action with the lead if it came into contact with the electrolytic solu-tions. But even when united into one continuous leaden trough, the metal should be completely pro-tected in every part by a good layer of acid-resisting varnish, to prevent the de-composition of the liquid by

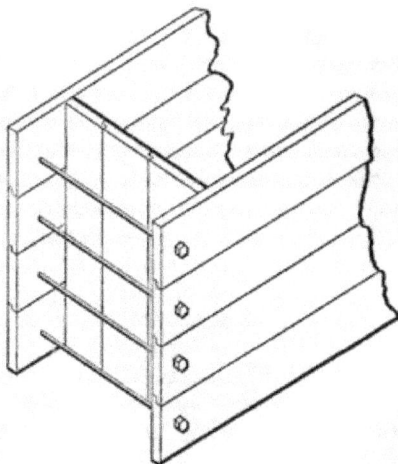

Fig. 57.—Plating-vat.

the lead through simple exchange. Iron tanks also are very largely used, and, indeed, almost universally so for hot solu-tions. These also, being constructed of a highly electro-positive material, must be carefully preserved from attack by the solu-tions, either by varnish, or better, by a good coating of enamel, which, since it is a fused complex silicate, forms practically a tank of glass, so long as it remains intact; but as soon as the enamel is chipped and the surface of the iron is laid bare, its use must be discontinued until a new coating can be given.

Wood is abundantly used, and, being a cheap material, easily worked, is especially well adapted for vats of unusual shape

which may have to be constructed for one particular class of
temporary work.   These tanks are best secured at the ends by
bolts and nuts, as shown in fig. 57, which serve to hold the sides
firmly pressed against the end pieces.   As wood alone is very
absorbent, they should be lined with gutta-percha or any other
water-proof material, and must be carefully watched so that
they may be re-lined as soon as leakage into the wood is
observed.   Wooden vats are sometimes lined with thin lead
sheet autogenously soldered, and this inner case may be varnished, or may be again lined with varnished wood.   A
mixture of 10 oz. of gutta-percha with 3 oz. of pitch and $1\frac{1}{2}$ oz.
each of stearin and linseed oil, melted together and well incorporated, has been found to afford a good protective covering
to lead or wood.

The tanks for hot solutions are best made of enamelled iron,
and may be set over a small fire-grate in which charcoal is
burned, or better—because the heat is more under control—
over a series of Bunsen's burners of the ordinary upright form,
or in the shape of horizontal rings; or they may each be surrounded with an outer jacket, the intervening space being filled
with waste steam, which is often available in large works and
may be economically applied.   Iron tanks are often made of
thin metal, and, if of large size, should be supported by strong
iron bearing bars beneath, to prevent them bulging when the
weight of the liquid is applied.

Vat-Connections.—Of whatever material the vat is constructed
it should be provided with an insulated rim around the top, to
carry the wires which conduct the current to the objects in
the bath.   This rim is best made of well-painted wood fitting
on to the top of the bath, and the outer portion should be at a
higher level than the inner,
as indicated in fig. 58,
which with fig. 59 illus-
trate the general arrangement as adapted to an iron
vat.   Around three sides
of the raised portion there
runs a short brass or copper
rod, A, ending in a binding-

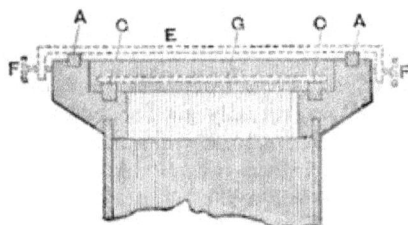

Fig. 58.—Rim of plating-vat.

screw, B, attached to the
positive pole of the battery.
Around the corresponding three sides of the inner or lower level
platform is a similar rod, C, isolated completely from A and
from the bath by the woodwork of the frame, and terminating

in the binding-screw, D, attached to the negative (zinc) pole of the battery. Resting on the two sides of the rod, A, may be placed any number of cross-rods of brass or copper, E, which can

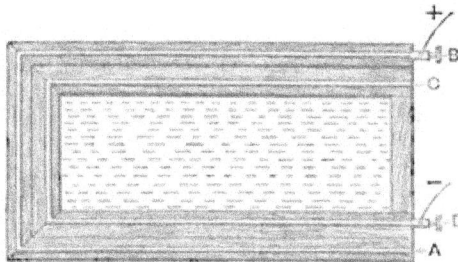

Fig. 59.—Rim of plating-vat.

be held firmly in position by the screws, F F; from these are suspended the anodes, which are thus placed in direct connection with the battery. Lying upon the lower rods, C, similar cross-bars serve to support the cathodes or objects to be coated. Any reasonable number of electrodes may be thus suspended in the same bath, the current flowing always from the anodes to the cathodes in parallel arc. When both sides of an object are to be coated, the anodes and cathodes must be placed alternately; such an arrangement has been shown in fig. 54, where A A represent anodes and C C cathodes; here both sides of the anode are used and dissolved, as the current flows from them in either direction to the cathode next adjoining. But where a deposit is required on one side only of a plate, as in printers' electrotyping, one anode may be placed between each pair of cathodes set back to back (fig. 60), so that both sides of the former are used, but only that side of the latter which is nearest to its corresponding anode.

Connection of Electrodes.— Other methods of connecting the electrodes with the battery are used, but that just described presents many advantages; thus the distance be-

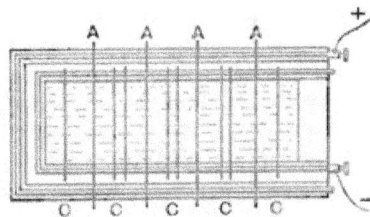

Fig. 60.—Arrangement of plates for electrotyping.

tween an anode and cathode may be adjusted at will, and either may be removed from the solution, examined and returned without disturbing the remainder, and without any manipulation of

binding-screws; the anode may be instantaneously transferred
to a cathode rod at the beginning or end of an operation, if
desired, to control the current (see p. 206); and the anode
supporting rods are held firmly in position by the external
screws—a similar arrangement, but of *internal* screws, may be
applied to the cathode rods if required.

Suspension of Anode Plates.—The anode plates are suspended
from the cross-rods by suitable hooks. The anodes may conveniently have a perforation in each of the upper corners,
through which the suspending hooks are passed; but as they
are liable to dissolve irregularly, even when every precaution
is used to ensure uniformity of liquid, the lower corners should
also be perforated, so that the plates may be suspended alternately from opposite sides. They are sometimes hung so that
they project above the surface of the liquid, and the supporting
wires, not being immersed, are, therefore, not liable to corrosion
and ultimate destruction by solution; but the plates themselves
will then have a shorter life, for they are most vigorously
attacked at the line of uppermost contact with the liquid, and

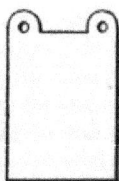

Fig. 61. — Mode
of suspending
anode plate.

will be worn away at this point while the remainder of the plate is still sound; moreover,
the portion of metal above the water line is
practically wasted. They are frequently made,
therefore, with projecting perforated lugs, either
at each corner, or only at two, as in fig. 61,
these alone project above the solution and,
while protecting the supporting wire from destruction, minimise the amount of useless anode
surface outside the liquid. To obviate the
destruction of the suspending hook, without
permitting any portion of the anode to remain above the bath,
some operators use but one side of the anode to face the objects
which are being coated. On the back reversed hooks are fastened
by which the plate is suspended, as in fig. 62; the hooks are
thus protected from dissolving by the anode which intervenes
between them and the cathode, except in the space between
the top of the plate and the surface of the liquid, which space
should, therefore, be made as short as possible.

Agitators.—The necessity for keeping the electrolyte in motion
to ensure uniformity in composition when a thick deposit is
required has been dealt with already; the manner of effecting it
must depend upon the appliances at command. Stirring by
hand is frequently relied upon, but it is liable to be accidentally
omitted, and being necessarily intermittent allows time for

partial separation to occur between two consecutive stirrings.
Mechanical agitation, which is more certain in its effect, may be
applied by such devices as working a small screw-propeller
slowly at one end of the bath; or
by blowing air into the solution
constantly, through a tube pass-

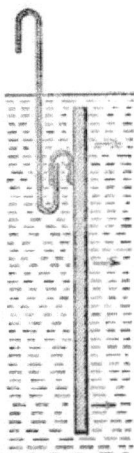

Fig. 62.—Mode of suspending anode.   Fig. 63.—Von Hübl's agitator.

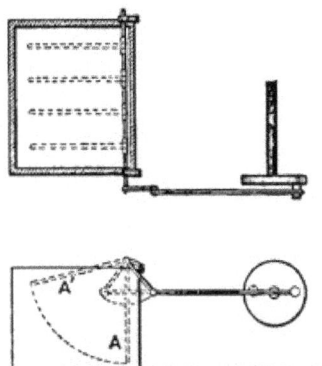

ing to the bottom of the vat, by means of a fan-blower, or by a
beating arrangement such as that used by v. Hübl in the electro-
type-baths of the Austrian Military Geographical Institute.

In this system a glass rod, A (fig. 63), is fastened to a crank
connected with an eccentric wheel or with any suitable device
for imparting a reciprocating, or to and fro, motion, so that at
each reciprocation the rod is moved through the arc of a circle,
from its original position at A to that represented by the dotted
line A', and is then returned to its first place at A.   Such a rod
is placed between each pair of anodes and cathodes, all the rods
or beaters being attached to the same crank-shaft which runs the
whole length of the vat, so that they may all be actuated by the
same mechanism.   The motion of the beaters need not be very
rapid—from 10 to 30 strokes a minute amply sufficing.   The use
of this or any similar device presupposes the existence of steam-
or water-power and machinery in the shops; where this is not
available, manual power must be substituted in connection with
any suitable mixing appliance, which must be set in motion
at frequent intervals.   Whatever motion is given must be
sufficiently vigorous to ensure thorough mixture of the solution,
but without disturbing the relative positions of anode and
cathode; and the mechanism must be so applied that it in no way
lessens the facilities for examining the progress of deposition.

Motion of Cathodes.—Silver and some other metals require a gentle motion to be imparted to the objects upon which they are being deposited. This may be done by enclosing the suspending rods of the objects within a wooden frame which is caused to move to and fro above the solution by means of an attachment to an eccentric. The frame may be suspended above the vat, or it may be caused to slide upon the edge of the bath.

Figs. 64, 65, and 66, illustrate a convenient method of fixing the sliding frame; it is supported on wheels placed at the corners, each wheel rolling upon an inclined plane, E (fig. 66), which may be set at any angle by means of a screw. The rod R thus imparts the necessary backwards and forwards motion to the system; so that, at

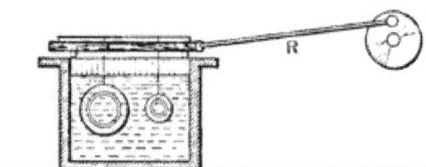

Fig. 64.—Mode of attaching sliding-frame.

every stroke, a double action occurs, and the frame with the objects suspended from it moves in a horizontal direction by virtue of the pull of the rod R, and in a vertical direction on account of the inclined plane, the extent of the latter motion being controlled by the screw which determines its angle of inclination. The frame with the objects should be caused to slide backwards or forwards once in about two or three minutes, through a distance of 2 or 3 inches.

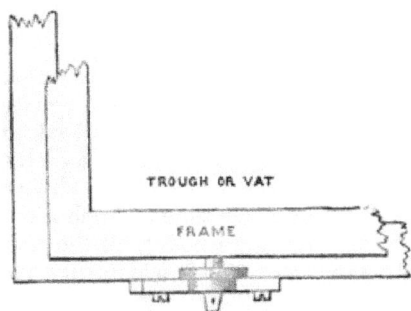

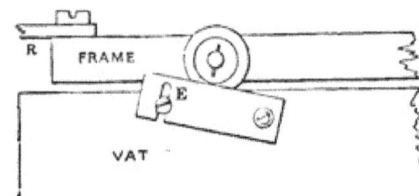

Figs. 65 and 66.—Mode of attaching sliding-frame.

Plating-Balances.— In plating with precious metals it is frequently required to deposit only a given weight upon the various articles; and although this may be approximately accomplished by calculating the time required to deposit a given weight of metal with a known current-strength and upon a known superficial area,* with the aid of the table given on p. 347, it is safer when absolute accuracy is required to use a plating-balance, by which the weight of metal deposited may be determined as the operation proceeds. In this instrument, a metal frame for carrying the cathode-rods is substituted for one scale-pan of a large balance of the ordinary description. The frame is supported from the beam of the balance by metallic connection, while the pillar of the scale is connected with the negative pole of the battery, so that electric communication is made through the parts of the balance. Having attached the objects to the frame and immersed them in the solution, they are counterpoised by placing weights in the opposite pan of the balance until equilibrium is restored, and the frame and the objects are suspended freely, the former, of course, above and the latter in the solution ; an extra weight, equal to that of the metal which is to be deposited upon all the objects in the aggregate, is now added to those already in the scale-pan, and the current is switched on until the deposited metal, just over-balancing the added weights in the pan, turns the scale in the opposite direction—an action which may be indicated automatically by causing the pointer or beam of the balance to release the hammer of a small gong, or to make contact with an electric bell.

Roseleur's Plating-Balance.—Roseleur has introduced a more elaborate balance by which the current is automatically cut off as soon as the beam of the scale is turned, so that the electrolytic action ceases directly the required amount of metal has been deposited ; this was effected by making electrical contact, not with the pillar, but by attaching a platinum wire to the arm of the balance which supports the weights, and arranging underneath it a cup of mercury connected with the negative pole of the generator, into which cup it dips to such a distance, that, as soon as the arm is raised by the reversal of the beam, the wire is lifted out of the mercury, and connection between cathode and battery is permanently broken. All the knife-edges of this balance work under mercury, in order to prevent overheating of these parts by the current flowing through them, and at the same time to lessen friction and obviate the corrosive action of acid fumes.

In the balance diagrammatically indicated in fig. 67, the current

* When it is only desired to deposit a total weight of metal, and not a certain weight per square inch, the superficial area of the objects may be neglected ; all that is required to be known is the mean strength of current applied, in ampères.

does not traverse the beam at all, but enters by a contact screw
attached to the supporting rod of the cathode-frame. The objects
are suspended in the bath
from the flat cathode-rod, C,
in the usual way; then when
these have been counterpoised
by introducing weights into
the pan, P, and the extra
weights representing the total
mass of metal to be deposited
have also been added, the
beam will turn, so that the
arm, A', rests on the stop, R',
which is rigidly attached to
the pillar of the balance to
prevent an excessive amount
of play; at the same time the
point of the screw, S, in the
supporting rod of the cathode-
frame, should just make good
contact with the block, M, at

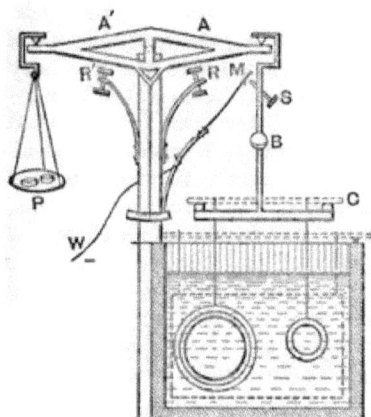

Fig. 67.—Balance.

the end of a spring, attached to a suitable fixed support, and connected with the negative pole of the battery; the spring must,
of course, be insulated from the pillar and all parts of the balance.
Through this connection the current passes from the bath to the
return-wire of the battery; both S and M should, therefore, be
tipped with platinum, which remains untarnished under all conditions, and, therefore, secures good contact. If the extremity
of S do not actually touch M, or if it press so hard upon it that
the spring is bent, adjustment may readily be made by turning
the milled head of the screw: the adjustment should be so made
that when the balance is in exact equilibrium the points are just
touching; so that when the beam rests upon the stop, R', the
spring becomes slightly bent and ensures perfect contact; but
when it rests upon the corresponding stop, R, contact is entirely
broken, and a space of at least the $\frac{1}{16}$ of an inch separates the
two platinum surfaces. The current now flows through the
circuit from the positive pole of the battery to the anodes, which
rest as usual upon the sides of the vat, thence through the solution to the cathodes upon which it deposits the precious metal;
and from the cathodes it traverses the supporting-rod, the screw,
S, and the spring, M, and returns to the negative pole by the
wire, W. As soon as the weight of metal precipitated is equal to
that added to the pan, P, the balance comes to equilibrium and

remains poised between the stops R and R'; but the current is still flowing because S is not yet withdrawn from R'; then, as soon as a slight additional amount of metal is deposited, the beam comes over and rests upon the stop, R, and, contact being broken, no further electrolytic action can ensue.   The cathode supporting-rod is made in two pieces joined by a ball-and-socket joint, B; any disturbance of the knife-edge of the balance caused by the necessary slow reciprocating motion imparted to the frame is thus obviated.   The motion is imparted by an inverted fork, running between horizontal guide-rollers, which spans the edge of the frame at the central point on one of its sides, the fork being, of course, attached to an eccentric rod; the length of the fork may be so arranged that as soon as the balance rests upon the stop, R, the frame falls out of reach of its action.

Weight-Corrections.—In using any of these balances, it must be remembered that the weight of a substance in water is less than its weight in air, and that the plated article will appear to weigh more after removal from the solution than it did when immersed; the actual weight of silver deposited by balance is, therefore, in excess of that indicated by the weights in the pan. It is quite possible to rectify this error in making the needful calculation.   The initial weight of the objects to be plated may be left out of account, because it is counterpoised while they are in the solution at the beginning of the operation, and they remain in the same liquid to the end.   But the weight of silver (or gold) deposited upon them will be less while it is in the solution than it would be outside, by the weight of an equal volume of the liquid.   For example, the specific gravity of silver may be taken at 10·6; that is to say, 1 cubic inch of silver weighs 10·6 times as much as 1 cubic inch of water; thus 10·6 ounces of silver, weighed in the air in the usual manner, would show only 10·6 − 1·0 = 9·6 ounces if it were weighed while immersed in water.   But the specific gravity of the solution is more than 1 as compared with water; regarding it as 1·1, the weight of the 10·6 oz. of silver becomes 10·6 − 1·1, or only 9·5 ounces, if counterpoised while suspended in this liquid, and this is equivalent to a loss of over 10 per cent.   Therefore, strictly speaking, to obtain an aggregate deposit of 10 ounces of silver upon a batch of articles, only 9 ounces need be placed upon the scale-pan.   For gold, similar calculations may be made, but the loss is not so great owing to the higher specific gravity of gold (= 19·3), so that the ratio of its weight in air to that in water is 19·3 : 18·3.   In the same way, for other metals the weight in the pan should be divided by the fraction :—

$$= \frac{\text{specific gravity of the metal}}{\text{specific gravity of metal} - \text{specific gravity of solution}}.$$

We are not aware that these calculations are often made in practice, and the thickness of silver deposited must, therefore, always be perceptibly greater than that intended—certainly an error on the right side from the consumer's point of view. A certain loss may be incurred in subsequent polishing-processes, but this should not be greater than would be compensated for by the small excess of metal, which is required to overcome the friction of the balance and bring the beam over to the opposite side.

Should any articles or anodes accidentally fall into the plating-vat, they should be carefully picked out by means of a long wire, bent at one end into hook-shape, or by a pair of light tongs, which may be made of brass, previously coated with the metal which is being deposited, or with some more electro-negative metal; the bare arm should not be introduced, because of the risk of blood-poisoning which may be caused by the contact of recent wounds with the plating-liquid.

For special classes of work special vats and special arrangements of all kinds may have to be made, and herein lies the

Fig. 68.—Roseleur's wire-gilding arrangement.

scope for the inventive skill of the operator. But, with a knowledge of the principles laid down in works on electro-metallurgy, there should be no serious difficulty in dealing with the various problems which may be presented.

Roseleur's Wire-Gilding Process.—Among these special arrangements, the method adopted by Roseleur for gilding wire by a single continuous process is especially interesting; a diagram of the plant is given in fig. 68. The tin-coated and well-cleaned wire is slowly uncoiled from a reel, R, and passed to the drum, D, on which the finished wire is ultimately wound, and whose rotation causes the wire to travel through the various stages of the process.   First it is passed into the hot gold-bath, E, heated by the furnace, F, and is maintained beneath the liquid by the glass rollers, G G; here, by the current supplied through the platinum-wire anodes, A A, and passed through the wire to the connection with the battery at B, the gold is deposited upon it. Passing thence, the gilt wire is guided by two wooden rollers, W, to a cleansing-bath of potassium cyanide solution; from this it is led into a vessel of clean wash-water, V, and finally between the calico-covered draining rollers, C, to the drying and anneal-ing tube, T, which is maintained at a dull-red heat by a charcoal-furnace.   Several parallel wires may, of course, be passed through the same process simultaneously.

Wire gauze, or fabrics of any kind, provided that they con-duct electricity, may be similarly coated by passing them beneath a roller in the depositing trough, and winding them finally upon a reel or drum.

Wagener and Netto's Doctor.—A device invented by Wagener and Netto for coating surfaces which are too extensive to immerse in a solution, may well be noted as an application of ingenuity to the solution of a difficult problem, although it is really only a modification of the apparatus long since known as a "doctor," which is used for gilding the lips of ewers and the like (see p. 221).   To a hollow wooden handle, H (fig. 69), is attached a circular anode, A, of the required metal perforated in the centre, and connected by a wire, W, with the posi-tive pole of the battery; in close contact with the lower surface of A is a flannel pad, E, held in place by the brass tube, T, which passes through the length of the handle and, being connected with the india-rubber tube, R, conveys the liquid electrolyte from any convenient receptacle to the pad, E.   This pad must be kept constantly wet, and the flow of solution is regulated by the clip, C, upon the india-rubber tube.   The surface to be coated is connected with the negative pole of the battery; now, by brushing the apparatus over the surface, electrolytic connection is made

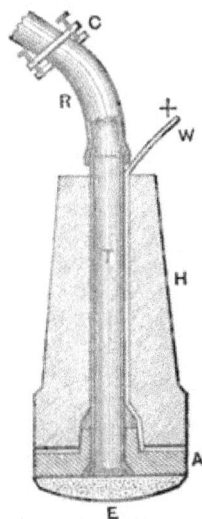

Fig. 69. — Wagener & Netto's "doctor."

between it and the anode, A, through the solution with which the pad, E, is wetted ; thus the electrolyte is decomposed and deposits its metal upon the required object, the thickness of film being regulated by the time during which the handle is held over each portion in succession.   As the electrolyte is used up, fresh liquid is supplied through the central tube.   For very irregular surfaces a long-haired brush, with short anode-wires admixed with the hairs, may be substituted for the sheet-anode and flannel-pad.   Care must be taken, of course, as in all electro-depositing work, that the surfaces are perfectly clean before attempting to coat them.

CHAPTER VI.

THE CLEANSING AND PREPARATION OF WORK FOR THE DEPOSITING-VAT, AND SUBSEQUENT POLISHING OF PLATED GOODS.

In this chapter it is proposed to treat of those adjunctory processes and arrangements which, not being purely electrolytic, and being common, moreover, to all branches of deposition, are less conveniently dealt with in the chapters devoted to electro-plating with the metals individually.

Objects must be Clean.—We have constantly urged that too much stress cannot be laid upon the necessity for absolute cleanliness in all operations, but this is especially to be observed in the preparation of the plate or object to receive the deposit, because the merest speck of tarnish, oxide, or grease—such as may result from merely fingering it—suffices to prevent the adhesion of the coating-metal at the points affected; the whole work may be ruined, the deposit may have to be removed, and the precipitation repeated from the beginning. Cleansing pro-cesses are, therefore, essential in every case; frequently, as when a bright deposit of a hard metal (*e.g.*, nickel) is sought for, it is necessary also to give the highest possible polish to the articles before immersing them in the depositing-vat. There are two systems of cleansing, chemical and mechanical, which must be varied to suit the metal which is to be coated.

Removal of Grease.—This is usually the first operation to be performed. Large amounts of oily or fatty matter should be removed by rinsing thoroughly in two washes of benzene. The benzene must be kept in tightly-closed vessels maintained in a cool position, as far as possible from any furnace or from arti-ficial lights (excepting, of course, incandescent electric lamps) on account of its rapid evaporativeness and dangerously-inflammable nature. Benzene and benzoline exert merely a solvent action upon the grease, and so extract the greater portion of it. On removal from this liquid the goods should be well rubbed with a soft cotton cloth if they were originally very dirty, and may often receive a further dip into fresher benzene with advantage. The last trace of grease is removed by a more chemical treatment,

such as immersion in boiling caustic potash solution. If the pieces
were not in bad condition at the outset, the mere application of
the potash may suffice, indeed it may be used alone in any case,
but the time required for the treatment must then be extended
in proportion to the amount of fatty matter to be removed.

The organic dirt is sometimes removed by heating the objects
to dull redness on charcoal, or, pre-
ferably, in a *muffle*-furnace, which
consists of a small fire-clay oven
(fig. 70), about 6 to 12 inches long
by 4 or 5 inches wide and high, in
which the objects are placed, so
that they do not actually come into
contact with the fuel, the burning
charcoal or coke being built up around the muffle in a suitable
furnace, and there is, therefore, no liability of their becoming
injured by the impurities from the heating agent. All grease
and other organic matter is thus completely destroyed, but the
surface becomes covered with a film of oxide which must be
thoroughly removed in the subsequent acid dip; this usually
imparts a dull or frosted surface to the metal, so that polishing
should be resorted to before plating if a bright surface is
ultimately required, especially when the covering metal is so
hard that it will not itself admit of polishing.   After this
polishing the objects must be rapidly passed through the potash-
vat, and again through the acid-baths, to ensure perfect cleanliness
prior to the actual plating operation.   The heating process being
really one of annealing, rolled, hammered or worked metal
becomes softened and so far altered in character that many
articles, such as spoons, forks, or dessert-knives, which are
purposely left unannealed, and, therefore, hard, are rendered
practically useless ; this alteration is indicated by the loss of the
characteristic metallic ring, that should be emitted by striking the
object in its hardened condition.   It is, of course, unnecessary to
remark that bodies which have a low fusing point, such as tin,
lead, pewter, Britannia metal, and the like, cannot be submitted
to the fire treatment.   This method is not, therefore, of universal
application, and, indeed, is rarely, if ever, to be preferred to the
other processes here described.

The potash-tank is best constructed of wrought-iron, which
may be heated by setting it over a gas-stove or charcoal fire, but
more conveniently, when a supply of steam is available, by
forcing the steam through a spiral or series of iron pipes placed
within the vessel and close to the bottom.   As the final stages

Fig. 70.—Muffle-furnace.

of the cleansing process should be conducted as close to the depositing-vats as possible, so that no time may be wasted between the operations, it is frequently convenient to utilise steam, which has been previously employed in a steam-jacket or coil of pipes, for heating the latter vessels whenever the plating-solution is to be worked hot. In this case the pipes in the potash-tub may be perforated, so that the steam and condensed water may pass into the liquid itself, which is thus maintained in constant motion, while the amount of water lost by gradual evaporation is compensated for. Such a system of heating compares favourably with the direct application of solid, liquid, or gaseous fuel, inasmuch as there is a greater choice of materials from which to construct the containing vessel. When external heating arrangements are adopted, the vat must be of fairly thin metal, to minimise the loss of heat-energy, and must be carefully supported to avoid strains from the weight of the liquid; but with internal steam-heating by coils of pipes, whether closed or perforated, any material which will resist the chemical action of the hot alkali is available, and the tank may have any convenient size, thickness, or shape. An extension of the steaming system, much to be recommended, is to continue the solid-walled pipe from the potash-vat into a second vessel, there discharging the steam from the orifices of a perforated pipe into the water, which is used for rinsing the pieces after removal from the alkaline solution.

The cleansing liquid is a 5 or 10 per cent. solution of caustic potash or caustic soda in water; if these be not available, sodium carbonate (washing soda) may be substituted, but its effects are not comparable with those of the caustic alkali, and it demands a far more protracted steeping of the articles to be cleansed. Since the caustic alkalies gradually become carbonated by the absorption of carbonic acid from the air, and since they also become gradually saturated with greasy matter, forming soapy substances by combination with them, fresh potash (or soda) must frequently be added to restore the strength of the bath; this precaution must be most carefully observed, as the bath soon becomes useless for its work, and the process then begins to give unsatisfactory results. To check this formation of carbonates by exposure to the air, the potash-vats should be closed with a tightly-fitting cover whenever they are not in use, and are not being heated. The caustic alkalies readily attack tin and lead and allied metals, and for this reason a prolonged potash dip must never be given to any objects made entirely, or in large part, of these metals (pewter, Britannia metal, and similar alloys), or to any which are joined by tinman's soft solder, which is an alloy of tin and lead ; if applied at all to these bodies, the dip must be momentary ; but whenever practicable it should be avoided altogether, inasmuch as a rapid dip does not afford sufficient security for that absolute cleanliness and freedom from grease which is quite indispensable. For such metals, and for any which it is undesirable to heat to the temperature of the boiling alkali bath, a very thorough treatment with benzene or any simple solvent of fat may suffice ; or the pieces may be carefully rubbed by means of a brush with a paste of very finely-crushed whiting and water, taking care that neither the fingers nor any greasy material be allowed to come into contact with the cleaned surfaces.

In using the potash dip, the small pieces should be slung upon copper wires if convenient (fig. 71), or held in perforated bowls of glazed earthenware (fig. 72), or in a copper wire sieve (larger articles are suspended singly from suitable hooks), and must be allowed to remain in the bath, with frequent agitation, for a space of time ranging from half-a-minute to ten minutes or a quarter of an hour, according to the amount of impurity to be removed. On completion of this treatment, the articles are immediately plunged into a large volume of hot water ; and, after this

Fig. 71.—Copper dipping wire.   Fig. 72.—Earthenware dipper.

preliminary rinsing, are rapidly and thoroughly washed by passing through successive wash-waters, which may be used cold ; they should then be transferred at once to the acid-baths, and thence to the plating-vats. From the moment they enter the potash-tank, they must on no account be touched with the fingers or with any surface that may be oily or greasy. The reason is not far to seek. Oil and water are not miscible, and if even the smallest patch be ever so slightly greased water will be repelled from that point, so that the acids cannot effect the work required of them, nor will the electrolytic solution be able to come into contact with the metal surface, with the result that this portion will be left uncoated, or at best covered with a non-adherent deposit which forms on the greasy matter.

The object being removed from the potash-vat, and now unstained with grease, the liquid in the plating-baths can make the most intimate contact with its surface in every part; and, therefore, when the current is passed, the required metal may be deposited uniformly over the whole article; but it would even then be found to have little or no adhesion to the surface covered, and would strip off when subjected to a very moderate amount of rubbing—most probably during the final polishing process. All objects which have been exposed to the air for any appreciable time after polishing are covered with a film of tarnish or rust, which is insoluble in potash and may be so thin as to be imperceptible, but which, by interposition between the two surfaces, prevents true metallic contact and consequently destroys adhesion. This film must, therefore, be most carefully detached, which is best accomplished by immersion in a bath of acid, or of a solvent suitable to the metal under treatment. The acid dip constitutes the final preparation for the plating-bath, and should leave a chemically-clean surface for the reception of the coating metal. After washing, tarnishing again takes place by unnecessary exposure to the air; and it is thus important that the object should be transferred with the highest rapidity from the acid-bath to the wash-waters, and from these to the electrolytic solution.

Copper, Brass, German Silver, and other alloys, of which the chief constituent is copper, are, as already indicated, cleaned and brightened by dipping into acid after removal from the potash solution. The common nitric acid of commerce, which emits red fumes and is known in the trade as *nitrous acid,* is very generally used for the acid dip; it consists of nitric acid in which lower oxides of nitrogen are dissolved, and the waste nitric acids from used Grove's or Bunsen's battery cells are not unsuitable for the work. It is used undiluted, the articles to be dipped are slung on wires or in the baskets already described; baskets of platinum-wire mesh are sometimes used for small articles, and as they are quite unattacked by the acid, and therefore neither weaken nor contaminate the liquid, they are to be preferred to all others; the prime cost alone is against them, but they soon repay this expenditure by their indestructibility and by the greater comfort and saving in their use. Since the acid exerts a powerful action upon the metal of which the articles are made, and the film of oxide is instantaneously

cleared from the surface, the plunge into acid must be but
momentary; the objects are then rapidly transferred to a vessel
containing a large volume of water, and are afterwards washed
again and again, the last time in quite clean water. It is false
economy in electro-plating to stint wash-waters at any stage.
Another acid dip, sometimes used in preference to the above,
is made by mixing together little by little and with the utmost
care, equal volumes of strong sulphuric acid and water, and
adding to each gallon of the mixture one pint of the common
aqua fortis (nitric acid) of commerce. The subsequent manipulations are similar to those already described. A dead or dull
surface may be given by plunging the article, previously dipped
in nitric acid, into Roseleur's mixture of 200 parts of yellow
nitric acid, 100 of oil of vitriol, and 1 part of common salt,
together with from 1 to 5 parts of zinc sulphate (white vitriol),
which should be made up the day before it is required for use.
After remaining in this liquid for several minutes (from five
to fifteen or even twenty) the articles are removed, plunged
momentarily into a bright-dipping bath, to restore a certain
amount of the brilliancy which is generally too thoroughly
removed in the dead dip, and are then rinsed as usual.

A bright lustre is given by placing for a few seconds in a
mixture of the yellow nitric acid and oil of vitriol in equal
parts, with 0·5 per cent. of common salt; like the last this
mixture should be made up the day before it is required, or
earlier, in order that it may be quite cold when required.
Instead of this solution, a mixture of 1½ to 2 parts of sulphuric
acid with 1 part of an old nitric acid dip is frequently used;
after mixing these acids, they should be put aside to cool,
and decanted from any crystals of copper sulphate which may
be formed by the action of the vitriol upon the copper contained
in the old *aqua fortis* (due to the partial solution of the objects
that have been immersed in it). Small proportions of hydrochloric acid or of sodium chloride, which by contact with the
vitriol liberates hydrochloric acid, or of lamp-black are sometimes added to this solution.

Potassium cyanide, dissolved in ten times its weight of water,
is often used instead of the acid dip for brass, especially when
it is essential that the original polish upon the article should
not be destroyed, as in the preparation of the objects for nickel-
plating. A longer immersion in this liquid is to be recommended, because the metallic oxides are far less readily soluble
in this than in the acid dips. In all cases the final cleansing
in water must be observed.

Iron and steel articles, which have been thoroughly polished, are dipped first into the boiling potash solution, and then, when thoroughly cleansed from grease, into a pickle consisting of water containing 10 per cent. of sulphuric acid or 25 per cent. of hydrochloric acid, as the nitric acid liquids used for copper and its alloys would have too powerful an action upon the more oxidisable metal iron. The pieces must not be allowed to remain too long even in the baths recommended, or they will show traces of the action in the pitting and unequal solution of the surface. When the metal is covered with oxide or scale, as after forging or heating to redness, more vigorous measures are needed to effect its removal; the objects are first suspended for two or three hours in a bath of dilute sulphuric acid (2 or 3 per cent. of acid for wrought-iron or steel, about 1 per cent. for cast-iron), which dissolves a little of the oxide and loosens much of the remainder, so that after washing well with water it may, for the most part, be detached by scouring with very fine sand. A second dip into the acid usually removes the last portions of scale; but, if necessary, the process must be repeated until the pieces are perfectly clean.

Zinc is first passed through the potash-bath, which exerts a distinct solvent action upon it, so that the process must be expedited, and is next dipped for a few minutes into water containing 10 per cent. of sulphuric acid; the mixture of acids used for bright-dipping brass (50 sulphuric acid, 50 nitric acid, and 0·5 salt) is sometimes used, but as it violently attacks the zinc, the operation of dipping must be rapidly effected, and the subsequent washing must be immediate and thorough. After treatment by either of these processes, scouring with fine sand and clean water must be resorted to, which has the effect, *inter alia*, of obliterating the black lines that indicate the position of solder after dipping in the acids.

Lead, tin, Britannia metal, and the like are very rapidly passed through the potash-bath, or, as already described, are scoured with whiting and water subsequent to cleansing in benzene, if this be necessary by reason of extreme greasiness. They should be polished finally by rubbing with lime, even after the potash dip; they then require only to be well rinsed in water to be ready for electrolysis.

All acid pickles used for different classes of work should be kept distinct from each other, so that one metal may not be dipped into a solution containing a more electro-negative metal, which would deposit upon it by chemical exchange. For example, zinc or iron must not be immersed in a pickle which is used for cleansing copper articles, because a certain amount of copper gradually dissolves into the liquid as successive objects are dipped, and this copper tends to deposit upon the more electro-positive metals afterwards brought into contact with it.

Quicking.—Many articles are "quicked" before being sub-jected to the operation of depositing other metals, especially silver and gold, upon their surfaces. This simply consists in giving them a superficial amalgamation by the deposition of a thin film of mercury, in order that many metals, which alone would deposit the coating-metal from the plating-liquors by simple immersion, may be rendered practically incapable of so doing; the resulting deposit being more adhesive and of better quality in consequence.

But quicking is frequently resorted to in order to increase the adhesiveness of deposited metals on objects which would have no action on the bath; for the mercury, being but little liable to tarnish by oxidation, retains a bright surface when exposed to the air for a period which would suffice to produce a film of oxide upon an unquicked surface, and thus prevent adhesion. Moreover, the solvent action of the mercury on both surfaces (especially on gold and silver) tends to unite the two metals in the most intimate contact, and may even, by interamalgama-tion, form a superficial alloy which would thus make adhesion perfect.

The principal metals so treated are copper, brass, German silver, and the like, previous to gold- and silver-plating, and zinc prior to nickeling. The quicking-solutions more commonly used are—the per-nitrate or proto-nitrate of mercury, the strength ranging from 1 to 2 ounces per gallon (Roseleur recommends 1 part of mercuric (per-) nitrate and 2 of sulphuric acid to 1000 parts of water); or the cyanide of mercury, which is made either by adding a solution of potassium cyanide to one of a mercury salt (nitrate, chloride, or sulphate), until no further precipitate is produced, allowing this cyanide of mercury to subside, washing it two or three times with water by alter-nately stirring it up, allowing it to settle and pouring off the clear liquor from the precipitate, then dissolving it in a further quantity of potassium cyanide and diluting with water; or by dissolving mercuric (per-) oxide directly in potassium cyanide solution.

The objects are merely dipped into these solutions, when metal is superficially dissolved from them and mercury is deposited in its place by simple chemical exchange. The duration of the dip, never much more than momentary, is governed by the amount of mercury to be deposited, which in turn depends upon the thickness of the object to be plated and that of the coat to be applied. Usually a thick object and a thick coating demand, or at least permit, a heavier mercury deposit than thinner ones, which are more liable to become brittle, and for which a mere momentary immersion will suffice. The character of the basis-metal* also influences the time required for mercury-deposition, inasmuch as the electro-positive metals have a more rapid action than those which are more electro-negative. Zinc, especially, requires careful quicking, as it not only deposits the mercury with rapidity, but is very readily penetrated by it, and is rendered brittle in consequence. The quicking-bath should be of such a strength that copper plunged into it becomes immediately covered with a silvery-white metallic film. The use of old or dilute solutions should be discontinued when they begin to yield a dark or almost black deposit of mercury, which is worse than useless, because the electro-deposited metal refuses to adhere to it. If the liquid be too strong, or contain too large an excess of free nitric acid, a similar result obtains; it is, however, easy to decide to which cause failure is to be attributed.

If the pieces have not been properly cleansed, the quicking-solution will give an irregular, patchy or discoloured film, instead of a clear silver-like uniform coat, owing to the presence of foreign bodies such as grease or oxide.

Mechanical Treatment.—Scouring with sand or pumice is best conducted on a wooden board placed above a tub containing the water or liquid to be used. The scouring-brush should be made with moderately-hard bristles (hog's hair is generally preferred), and is used by plunging it into the water, withdrawing it, shaking gently to remove the excess of the liquid, dipping in the powdered pumice-stone, and at once rubbing it over the whole surface of the object. To yield good results, the brush must be constantly charged with the powder, but while keeping it thoroughly moist, excess of water is to be avoided; the dipping into water and powder may thus have to be frequently repeated if the object be at all large. When the scouring is not to be followed by a potash dip, the pieces should not be touched with the bare hands,

---

* The word *basis-metal* is here applied to the metal which forms the object or base upon which an electro-deposit is ultimately to be given; the term base-metal, though more euphonious, has a second signification which might prove to be misleading.

and both brushes and powder must be examined to see that there is no trace of greasy matter attached to them.

In preparing metal for the chemical treatment previous to actual electro-deposition, the pieces should be polished at least in part; as a rule, however, it will be found that the coating-metal adheres less satisfactorily to a surface which is perfectly polished than to one which is in a slight degree, it may be almost imperceptibly, roughened. Nevertheless, the polishing must be so far completed that all marks of the file or tool are obliterated, and the whole surface has only a regular, and hence almost invisible, roughness; for it is difficult to remove file-marks or scratches afterwards from the plated article without cutting through the coating, or at least rendering it dangerously thin.

Deep irregularities must first be removed, and this may necessitate the use of a file; the marks of the latter must now be erased by rubbing with some material such as emery; and this in turn leaves finer markings, still too coarse to be left untouched, and which necessitate the use of polishing tools. Of these the most successful are those which are caused to revolve rapidly in one plane by suitable mechanism, and which not only economise time but have a perfect regularity of action and produce true parallelism of the fine lines scratched by them. It is well known that the best surface is always obtained when the polishing tool is passed over it uniformly in the same direction, and that any motion which produces cross-lines, no matter how fine they may be, and thus gives rise to cross-reflections of light, destroys the evenness of the appearance. Hand-polishing is especially liable to produce these cross-lines, and thus entails a greater expenditure of time and care.

Polishing-Lathe and Dolly.—The lathe-action is generally used for polishing; discs or *bobs* of stout leather, usually hippopotamus- or walrus-hide, about half-an-inch to one inch in thickness, and four or six inches in diameter, are rotated rapidly on the lathe-spindle by means of a treadle like that of a lathe or of a grindstone, and while thus in motion the workman with one hand firmly presses the object to be polished against the lower side of the leather bob, while with the other he allows a gentle stream of fine sand to fall upon the top of the disc as it revolves towards him from above downwards. Trent sand is usually deemed most suitable for the work; it may be used repeatedly. The pieces may be first treated with fresh rough sand to obliterate the deeper markings, and afterwards with worn sand, which has been used many times. For this purpose the bobs should make 1,500 or 2,000 revolutions per minute, but this is a high rate of rotation to be maintained by a treadle action, and is, therefore, more satisfactorily communicated by steam-power. Fig. 73 shows a bench power-spindle suitable for this purpose; the base of the stand is bolted firmly to a strong table or bench; the spindle, S, has a screw at either end, to which the bobs, B B', are firmly attached; and between the

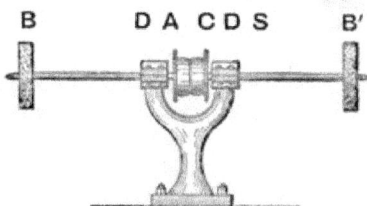

Fig. 73.—Bench power-spindle.

forked arms of the stand, which carry the bearings, D D, of the rotating spindle, are fast-and-loose pulleys, one of which is keyed firmly to S at A, so that, when connected with a large pulley on the main shafting in the shop by means of a leather belt, it acts as a driving pulley and imparts the required motion to the bobs; the other is free to turn loosely upon S at C, so that when the belt is shifted to it from the driving pulley, it alone rotates, and the spindle remains at rest. A lever must be conveniently placed to shift the belt from the one pulley to the other at a moment's notice. Two workmen may use such a tool simultaneously; one standing at B applies the coarse sand only, and when the object is thus sufficiently treated, hands it on to the second operator at B', who finishes it with the finer sand. Even now the metal is not absolutely bright, but requires a final polish with a finer material, which may be given by bobbing the article with a little fresh finely-crushed quicklime (Sheffield lime is particularly well-suited to the work) mixed with a little oil. The lime should be thoroughly caustic, and as it rapidly absorbs both carbonic acid and moisture from the air, it must be stored in air-tight closed boxes as soon as possible after burning, and should only be removed from these a little at a time as required for use. The last polish of all is given by a small quantity of lime applied without oil by the projecting edges of a series of calico rings clamped one upon another in a wooden holder with a central hole, by which it is screwed to the lathe-spindle in place of one of the bobs. This instrument is known as a *dolly*.

Scratch-brushing consists in submitting the surfaces of articles to the polishing action of a number of fine wires set on end. The wire selected for this purpose must be harder than the metal to be treated by it, or it will have little or no action, and may even cover its surface with a thin film of metal worn off from the wires by attrition; when, for example, nickel is scratch-brushed with brass wire the surface becomes quite yellow in tint; it must not only be relatively hard, but must also be actually and intrinsically rigid and stiff, so that the points shall not be readily bent over out of shape when in use. Hard-drawn thin brass wire, which may be made partially or wholly soft if required, by suitable annealing, is the usual material for scratch-brushes, but occasionally steel, or even spun-glass, may be employed for treating extremely hard surfaces.

Hand scratch-brushes are about 6 or 8 inches long, and are made by firmly binding a large number of wires in the middle so that they form a compact bundle, with the ends free for the space of half-an-inch to an inch. One end is then dipped into soldering fluid (hydrochloric acid, in which as much zinc as possible has been dissolved) and then into a ladle of melted soft solder, to firmly unite the various wires and so form a solid brush; the whole is then mounted on a wooden handle for convenience, with the free

Fig. 74.—Short wire brush.

ends of the wires extending beyond the handle, as in fig. 74. Occasionally a double length of the wires is taken, and they are simply bent over upon themselves and bound round the centre, leav-

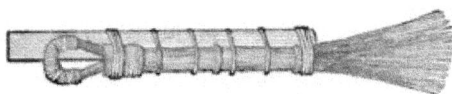

Fig. 75.—Long wire brush.

ing a loop at one end, and are afterwards mounted as before on a wooden handle (fig. 75); but it is less easy to produce regularity in the laying of the wires by this means. Either of these brushes may be used upon small surfaces; for large areas, the wires are

Fig. 76.—Wire scrubbing-brush.

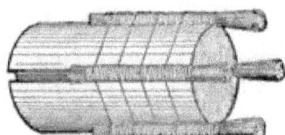

Fig. 77.—Machine brush.

mounted in handles in the form of a scrubbing-brush, as shown in fig. 76.

When the motion is to be applied to the brushes by machine power—and this method is to be preferred to hand labour—a number (4, 6, 8 or more) of hand-brushes may be mounted on

the periphery of a wooden drum, as indicated in fig. 77 ; or a
better form is made by setting the wires radially in a circular
handle, so as to form a disc of from 5 to 6
inches in diameter (fig. 78). Both forms of
circular brush are mounted on the spindle of
the lathe, or on that driven by machine
power, shown in fig. 73.

For surfacing the interiors of vessels, a
brush of the shape indicated in fig. 79 is
useful, or on an emergency an old hand
scratch-brush may suffice, the wires of which
have become turned over at the points.

Fig. 78.—Rotary
machine brush.

In using any of these brushes, a liquid
lubricating medium is employed; this is most
generally stale beer, but many other liquids,
such as crude tartar dissolved in water, diluted vinegar, or decoction of soap-wort, are
supposed by some operators to produce a
better effect, and are, accordingly, substituted
for it. The brushes are useless when the
ends of the wires have turned over upon
themselves; if they cannot then be straightened by means of a wooden mallet, the extreme
tips must be cut off by a sharp metal chisel.
The circular lathe-brush should be mounted

Fig. 79.—Wire
brush for in-
ner surfaces.

upon the spindle, sometimes on one side, sometimes on the other,
so that the direction of rotation is reversed, and the wires strike
the object alternately with different sides of their surfaces ; thus
the latter are not unduly bent in one direction. It is essential
that the wires be kept in good order, and an occasional dip into
potash to remove grease, or into the acid dip to remove oxide,
may have to be resorted to.

The hand-brush is used by holding it in the same manner as,
and imparting to it somewhat the motion of, a paint-brush. The
lathe-brush is mounted upon a spindle, and should be arranged
with a small reservoir above to contain the lubricating fluid, a
small pipe with a tap serving to conduct the solution from this
to a point immediately above the rotating brush, upon which
the drops gradually fall ; the piece is held firmly underneath the
brush, but slightly on the side nearer the operator, so as to meet
the wires as they descend. Around the brush is a metal screen
to prevent splashings produced by the rapid rotation, and
beneath it is a tray with an overflow pipe conducting to a
receptacle placed below, to retain the waste solution for use again.

9

The operation of scratch-brushing is had recourse to after deposition, in order to brighten the dull deposit; sometimes even at intervals during the process to secure a good coating; sometimes beforehand to brighten the object finally before immersion in the plating-vat. Whenever it is used prior to or during deposition it is obvious that every trace of the lubricating liquid must be washed away before placing, or replacing, the article in the bath.

Burnishing.—This is a process applied to finally polishing silver and some other deposited metals, and consists in rubbing the whole surface under considerable pressure by a very hard and, at the same time, highly-polished surface; it may be effected after scratch-brushing the articles, or is often used as a substitute for this latter operation. The burnishing tools are usually made of steel for the first or *grounding* process, and of a very hard stone, such as agate or blood-stone, for finishing.

These tools must be kept in the highest degree polished by rubbing them vigorously with very finely-crushed crocus or rouge-powder on a strip of leather, fastened upon a piece of wood which is placed in a convenient position upon the working bench. The burnishers are of various shapes to suit the requirements of different kinds of work, the first rough burnishing being often accomplished by instruments with comparatively sharp edges, while the finishing stages are accomplished with rounded ones. The annexed sketch (fig. 80) illustrates a few of the patterns commonly employed. Soap suds may be used to lubricate and moisten the burnishers.

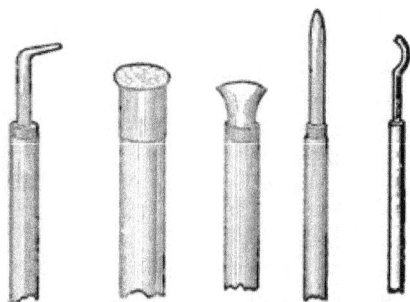

Fig. 80.—Burnishers.

Silver-plated goods may be readily polished by submitting them to the action of the lathe-bobs, such as those already described, or of wood covered with leather, or of brushes, upon which is maintained a small quantity of tripoli-powder mixed with a few drops of oil. The last polish is given, either by the application of rouge by constant rubbing with the fleshy portions of the hand, or by a dolly in which swan's-down is substituted for the calico between the wooden clamps (see p. 127), using with it the finest possible paste of rouge-powder, entirely free from gritty matter, which would destroy rather than improve the existing polish.

The results of scratch-brushing and burnishing are quite different, and each system has its own special advantage. Electro-deposited metal is always crystalline, however close the texture may be ; and being thus made up of an aggregation of minute crystals, the light falling upon it is not evenly reflected, but is more or less scattered by the varying facets of the crystalline surface; and thus, although metallic, it has a *dead lustre*. Scratch-brushing followed by buffing, or bobbing, has the effect of very slightly flattening the projecting portions downwards upon the surface, but mainly of grinding them off until they are level with the lowest portions, and so a perfectly even and uniform surface being produced, light is reflected as it were from a mirror; but no practical alteration of the physical condition of the coating results. Burnishing, on the contrary, scarcely effects any grinding of the irregularities, but rather produces the level surface by flattening the raised portions into adjacent cavities, so that the pressure exerted tends to fill up any pores or inter-crystalline spaces, and so to yield a more solid coat. Thus burnishing produces a denser, more durable, and more solid covering, but the colour and general appearance is somewhat less satisfactory, possibly because the irregularities are merely rounded off and not entirely effaced, so that the surface is not so absolutely true as that yielded by good buffing and dollying.

Steel, which is too hard to be polished by the methods given above, should be rough-polished with the emery-wheels, then *glazed* by the action of bobs of wood covered with leather, to which a mixture of the finest emery-powder with oil is applied. A strip of a soft alloy of lead and tin is sometimes substituted for the leather upon the bob. The finishing polish is administered with the best crocus-powder.

For *nickel deposits* the object should be so thoroughly polished before plating that the minimum of work shall be necessary afterwards, owing to the extreme hardness of the deposited metal. Careful treatment with the Sheffield lime bob above described should then suffice to bring up the finest possible surface upon the pieces.

CHAPTER VII.

THE ELECTRO-DEPOSITION OF COPPER.

It is frequently necessary to give a coating of copper to metals ;
chiefly to those, such as iron, which are more electro-positive,
occasionally with the object of imparting to them the external
characteristics of copper, but more often in order to enable them
to receive a good deposit of a less electro-positive metal. But
by far the most extensive application of electro-plating with
copper is to be found in electrotyping, or obtaining *facsimile*
copies of various objects for the use of the printer or sculptor.
The (acid) copper solutions present fewer difficulties in management than perhaps those of any metal, permitting at once a wider
range of current-strength and a greater variation of bath-com-
position.

COATING BY SIMPLE IMMERSION.

Iron.—Iron is practically the only metal that is coated with
copper by simple immersion, and only small articles of this body
are usually so treated. An acid solution of copper sulphate,
made by dissolving about 2 ounces of the blue salt in a gallon of
water, and adding about $1\frac{1}{2}$ to 2 ounces of sulphuric acid, may
be employed with advantage, but considerable latitude is permissible in the proportions adopted. The deposition of the
copper upon the surface of the iron is almost instantaneous, and,
indeed, a long exposure in the solution produces a slimy precipitate which has almost no adhesion to the basis-metal ; such a
deposit, Roseleur recommends, should be mechanically consolidated and attached by rolling, if the metal be in the form of
sheet, or by passing through the dies of a wiredrawer's plate, if
it be in the condition of wire. Before dipping any iron or steel
article into the copper solution, it must be thoroughly cleansed
by plunging it consecutively into the caustic alkali liquids and
the suitable acid dips, described in the last Chapter ; then,
when thoroughly brightened by the acid, it is immersed in the
copper-bath.

Steel Pens.—Steel pens may be coppered superficially by
treatment in the liquid already described, but are more satisfactorily coated by thoroughly stirring them, after cleansing,

in sawdust moistened with a solution of half an ounce of copper sulphate with a like weight of sulphuric acid per gallon of water. The mixture is usually effected in a barrel or drum mounted upon a horizontal axis. The long hexagonal drum outlined in fig. 81 is a convenient arrangement for this purpose. It is mounted so that it may be turned on its horizontal axis, the pins at either end resting in bearings upon the upright supports ; one side is hinged, so that it may be opened to admit or discharge the damp sawdust and pens, and when closed is held in position by a suitable catch. An improvement upon this form may be made by substituting short lengths of tube for the pins at the ends of the drum, and instead of causing them to rotate within bearings, passing a fixed rod completely through the drum, so that the tubes turn upon this rod which is held firmly by the uprights, and which carries fixed arms within the drum. Thus, on rotating the latter, the contents are turned over, and are more thoroughly mixed together by the arms or beaters stationary within it. The fixed rod and beaters should be made of brass or of iron completely sheathed in copper. In using the apparatus, it is first half-filled with the moistened sawdust,

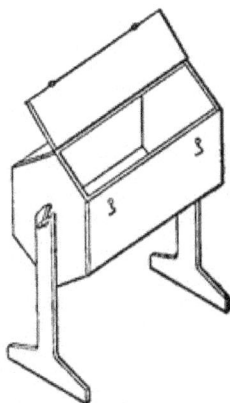

Fig. 81.—Mixing-drum.

then the pens are introduced ; the lid is closed and fastened in place, and the drum is rapidly rotated on its axis for a few minutes, until it is judged that every pen has been thoroughly coated, when it is stopped with the door at the lowest point, and this being opened allows the contents to fall upon the floor. The mixture is now placed on brass sieves, the mesh of which is of such size that it passes the sawdust through, but retains even the smallest articles that have been treated ; the sieves containing the latter are now plunged twice or thrice into fresh water, and the washed pens are transferred to a second rotating drum, in which they are dried by contact with hot, clean, and dry sawdust, which is subsequently separated from the finished nibs by means of sieves.

Other solutions for coppering by simple immersion have been recommended, and notably those of Kopp, who coats iron in cupric chloride containing a little nitric or hydrochloric acids; and Puscher, who treats brass by exposing it to a solution of copper sulphate and ammonium chloride.

Obviously in all these processes the deposition of the copper is due to an exchange of a more electro-positive metal (*e.g.*, iron) for the copper contained in the solution; thus the latter gradually accumulates a large quantity of iron, while it loses a corresponding amount of copper (56 of iron being equivalent to 63·5 of copper); and for this reason the bath must be watched to ensure that it is maintained at approximately the right strength.

The simple immersion-process is not strictly electrolytic, but merges into a *single-cell* process when, as by Weil's method, a piece of zinc is placed in contact with the metal to be coated, to facilitate the deposition of the required metal from a solution which is tardy or inactive.

SINGLE-CELL PROCESS.

Weil's Process.—Weil's process, which he has used for coppering cast-iron pieces, even of large size, consists in dissolving 5½ ounces of copper sulphate, 13 ozs. of soda-lime (containing 50 per cent. of caustic soda), and 24 ozs. of potassium-sodium tartrate in each gallon of water, and in submitting each piece of iron, with a fragment of zinc attached to it, to the action of this solution. The zinc, being in metallic connection with the iron, sets up a current as it dissolves in the liquid, and deposits the copper, therefore, not on itself but upon the iron, to which it is electro-positive; the duration of the immersion may range from a few hours to several days, as the deposition proceeds very slowly.

The single-cell process was actually the source of all the others, for by its aid the art of electrotyping was first accomplished. In its simplest form it is well represented by the arrangement of Weil's which we have just described, but a porous cell is almost everywhere used to contain the zinc, so that it shall not be immersed in the copper liquid. Fig. 82 illustrates a depositing-apparatus of this type; the outer jar, *a*, which may be made of glass or earthenware, is filled to about two-thirds of its height with a nearly saturated solution of copper sulphate, and contains an inner cell of porous earthenware, *b*, closed at the bottom, within which is a plate of zinc, *z*, standing in a moderately-strong solution of common salt or sal-ammoniac, or, preferably, of dilute sulphuric acid. In the latter case the zinc must be amalgamated; the liquids in the two cells should stand at the same level. The zinc plate should project above the porous pot, and have soldered to it a piece of copper wire, which serves to connect it with the object to be electrotyped, *c*. Thus a species of Daniell-cell is formed, in which the zinc, dissolving in the acid liquid of the porous cell, deposits copper upon the conductor in the outer jar; and crystals of copper sulphate should be suspended in the liquid at the upper portion of the outer cell, to replace the metal deposited from the solution upon the negative plate, just as they are in the ordinary battery-cell.

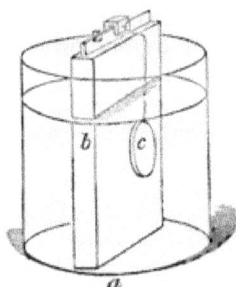

Fig. 82.—Single-cell depositing apparatus for one object.

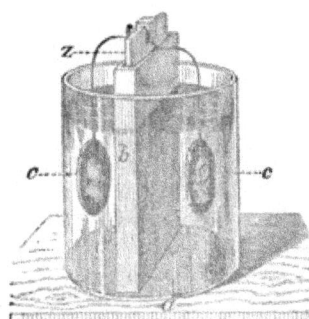

Fig. 83.—Single-cell depositing apparatus for two objects.

Several objects may be coated simultaneously without detriment to the working of the cell; all must of course be attached to the zinc, and they may be suspended around the central zinc rod, as indicated in fig. 83; the arrangement here figured on a small scale may be made of any required size by substituting wooden vats for the glass containing-vessel, and using porous cells and zinc plates of corresponding dimensions. In all the methods of deposition which we have been considering, one face only of the object is turned towards the zinc, and that face alone will receive a deposit of copper; this is suitable enough when it is only required to produce an electrotype from a coin, the two sides of which are separately treated; but when an object is to be completely covered with copper, prior to receiving a coating of a different metal, some such arrangement as that sketched in plan in fig. 84 is to be recommended. The object is suspended in the centre of the tub containing the copper solution from two cross-rods which rest on a circular wire connecting all the zincs in their separate porous cells, these being arranged round the circumference of the containing-vessel. In this manner the object to be coppered is completely surrounded with zincs,

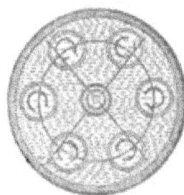

Fig. 84.—Arrangement for electrotyping all surfaces at once.

and the deposition proceeds with equal regularity on all portions. Porous diaphragms may be made of parchment-paper or
of plaster of Paris, but are less satisfactory in use, and should
only be adopted as a temporary substitute for the unglazed
earthenware cells upon emergency.

The single cell would seldom be used in practice when a
separate battery-plant could be obtained, because it is more
clumsy in its arrangements, the process is less under control, and
the solution gradually becomes exhausted of copper unless well
tended.

DEPOSITION BY BATTERY; OR SEPARATE-CURRENT PROCESS.

The principle of this process has already been fully explained ;
a current of electricity is passed from a copper plate (anode) to
the object which is to be coated (cathode), both being immersed
in a solution containing copper ; a quantity of copper, depending
entirely on the strength of the current, is thus dissolved from
the anode, and an equal amount is deposited upon the cathode.
Such details as strength of current, duration of process, composition of bath, and disposition of plant must be determined
by the character of the work under treatment. In the remainder
of the chapter it is proposed to treat first of the electro-deposition
of copper generally, then as applied to the covering of iron,
brass, or other metals for protective, ornamental, or other
purposes; while the electrotyping of printers' plates and art-
electrotypy will be dealt with in a separate chapter.

The Battery.—The battery employed is very frequently that
of Smee, which is a favourite with printers' electrotypers; the
Daniell and bichromate, or modifications of them, are, however,
also largely used. For the acid copper-baths a comparatively
weak current of low electro-motive force is required, and any
attempt to hasten the deposit by increasing the battery-power
will result in defeat, owing to the production of brittle and
crystalline or spongy copper. The alkaline bath requires a
higher electro-motive force, such as would be provided by two,
or even three, Bunsen- or bichromate-cells in series; but the
volume of current must not be excessive, on account of the
lower solubility of the anodes in the solution, which would lead
to a portion of the oxygen deposited at the anode escaping
without combining with copper, and this in turn would result in
a lower rate of solution than of deposition, and so to a gradual
impoverishing of the liquid. The number of cells to be used
must depend upon the quantity of the work ; with the acid

copper-solution they will all be arranged in parallel arc, and will be increased in number as the area of cathode surface is multiplied. A dynamo may be substituted for the battery, but it must have a very low electro-motive force, and will need an intelligent interposition of resistance-coils, unless the work is very extensive.

The Solutions.—For coating metals which are less electro-positive than copper, and for the production of electrotype-plates, a simple solution of $1\frac{1}{2}$ pound of copper sulphate and $\frac{1}{2}$ pound of concentrated sulphuric acid in each gallon of water, will be found to give excellent results with a current of about 1 ampère per square decimetre (= 0·064 ampère per square inch, or 9·3 ampères per square foot) of cathode surface. The bath should be made up by placing the weighed quantity of crystallised copper salt in a suitable vessel, and pouring upon it about four or five pints of boiling distilled or rain-water, and stirring until the crystals have quite dissolved. If the solution be not now perfectly clear, owing to the presence of insoluble impurities in the copper sulphate, it must be filtered by passing it through a cone of blotting-paper fitted into a glass funnel (see p. 54), which will remove all mechanical impurities. The remainder of the water, necessary to make up the solution to the volume of one gallon, is now added cold; and when the mixture is thoroughly cool, the sulphuric acid is cautiously added in a gentle stream, while the liquid is briskly stirred with a glass rod, or if glass be not at hand, with a clean wooden stick or a length of copper rod. Iron must on no account be used, nor may iron or zinc containing-vessels be employed to hold copper solutions, because these metals deposit a portion of the copper and contaminate the liquid by passing into solution themselves. Iron vessels, protected internally by a sound coating of enamel, may, of course, be used, but glass, glazed stoneware, or even wood are preferable, as the enamel is always liable to become chipped.

For treating metals such as zinc and iron which, being more electro-positive than copper, would take a non-adhesive deposit in the acid solution we have just described, recourse must be had to a special bath. In the following table are given the percentage compositions of a number of different copper solutions which have been advocated by various authorities. It will be seen that the majority of these take advantage of the solubility of copper cyanide in a solution of potassium cyanide, while the remainder, for the most part, use copper oxide dissolved in alkaline liquids containing salts of tartaric acid. The chief

TABLE VIII.—SHOWING THE COMPOSITION OF COPPER BATHS, RECOMMENDED BY VARIOUS AUTHORITIES.

No.	Authority	Special Application of Bath.	Most suitable Temperature (F.)	1 Copper Acetate.	2 Copper Carbonate.	3 Copper Cyanide.	4 Copper Sulphate.	5 Potassium Cyanide.	6 Sodium Sulphite.	7 Sod. Bisulphite.	8 Ammonia (.880).	9 Caustic Soda.	10 Sodium Carbonate.	11 Potassium Carbonate.	12 Potassium Hy. tartrate.	13 Sodium-Potassium Tartrate.	14 Sulphuric Acid.	15 Pure Water.	Special Method of Preparation.
1	Acid bath Déplerre	Electrotype	Cold				150										50	1000	Dissolve 5 in 800 of 15; 1 and 8 in 200 of 15; mix.
2	Déplerre	Zinc rollers	"	14				20			8							1000	Boil 12 with 15; saturate with 2; then add 11.
3	Eisner	Iron & zinc	"		q. s.									little	100			1000	Dissolve 200 of 5 in 15; saturate this with 3; add rest of 5.
4	Gore	"	Hot or cold			q. s.		22·5										1000	Dissolve 5 and 6 in 800 of 15; proceed as above (Déplerre).
5	Japing	"	"	14				20	12		8							1000	Dissolve 4 in 450 of 15; mix with 5 dissolved in rest of 15.
6	"	"	"				76	76 to 108										1000	Dissolve 5, 6, and 10 in 800 of 15; mix with 1 and 8 in 200 of 15. Vide this vol., p. 138.
7	Kasalowsky	...	Warm	20				28	8	20	12		20					1000	
8	Roseleur	Iron & steel	Hot or cold	20				20		20	11		20					1000	
9	"	"	Cold	18				16		20	14		40					1000	
10	"	"	Hot	20				22		8	12		20					1000	
11	"	{Cast-iron, tin, zinc}	Hot or cold	11				16		12	8							1000	
12	Watt	Zinc	Nearly boiling 108°—130°	18			50	22		4	6							1000	Dissolve 4 in 250 of 15, add excess of 8, and 250 of 15, add slowly (stirring) 5 in 500 of 15.
13	"	"	"					125			excess							1000	Dissolve 4 in water, add 8 and 11, then 5, and decant.
14	"	Iron, zinc, &c.	...				12·5	37·5			12·5		25	25				1000	Dissolve 7 and 10 in 500 of 15 (=a); dissolve 5 in rest of 15 (=b); dissolve 1 in 8 (=c). Mix a and c; add b; boil and filter.
15	Weiss	Zinc	...	20				25		20	12		25					1000	
16	"	{Zinc, tin, lead, iron}	...				30	25		20			65					1000	
17	Weil	"	...				35					80				150		1000	

NOTE TO TABLE.—Cyanide of Potassium containing 95 per cent. of pure salt is understood; if a less pure article be used, proportionately more must be added.

The black figures in the last column refer to the numbers of the vertical columns denoting the various ingredients.   q.-s. = *quantum sufficit.*

variations are due to the substitution of copper acetate or of verdigris for the sulphate, and of the single tartrate of potash or soda for the double tartrate (of the two metals together), and in the addition of varying proportions of other substances which play a minor part in the action of the bath.

Of all these liquids, the bath of Roseleur is, perhaps, the most generally useful, as it is equally applicable to all metals, and may be worked at any desired temperature. It is best prepared by working up $3\frac{1}{4}$ avoirdupois ounces of copper acetate into a thick paste with a small proportion of water, so that every particle may be thoroughly wetted; then an equal weight of sodium carbonate in crystals should be added, together with about $1\frac{1}{2}$ pints of water, and the whole mass should be well stirred together for a few minutes. An exchange will thus take place between the sodium carbonate and the copper acetate, with the result that the water will contain in suspension insoluble copper carbonate as a pale green-coloured precipitate, and in solution soluble sodium acetate; $3\frac{1}{4}$ ounces of sodium bisulphite are now added in a second $1\frac{1}{4}$ pints of water; and finally the remaining 5 pints of water, containing $3\frac{1}{4}$ ounces of potassium cyanide, is introduced. The pale yellow-coloured precipitate produced on the addition of the bisulphite will be gradually dissolved on stirring with the cyanide, and should completely disappear after a few minutes, leaving a solution which is practically colourless. Should it not be so, a little more potassium cyanide solution must be added by degrees, until the decolorisation is perfect. Failure in the first case is probably due to the use of more than usually impure potassium cyanide. If necessary, the bath should be finally filtered.

The alkaline baths require more careful watching than the acid liquors, owing to the comparative insolubility of the anode; if, on examination after use, the strength is found to be greatly diminished, it may be restored by adding to it a sufficient quantity of a copper cyanide solution prepared as follows:—A solution of sodium carbonate is added to one of copper sulphate until no further precipitate or cloudiness is observed on the addition of another drop of the soda solution. The mixture is allowed to stand until the copper carbonate which forms has subsided; the clear liquor is now poured off, and potassium cyanide solution is added until the whole of the slimy powder is redissolved to a colourless fluid. If, on the other hand, the use of too large anodes has charged the bath too highly with copper, it will have a bluish colour, which must be discharged by the addition of a sufficient quantity of potassium cyanide. The appearance of a white or pale brown precipitate on the anode indicates the necessity for a further addition of copper salt, which Roseleur recommends should in this case consist of the ammonium-copper acetate, formed by adding excess of ammonia to a solution of copper acetate, so that a perfectly clear intense blue liquid results; this liquid should be added little by little, until the blue shade produced by it in the solution disappears only with difficulty; any excess of copper may of course be neutralised by adding potassium cyanide.

The acid baths are almost invariably used cold, for if allowed to become warm the resistance is diminished and the current may then be too intense to give a good deposit; the others are more frequently warmed—some even to boiling, as indicated in the table given above. The necessity for stirring the solution and the methods by which it is accomplished have already been described on pp. 108, 109.

The Anodes.—These should be of the purest quality only; for electrotyping in all its branches it is well to use only electrotype-copper—that is, metal which has already been electro-deposited from a pure solution. If this cannot be obtained when required, the best rolled copper should be used, as it is likely to contain fewer impurities than the cast metal. The impurities in commercial copper are for the most part insoluble in the bath, and, therefore, gradually cover the anodes with a dark grey or black slime, which accumulates on the surface as the copper dissolves, and by degrees becomes detached and sinks to the bottom of the bath; but remaining for a time suspended in the solution, especially if the bath is agitated as it should be, it tends partly to become attached to the cathode, and so to give rise to difficulties in deposition. The composition of the slime will manifestly depend upon the brand of copper used and the impurities which it contains; an analysis of a specimen examined by Max Duke, of Leuchtenberg, in 1848, gave—33 per cent. of tin; 25 of oxygen; $9\frac{1}{4}$ of antimony, and an equal percentage of copper; $7\frac{1}{4}$ of arsenic; $4\frac{1}{2}$ of silver; $2\frac{1}{2}$ of sulphur; $2\frac{1}{4}$ of nickel; 2 of silica; $1\frac{1}{4}$ of selenion; and 1 of gold, with smaller proportions of cobalt, vanadium, platinum, iron, and lead. Other samples would contain bismuth in addition to the above. When this slime appears upon the anodes, they must from time to time be removed from the bath, rapidly washed with water, and returned.

The anodes in the case of copper should present approximately the same surface as the cathodes; any considerable departure from this rule will, on the one hand, cause an excess of copper to pass into the bath, especially if a very acid solution be employed;

or on the other, it will give rise to a decrease in the strength of the solution.

The Character of the Copper Deposited.—The copper deposited electrolytically may vary, according to the conditions of working, from a loose black powdery deposit, such as results from using a strong current or weak solutions, or whenever hydrogen is deposited simultaneously with the copper, to a minutely crystalline, highly tenacious reguline metal, when the current is

TABLE IX.—SHOWING THE EFFECT OF VARYING CURRENT-STRENGTHS ON THE COPPER DEPOSITED FROM NEUTRAL COPPER SULPHATE SOLUTIONS (*von Hübl*).

Intensity of current in		Deposit from Solution with 5 per cent. Copper Sulphate.		Deposit from Solution with 20 per cent. Copper Sulphate.	
Ampères per sq. inch.	Ampères per sq. decimetre.	Appearance.	Character.	Appearance.	Character.
0·013	0·2	Very coarse crystals.	Very brittle.	Very coarse crystals.	Exceedingly brittle.
0·026	0·4	Coarse crystals.	Brittle.	Coarse crystals.	Very brittle.
0·052	0·8	Rather coarse crystals.	Fairly good.	Rather coarse crystals.	Very brittle.
0·193	3·0	*	*	Exceedingly fine grain.	Excellent.

TABLE X.—SHOWING THE EFFECT OF VARYING CURRENT-STRENGTHS ON THE COPPER DEPOSITED FROM COPPER SULPHATE SOLUTIONS CONTAINING 2 PER CENT. OF FREE SULPHURIC ACID (*von Hübl*).

Intensity of current in		Deposit from Acid Solution with 5 per cent. Copper Sulphate.		Deposit from Acid Solution with 20 per cent. Copper Sulphate.	
Ampères per sq. inch.	Ampères per sq. decimetre.	Appearance.	Character.	Appearance.	Character.
0·013	0·2	Small crystals.	Somewhat brittle.	Small crystals.	Somewhat brittle.
0·026	0·4	Small crystals.	Good.	Small crystals.	Good.
0·052	0·8	Very small crystals.	Good.	Very small crystals.	Good.
0·077	1·2	*	*	Exceedingly fine grain.	Excellent.
0·258	4·0	*	*	Exceedingly fine grain.	Excellent.

* *Under these conditions hydrogen is evolved with the copper, and the deposit of the latter is, therefore, worthless.*

well proportioned to a solution of the right strength.   The physical condition of the copper is a matter of prime importance to the electrotyper who, unlike the electroplater, is required to produce a thick deposit which must be sufficiently tough, pliant, and tenacious to withstand a considerable amount of rough usage, receiving at first no support from backing metal.   The experiments of v. Hübl, to whom reference has been made in an earlier chapter, are of extreme interest in consequence of the light which they throw upon the relation of strength and toughness of electrotype-copper to the conditions of the depositing-bath.

Several points are clearly brought out by these tables, notably the disadvantages attending the use of weak and neutral solutions and of very weak or very strong currents.   Thus, with a 5 per cent. solution, no current higher than 1 ampère per square decimetre may be used on account of the resulting deposition of hydrogen, while with a 20 per cent. solution the current-density may even rise to 3 or 4 ampères per square decimetre without ruining the deposit; further, the addition of acid to the weaker solution can alone render it fit for use, while even with the stronger solution it enables a smaller current to be employed successfully.

Very weak currents are clearly to be avoided, because they give a coarsely-crystalline and brittle deposit, while dilute solutions are more readily overburdened with current, so that the copper is spoiled.

In a second series of experiments, v. Hübl determined the maximum volume of current, which may be safely applied to the acid copper-solutions under certain varying conditions, to be as follows :—

TABLE XI.—SHOWING THE MAXIMUM CURRENT-DENSITY FOR ELECTROTYPING (von Hübl).

Composition of Solution.	With Solution Undisturbed, Ampères.		With Solution in Gentle Motion, Ampères.	
	Per Sq. Decimetre.	Per Sq. Inch.	Per Sq. Decimetre.	Per Sq. Inch.
15 per cent. copper sulphate, neutral, .	2·6 to 3·9	0·168 to 0·252	3·9 to 5·2	0·252 to 0·335
15 per cent. copper sulphate with from 2 to 8 per cent. sulphuric acid, . .	1·5 „ 2·3	0·097 „ 0·148	2·3 „ 3·0	0·148 „ 0·193
20 per cent. copper sulphate, neutral, .	3·4 „ 5·1	0·219 „ 0·329	5·1 „ 6·8	0·329 „ 0·439
20 per cent. copper sulphate with from 2 to 8 per cent. sulphuric acid, . .	2·0 „ 3·0	0·129 „ 0·193	3·0 „ 4·0	0·193 „ 0·258

In a third series, the breaking weight, elasticity, and stretching-power under tensile (or pulling) strain were determined: new acid baths containing from 10 to 20 per cent. of copper sulphate, and old baths with from 5 to 20 per cent., were alike submitted to the action of currents of different strengths. The conclusions which may be drawn from his results are :—

(1) That the actual strength of the deposited copper increases as the current-volume rises from 0·61 to 3 ampères per square decimetre.

(2) That the amount of extension at the moment of rupture, under tensile stress, diminishes as the intensity of the current is increased between these limits.

(3) That the highest limit of elasticity is exhibited by copper deposited by a current of from 1 to 1·5 ampères per square decimetre.

(4) That the elasticity alone is practically affected by variations in the strength of the solution between 10 and 20 per cent. of copper sulphate, the limit of elasticity of the deposited metal increasing with the percentage of copper in the bath.

Thus, to quote actual figures, in a new 20 per cent. bath containing 4 per cent. of acid, a current of 0·61 ampère per square decimetre yielded a copper which broke under a strain of 18·0 tons per square inch after stretching 27 per cent. of its original length; while a current of 2·22 ampères per square decimetre deposited a metal which required 23·6 tons per square inch to fracture it, but which only extended 16 per cent. In an old solution of equal strength, a current of 1 ampère per square decimetre produced a deposit which gave way at 17·2 tons per square inch, with 26·4 per cent. elongation; while the copper thrown down by a current of 2·5 ampères ruptured at 18·7 tons per square inch, at 25·1 per cent. increase of length; and that precipitated by 4 ampères per square decimetre yielded at only 15·3 tons per square inch under a stretch of 19·4 per cent.

The highest breaking strain in the whole series of trials was that of 23·6 tons per square inch above recorded; the greatest extension was 33 per cent., shown by a metal deposited by a current of 1 ampère per square decimetre from a solution containing 15 per cent. of copper sulphate and 3 per cent. of sulphuric acid; but the breaking weight of this specimen was only 17·3 tons per square inch.

Some of these samples of copper may be said to compare not unfavourably even with good rolled copper, such as would ordinarily be used for engraved plates, which are required to withstand severe strain in the various processes to which they are to be applied.

Electro-plating with Copper.—The objects to be coated are first thoroughly cleansed by the methods indicated in the last chapter. Cast-iron articles are especially liable to contain particles of sand and other non-metallic impurities embedded in the surface; these must be carefully looked for and removed before the treatment there described. On removal from the last acid dip, the pieces must be thoroughly washed and at once suspended in the copper-vat, by means of copper wires or hooks slung from the cross-bars attached to the negative (zinc) pole of the battery. The method of suspension will, however, depend upon the size and shape of the pieces, and upon the skill and ingenuity of the operator. Large objects may be hung individually from the cathode-rods, but the wires must be attached to portions of the object where they will not leave permanent marks upon the finished work; or they may be rested in wire-slings; or, if very heavy, they may be allowed to stand on the floor of the vat with conducting wires attached underneath; but in this latter case no deposit can of course form upon the bottom of the object, which must be treated subsequently, if necessary; light articles which are perforated in any part may be slung on copper wires. The methods of supporting the pieces are obviously innumerable; but it must always be remembered that whenever they are hung from wires, their position must be slightly shifted from time to time, to prevent the formation of wire-marks due to imperfect coating at the points of contact. The objects should be momentarily removed from the vat for examination soon after the deposition has commenced, for at this time imperfect cleansing will be clearly evident, so that if the coating be incomplete the cause may be at once detected and the defect remedied. Finger- or grease-marks, produced by handling after cleansing, can thus be removed by washing the pieces in water, immersing for a moment in the potash-vat, rinsing, dipping in the acid liquor, once more rinsing, and finally replacing in the copper-bath, when the clean surface being restored, the action proceeds in due course. Small articles may also be coated by placing them in a perforated porcelain ladle (fig. 72, p. 120), on the bottom of which rests a coil of copper wire attached to the negative pole of the battery, so that they form a continuous cathode by mutual contact; in this ladle they are plunged into the copper-vat and constantly agitated until the required thickness of metal is obtained; the anode should in this case take the form of a cylindrical sheet of copper surrounding the ladle, or it may be a disc of copper-plate, which is placed above the ladle after its introduction into the bath.

All portions of any object which are not to receive a coating of copper should be painted over with an insulating varnish (p. 344), which may be removed subsequently.

The duration of the process depends upon the thickness of deposit required; for most objects treated in this way—whether iron or zinc to be subsequently silvered, silver to be gilded, or iron which is to receive a coating, either protective or ornamental—a very thin wash will suffice, such as might be imparted in a period ranging from a few minutes to an hour. The alkaline baths that would generally be used for this class of work are slower in action than the acid baths.

On removing the work from the vat, it must be dipped into three or four successive wash-waters; as each piece is transferred to the washing-tank it carries with it a notable quantity of copper solution from the bath, so that the water in the first tub should not be frequently renewed, but should be allowed to remain until, many articles having been treated, it has become gradually charged with copper; it is then poured into a different vessel so that the dissolved metal may be recovered from it. Even the second tank receives a gradual addition of copper, and it is well to allow this also to accumulate until the solution in the first tank is discharged, when this liquor may be substituted for it. The water in the third and fourth tanks should be very frequently renewed, and, since they should contain no more than a trace of copper, may be discarded. On leaving the last wash-water the objects should be free from copper solution, and after drying are ready for the market. They may be dried in ovens heated to the temperature of boiling water, or in hot box-wood sawdust if they be small enough to render such a process practicable. When the articles are receiving the copper coat merely as a preliminary to the deposition of other metals, they should, if possible, be transferred directly from the final washing-tank of the coppering process to the depositing-vat of the metal next to be deposited. This system is to be recommended because, exposed to the air of towns, copper is most liable to receive a sulphur-tarnish which ordinary potash and acid dips might fail to remove completely.

The difficulties likely to arise in the process of deposition, and the methods by which they may be overcome have been treated of in the earlier portion of this chapter, dealing with the solutions and anodes.

Sundry Applications of Copper Deposition.—Thick deposits of copper for various purposes are now made by means of the dynamo-electric machine.

An interesting example of such a process is seen in Wilde's system of applying a thick copper coat to iron printing-rollers. The roller is first covered with a thin film of copper in the alkaline bath, and being thus protected it may afterwards be safely treated in the more rapidly-acting acid solution. For this purpose it is placed on end, mounted upon a vertical insulated axis in the acid copper-vat, and is connected at top and bottom with the negative pole of the dynamo; within a short distance off it is mounted a copper cylinder of about the same size, which is connected with the positive pole, and, therefore, plays the part of anode. The two cylinders are simultaneously rotated upon their axes at a moderate pace, and a gradual transference of copper from anode to cathode takes place. In this way an even thickness of deposit is obtained over the whole surface; uniform along the length of the roller, by reason of the motion imparted to the bath, which maintains an equal density of solution throughout; and uniform as to its diameter, because the rotation constantly brings fresh surfaces opposite the anode. Moreover, a high rate of deposition, or, in other words, rapid work, is permissible, because of the motion in the bath; for we have seen (p. 142) that a bath which is well agitated will take a stronger current than one which remains tranquil. Additional means for securing a thorough mixture may be employed by adapting a small propeller-blade to a shaft run by the same machinery that actuates the rollers.

Watt, in his treatise on Electro-Deposition, describes a plant which was used for a similar purpose by the Electro-Metallurgical Company of London. The rollers were dipped into potash solution contained in a steam-jacketed cylinder 15 feet high and 3 feet in diameter, and a preliminary wash of copper was imparted by an alkaline copper-bath in a similar tank. The anodes used in the acid bath consisted of copper billets 4 inches square and 12 feet long, while the tank itself was capable of holding 60,000 gallons of solution.

A third method which has been successfully utilised for coppering rollers consists in slowly rotating them on a horizontal axis about 12 or 18 inches below the surface of the bath, and between two curved copper anodes fixed on either side and nearly meeting beneath the cylinder.

Tubes of any size, shape, or section may be gradually built up in a similar manner, and, if the current be properly proportioned to the strength of the solution, should not be greatly inferior to ordinary solid-drawn copper tubes, and would, doubtless, be, on the average, stronger than the common brazed tubes so frequently employed. Straight tubes may be built up upon a well-polished iron mandrel or core, which may be subsequently withdrawn, the surface of the iron being black-leaded to prevent the adhesion of the two metals; bends and unusual shapes may, of course, be formed by employing a moulded core made of fusible materials, such as are described in the next chapter, which can afterwards be removed by the application of a gentle heat. When, however, the tubes are required to withstand a high fluid-pressure either from within or from without, the utmost care is required to guard against the formation of a largely-crystalline deposit, because, the grain of the metal being "open," it would be more or less porous, and would give rise to leakage of liquid through the tube walls themselves.

Elmore has recently sought to approximate an electro-deposited tube to one "drawn" in the usual way, by causing a burnisher to constantly traverse the length of the tube backwards and forwards, as it slowly rotates in the bath, so that the metal is continually consolidated, compressed, and brightened as it becomes deposited, and thus yields a copper possessing in a high degree both density and solidity. With the aid of this process a current-strength of about 16 ampères per square foot (1·7 amp. per sq. dm.) has been found to deposit copper, which would stand a stress of 26·5 tons per square inch, with an extension of 16·5 per cent., the limit of elasticity being reached with a load of 23·3 tons per square inch. These numbers are the mean of three closely concordant results of tests, made by Professor Kennedy, with samples measuring 4 inches long by from 1·245 to 1·272 inches wide, and from 0·131 to 0·135 inch thick. The action of the burnisher is said to increase the density of the metal to such an extent that its specific gravity may be as high as 9·2.

It is interesting to compare these figures with those given on p. 143, by which it was shown that von Hübl has succeeded in producing by simple deposition, and without the aid of a burnisher, a specimen of copper, which breaks under a stress of 23·6 tons. Nevertheless, the extra 3 tons of strength gained by Elmore's device (that is, on the supposition that the maximum limit of strength has actually been reached by von Hübl) undoubtedly enhances the value of the metal for seamless copper tubes by far more than a proportionate amount; and the application of the principle may be expected to develop, perhaps, in other directions than those at present contemplated.

An application of copper deposition which has suggested itself to the electrician, is the coating of an iron or steel wire with any desirable thickness of copper, for overhead telegraph-wires, so that the full advantages of the greater strength of the steel may be combined with the higher conductivity of the copper. To this end the American Postal Telegraph Company have arranged a series of copper-baths at their works, through which the iron wire is passed by means of rotating-drums until sufficient metal has been deposited. Twenty-five large dynamos were used to actuate 200 baths; and these were capable of depositing daily 500 lbs. of copper per mile upon 20 miles of steel wire, weighing 200 lbs. per mile; the time occupied by any point upon the surface of the wire in traversing the whole chain of baths being about 60 hours.

The number of uses for thick deposits of copper is endless, and very many other applications unenumerated here have been successfully carried into effect—as, to take a solitary example, the formation of copper rings upon small cylindrical shell (projectiles) to enable them to take the grooves of the gun. The process for producing these thick coatings is, as we have seen, very simple, but presents the extreme disadvantage of requiring a great expenditure of time to yield any considerable depth of deposit. The use of the dynamo is almost an essential to the work; the continual wear and tear of the batteries, and the worry of continual inspection and renewal, added to the current expenditure of costly zinc, rendering the application of the galvanic cell practically impossible.

CHAPTER VIII.

ELECTROTYPING.

The object of electrotyping is the production of an exact *facsimile* of any object having an irregular surface, whether it be an engraved steel- or copper-plate, a wood-cut, or a forme of set-up type, to be used for printing ; or a medal, medallion, statue, bust, or even a natural object, for art-purposes.

In all cases a reversed mould of the object is first obtained, and upon this the copper is electrolytically deposited to a sufficient thickness. It was very early discovered (p. 5) that metal deposited upon an uneven surface, and then wrenched apart from it, exhibited in reverse an absolutely-perfect reproduction of every irregularity or line upon the surface of the original, and that by depositing a second coat upon this new surface, a re-reversed image of the first surface was obtained which, as to minuteness of detail, defied distinction from the former. Hence impressions from a single printing-plate or block may be repeated an indefinite number of times by retaining the original plate merely as a means for producing any number of duplicate copies, and not taking press-copies directly from itself. Works of art from the chisel of the sculptor or the tools of the moulder may thus be similarly multiplied, or the process may even be substituted for that of the foundry in obtaining statues from the sculptor's model. So delicate, accurate, and refined is the copy, that the finest grain of wood, even the peculiar sheen of satin wood, is perfectly imitated, and it is even possible to reproduce the daguerreotype photographic image with its infinitely minute gradations of light and shade.

In all classes of electrotyping, then, the first step is to obtain a cast, *intaglio,* or negative-impression of the object, in which the projecting portions of the original become depressions, and *vice versâ.* Occasionally it is possible, and often it is preferable, to effect this by rendering the surface of the object somewhat dirty, and depositing a sufficient thickness of copper or silver upon it directly, to enable the two surfaces to be separated without fracture or distortion of the deposited plate. Thus a faithful negative is obtained which is not adhesive to the unclean surface

and may be readily detached; it is then ready to receive a coating upon itself, and this on removal is found to be a true copy
of the original in every respect, so that it may be substituted for
it—as, for example, in the printing-press. But it frequently
happens that the original subject would be spoiled if placed in
the depositing-bath, or that it has undercut surfaces, which
would prevent the deposited coat from being detached; in such
cases, and, indeed, more usually, casts of the object are first
taken in a suitable moulding material, which, having been
rendered a conductor of electricity, at least superficially, receives
the deposit in the form that is to be substituted for the original.

Moulding Materials.

For moulding, any material may be used which is capable of
taking accurate impressions of an object, and which is not so
brittle or so soft at ordinary temperatures as to be broken or
bent out of shape during subsequent processes; provided that it
is a conductor of electricity, or may be made so superficially;
and provided, also, that it neither injures the object to be
moulded from nor is affected in any way by the solutions in
which it will be placed. The principal materials employed are
gutta-percha, bees'-wax, plaster of Paris, fusible metal, gelatine, or
sealing-wax; or compositions in which these bodies are used, in
conjunction with others added to modify their properties in
some desired manner.

Gutta-percha and Compositions in which it is used.—Gutta-
percha is usually obtained in the market in the form of stout
sheet, into which it has been manufactured from the crude
masses imported from abroad. It should be of good quality,
otherwise the foreign impurities present in it are liable to ruin
a cast, if at least any of them should be present on the surface
which is to receive the impression. It is a non-conductor of
electricity, and possesses the property of becoming so plastic at
the temperature of boiling water that it may be pressed into the
finest lines of an uneven surface; on cooling *in situ* it again
becomes rigid, and thus preserves a faithful and permanent impression of them, while remaining sufficiently elastic to allow of
its being used for reproducing work that is slightly "undercut."
Heated to a temperature of 95° to 100° C. (203° to 212° F.) in
water, or better, in an oven warmed by a hot-water jacket, a
piece of gutta-percha may be pressed on to any surface which it
is desired to reproduce. The surface of the cast thus obtained

requires only to be rendered conductive, and is ready to receive the metal by electro-deposition. Gutta-percha contracts considerably on cooling; in order to ensure a good impression, it should, therefore, be allowed to cool in contact with the original object, so that no superficial contraction can possibly occur by reason of the intimate contact between the two irregular surfaces, which prevents the sliding of the one upon the other.

Where the very finest lines are to be reproduced with precision, as in the copying of engraved steel-plates, a mixture of gutta-percha with not more than about half its weight of fatty matter is frequently used, which is "thinner" than the gutta-percha alone, and may even be melted and simply poured over the plate. Such a mixture is :—

Gutta-percha,	66 parts by weight.
Lard,	33 ,, ,,
Russian tallow,	1 ,, ,,

The lard should be refined by melting in an earthen pot or pipkin and pouring it into hot water, when the fat remains fluid upon the surface of the water, while the mineral impurities sink to the bottom of the vessel; on cooling, the lard solidifies and may be removed from the water. The tallow should be similarly treated. Then the gutta-percha is cut into small strips, mixed with the requisite quantities of the other materials and melted slowly in a porcelain dish, or in an oven, which should be maintained at the lowest temperature compatible with complete fusion of the mixture. The mass is thoroughly incorporated by stirring, and is ready for use at once. These mixtures must not be overheated as the gutta-percha at high temperatures becomes brittle and useless.

The Russian tallow is often omitted from the mixture, and many operators use linseed or other oils in preference to lard. The addition of a little plumbago is to be recommended; it facilitates the production of the conductive film required for electro-deposition. Gore recommends, for moulding from medals and coins, a mixture of two parts of gutta-percha and one of marine glue, which must be thoroughly incorporated, and is then used in the plastic state like gutta-percha alone.

Gutta-percha is best adapted to copying plates, medallions, coins, or medals which are not too deeply undercut. To effect this, it is made plastic by heating to 100° C. (212° F.) as described above; a sufficiently large fragment is now carefully rolled and kneaded in the hand until it forms a quite uniform sphere without visible lines or seams; then, while still plastic, the ball is rested on the centre of the medal, and pressed into place with the thumb, while the fingers gradually flatten it and extend the material outwards and downwards, beginning at the centre, until it completely covers the whole surface; it is then placed under gentle pressure until cool, when it will be found that the surfaces will part without effort, and the gutta-percha intaglio will be ready for the next process. The medal should be well brushed with plumbago before moulding to prevent adhesion, and the fingers should be thoroughly moistened or rendered slightly oily or they will be liable to cling to the gutta-percha. Very large surfaces cannot satisfactorily be treated in this manner, and it is expedient in dealing with them to render plastic a sheet of gutta-percha, slightly larger than necessary to cover the plate; then placing it upon the surface, it is pressed into intimate contact throughout, beginning from the centre and working outwards as before, in order to prevent the entanglement of air-bubbles between the two surfaces, which would prove fatal to the sharpness of the impression. The plate and mould are finally allowed to cool under a gentle and even pressure.

When the fluid mixture is employed, the medal or plate is surrounded with a narrow rim of paper or thin metal placed on edge, beyond the limits of the part to be copied, and the melted material is poured slowly on to the centre and allowed to flow gently from this point over the whole surface, where it should be permitted to remain for several hours to ensure that it is completely set.

Either the gutta-percha alone, or the composition made from it, may be used over and over again, but after a long period of use, when so much plumbago has accumulated in it that it becomes hard and brittle, it is necessary to add a sufficient quantity of fresh gutta-percha (or composition) to regenerate it for use.

Bees'-wax and Compositions in which it is used.—Bees'-wax is very largely used in various compositions, and, indeed, is the usual basis of mixtures for electrotyping set-up printers' type. As a general rule, the ordinary yellow bees'-wax of commerce is sufficiently good for the purpose; but as the presence of some of the commoner adulterants increases its tendency to crack on cooling, especially in cold weather, some operators have preferred to melt the wax in a pipkin and to add to it about one-tenth of its weight of white lead (lead carbonate); after stirring, the excess of lead is allowed to separate and collect at the bottom, when the melted wax, still retaining a small quantity of lead, is poured into slabs and may be remelted at will. The presence of the lead has the twofold advantage of checking the liability to crack, and of increasing the specific gravity of the composition, so that it sinks readily in the depositing-vats. As a general rule, however, this precaution is unnecessary.

Usually the wax is not employed alone, but is mixed with stearine, Venice turpentine, or other bodies which tend to prevent the cracking of the moulds. For taking electrotypes of printers' formes in the usual way with the aid of pressure, an excellent mixture is made by melting together and stirring well :—

Bees'-wax, ...	85 parts by weight.
Venice turpentine, . . .	13 ,, ,,
Plumbago, ...	2 ,, ,,

The composition is then poured into flat trays, and when nearly set, the typographical plate is pressed upon its surface in a manner to be described hereafter (p. 168). Many other mixtures are, of course, used, different workers adopting their own formulæ. Thus, Urquhart recommends the same ingredients as above, but mixed in the proportion of $87\cdot3 : 9\cdot7 : 3$ for this class of work, or in the proportion of $80\cdot5$ of wax ; $27\cdot9$ of mutton-suet, and $1\cdot7$ of graphite for moulding when pressure may not be applied, on account of the fragility of the original object. Good mixtures may be made, according to the formula used by Volkmer, by melting together 70 parts of wax and 30 of stearine, or to that preferred by Watt, consisting of 70 of wax, 26 of stearine, and 4 of litharge or flake-white. For medal-work Walker was in the habit of using $69\frac{1}{2}$ parts of spermaceti with $15\frac{1}{4}$ each of mutton-suet and bees'-wax. For certain uses many operators have introduced rosin into the composition; Weiss employs a mixture of 62 of bees'-wax, $28\frac{1}{2}$ of mutton-suet, and $9\frac{1}{2}$ of rosin to produce a replica from a statue, the first cast of which has been made in an elastic composition attackable by water, as will be explained later ; and a mixture of 50 parts of wax, with 50 of rosin, and from $3\frac{3}{4}$ to 5 of a solution of phosphorus in carbon bisulphide (1 of phosphorus in 15 of bisulphide), was recommended by Parkes for the preparation of specially-conductive moulds.

In using melted wax to prepare matrices from medals and the like, it should not be poured on until it is nearly solidifying; then (having previously slightly oiled the surface of the object, surrounded it with a strip of paper or metal placed on edge, and rested the whole upon a gentle slope) the wax should be gently poured upon the lowest part and allowed to flow with an even motion over the whole surface ; as soon as it is set, the containing rim should be removed, because the composition adheres to it somewhat strongly, and the contraction due to cooling may cause the wax to fracture at the centre. Then, after standing for one or two hours, the impression may be safely detached from the original.

Plaster of Paris.—Only the finest quality of plaster should be used, which has not been overburnt, is very finely ground, and has been stored in a well-covered jar or bottle, so that it shall not have deteriorated by exposure to moist air. But, even at the best, it is not to be recommended in comparison with the materials enumerated above. In applying it, the object is first thoroughly well covered with an almost imperceptible film of oil, by means of a soft rag or a tuft of cotton-wool; then having surrounded it, if necessary, with a paper rim the plaster is made into a thick cream by sprinkling it into a sufficient body of water, and without loss of time, a small quantity is poured upon the surface of the object, and with great rapidity brushed into every corner or line by means of a moderately hard brush, to prevent the formation of air holes by entanglement of air in the liquid; then, immediately, the remainder of the paste is poured over it and allowed to remain until it has completely set, when it may be removed in the usual way. The plaster must be free from grit, and must be well mixed with sufficient water, or it will set too rapidly to allow of the treatment above recommended; but in any case the operations must be conducted very quickly, as the cream speedily assumes a pasty and semi-solid condition.

The matrix so prepared is porous and absorbent, and is not ready to be placed in the bath of copper-liquor until it has been water-proofed, which is usually effected by saturating the whole surface with wax or solid paraffin. The mould, if small, may be simply rested on its back in a shallow tray of melted wax, which must be insufficient to cover it completely, until the latter has penetrated through to the face by capillary action; it is then removed and cooled. Large moulds may be superficially covered by pouring the melted wax over the whole surface and then transferring them to an oven or hot closet, until the layer of wax just melts and becomes absorbed by the plaster.

Fusible Metal.—Certain alloys melt at a temperature below that of boiling-water, and may, therefore, be used to obtain casts from the most delicate objects, or even from organic bodies, such as animals, fruits, and flowers; but, although the casts obtained in this manner are very sharp and accurate, the process is rarely employed because of the expense entailed by the waste, however slight, entailed by the constant use of a costly metal. The principal so-called fusible metals are compounded as follows :—

TABLE XII.—Showing the Compositions of Certain Fusible Alloys.

Name of Alloy.	Percentage Composition.				Approximate Fusing-Point.
	Bismuth.	Lead.	Tin.	Cadmium.	
Newton's,	50	31·25	18·75	—	202° F.
Rose's,	50	25	25	—	201°
Lichtenberg's,	50	30	20	—	197°
Darcet's,	50	20	30	—	197°
Wood's,	50	25	12·5	12·5	141°
Lipowitz's,	50	27	13	10	140°

Any of these may be used, but those with the lower melting-point are naturally to be preferred. The metals should be thoroughly mixed by long stirring while still fluid; and they may even with advantage be poured into slabs and remelted several times with the same object in view; the necessity for perfect alloying will be recognised when it is remembered that the melting-point of any of the four components alone is in excess even of that required for Newton's alloy, the numbers being approximately, lead 533°, bismuth 515°, tin 442°, cadmium 608° F. The melted alloy may be poured upon the surface of the object to be copied; or, for the treatment of medals and coins, it may be run into a shallow tray; then, when nearly solidifying, any dross upon the surface is removed by skimming with a piece of card, and at once the medal is dropped upon the surface and pressed in this position until the fluid has completely set, which it should do almost immediately. No difficulty should be experienced in separating the two surfaces or in obtaining successful results after a few trials. The addition of mercury to reduce the fusing-point of the alloy, which is recommended by some operators, should not be practised, but it is especially to be avoided when gold or silver objects are being treated, as both these metals have a very strong tendency to amalgamate or take up mercury, and in so doing to alter in appearance. The special recommendation to the use of matrices of fusible alloy is that they are metallic, and are, therefore, conductors of electricity, requiring no treatment before placing in the bath, beyond rendering them slightly dirty on the surface to prevent adhesion of the deposit. The metal of old moulds may of course be remelted as often as desired.

Elastic Moulding Material.—When objects in high relief are deeply undercut, none of the moulding materials hitherto mentioned can be successfully applied, because it is not possible to detach the cast from the original without fracturing the one or the other. A special material, which shall possess so much elasticity that it may be drawn over projecting portions unhurt, and yet on its release return completely to its original form, must, therefore, be sought. Such a material may be made by mixing gelatine with sugar or treacle, the latter substances preventing the cracking of the substance on drying.

The mixture originally used by Parkes was made by soaking 80 parts of glue in cold water over night, or for several hours, until it becomes perfectly soft; then pouring off the water, it was melted in a hot-water jacketed arrangement, similar to an ordinary glue-pot, and when quite fluid 20 parts of common treacle were stirred in. Busts and statues may be readily copied with the aid of this mixture, but it has the inconvenience that it swells up when wetted, and cannot, therefore, be safely placed in the depositing-vat without special preparation. To render it capable of resisting the liquid, the hardening power of potassium bichromate on gelatine is utilised, the matrix being immersed for a short time in a 10 per cent. solution of the bichromate and then placed in direct sunlight until dry; or 2 per cent. of tannin may be added to the moulding-mixture before casting, to render it waterproof, in which case the bichromate process may be dispensed with. It is not, however, safe to rely upon these waterproofing mixtures, but when the elastic composition must of necessity be used, a reproduction of the original object should be made in wax-composition, and then from this a second matrix should be prepared in plaster of Paris, from which the wax may be melted away without injuring the delicacy of undercut portions. The manner of accomplishing this is described on p. 176.

Sealing-Wax.—Sealing-wax is only employed for very small work, such as the reproduction of seals or signets, and is, therefore, rarely required in practice. It is applied by melting a portion on to a sheet of card or metal, and after breathing lightly on the signet taking an impression as in sealing a letter.

Rendering the Mould Conductive.—Having obtained a mould in any of the above materials (with the exception of the fusible alloys) it is necessary to cover it with a film, which will enable the surface to conduct electricity sufficiently well to receive the first deposit of copper uniformly. If it were convenient, the covering of the surface with a superficial metallic layer would obviously be the most satisfactory method, and where it is desired to produce an especially fine reproduction of a delicate design, this is frequently done. Parkes' system of metallising the moulds consisted in preparing the matrix with the special material, containing phosphorus dissolved in carbon bisulphide, described on p. 153; then, on dipping this mould into a solution of silver nitrate, the phosphorus present on the surface reduces silver to the metallic state, and thus coats the whole with a superficial but continuous layer of metallic silver. Now on dipping the matrix into a solution of gold chloride, a partial exchange of gold for silver takes place, and the whole design is covered with an almost imperceptible film of metal, which is capable of conducting the current and depositing copper in the finest lines and interstices.

The more usual—practically the universal—method is to brush the best and most finely-ground plumbago thoroughly into every detail of the design. Plumbago or black-lead of the best quality is a sufficiently good conductor, but the lower grades are proportionately inferior in this respect. The utmost care must be taken to rub the plumbago with a fairly hard brush into every corner of the mould, for it must be remembered that any point which remains untouched can receive no deposit of metal, with the result that the electrotype-copper will prove defective or covered with pin holes. The plumbago is best applied by sprinkling the fine powder over the whole surface to be covered, and then rubbing it persistently in with circular strokes from the brush until the whole presents the uniform characteristic submetallic dead-black appearance of black-leaded articles.

By soaking 100 parts of the plumbago in 10 parts of silver nitrate dissolved in 200 of distilled water, drying and exposing the mixture to a red heat in a crucible well protected from the action of the air by a fire-clay cover, the silver nitrate is first absorbed by the plumbago, and by the ignition is converted into metallic silver ; and thus the black-lead becomes most intimately mixed with the best conductor known, in the finest state of subdivision. It may be similarly gilt by treating 100 parts with 2 parts of gold chloride dissolved in 200 of ether; simple exposure to light followed by gentle heating in an oven then suffices to metallise the gold. But these methods are costly and the materials can be used only once, for as soon as they are spread over the surface of the matrix, it is practically impossible to recover the precious metal; they are, therefore, rarely met with in practice.

Other metals may be obtained in a finely-divided state. Adams proposed to use powdered tin upon the surface of the moulds. This is prepared by briskly stirring melted tin with an iron rod just as it reaches its solidifying-point; so that the tin is broken up into globules, each of which is coated with a pellicle of oxide, that entirely prevents its union with contiguous particles. The whole of the metal is now poured on to a sieve of the finest muslin or wire mesh, the portion which passes through being retained for use, while the remainder is remelted and again treated in the same way. A still better plan is to stir up the globules with water, allow them to stand for one or two seconds in order to give the heavier particles time to subside, and decant the water, with the finest grains still in suspension, into a second vessel, where it is allowed finally to subside. The clear liquid is then poured away, and the tin powder is dried for use.

A process invented by Knight is used in America. The mould is well cleaned with water delivered from a rose jet, and is then flooded with a solution of copper sulphate. A small quantity of the finest iron filings is now sprinkled over the surface with the aid of a pepper-caster, with the result that a spongy deposit of copper is reduced over the whole area by exchange with the metallic iron, and may be made to form a continuous film by brushing the surface during the time of precipitation. Wahl has accidentally found that pure iron, obtained by reducing ferric oxide in hydrogen, does not give an adhesive copper, owing, as he thinks, to the extreme rapidity of the exchange, the comparatively large though still minute grains of the filings giving a slower reaction than the chemically reduced iron, so that the copper is reduced in actual contact with the surface of the matrix itself during the time of brushing.

PRINTERS' ELECTROTYPING.

The principal applications of electrolysis to the requirements of the printing-trade are to be found in the reproduction of engraved steel or copper plates, of wood-blocks or set-up type. For the last-named purpose, electrotyping has a keen rival in the process of stereotyping, which is more largely in favour in England for heavy newspaper or book-work; but for copying wood-blocks or engraved plates, electrotyping stands alone in the field, the other processes being inapplicable, or productive of inferior results.

ENGRAVED PLATES.

Steel Plates.—Steel plates carefully engraved have a very high value, and as they are slowly worn away by repeated use in the press, in spite of the hardness of the material, their faithful reproduction by electrolytic means effects an immense saving when a large number of impressions are required from one design. Thus, the steel plate may be preserved intact, and any number of copper replicas may be formed, each of which when steel-faced is as lasting and as clear as the original, so that impressions may be multiplied to any degree.

Steel plates, which are to be stored, are usually coated with wax to preserve them from oxidation, and the first step towards obtaining an electrotype is, therefore, the complete removal of the wax. This may be insured by boiling the plate for ten minutes in a strong solution of caustic potash (but not, of course, in the same bath that is used for cleaning plates previous to electro-deposition), then rinsing thoroughly with water, and finally rubbing with a soft rag moistened with benzene. The plate may now itself be made the cathode in an *alkaline* copper-bath, by which means a reversed plate or negative will be produced; or a suitable moulding material, preferably the mixture of gutta-percha, lard, and tallow recommended on p. 151, may be used to produce a non-metallic matrix, which is then rendered conductive as to its surface. The mould, whether made by casting or by electrolysis, is finally suspended as the cathode in an acid copper-vat, until a sufficient thickness of metal has been deposited to fulfil the requirements of the original plate.

In making metallic matrices from valuable steel plates, it is safer not to undergo the risk of spoiling them by suspending them at once even in an alkaline copper-bath, because the slightest trace of solvent action on the steel will be clearly shown by the rounding off of the finer lines, and by the consequent want of sharpness and brilliancy of texture in the resulting engravings. This may be avoided by the expedient of taking a silver matrix from the plate by electro-deposition (see p. 208), and then detaching the silver and depositing copper upon it so as to form a reproduction of the original plate; then, from this, in turn, a second matrix is obtainable, but this time in copper. The steel plate may not be required again, except to renew its replica in copper, in case of accident; and the silver plate having fulfilled its mission may be brought again into use at once by remelting, so that there need be no large stock of a costly material lying idle.

Copper Plates.—Whenever a copper plate is to be electrotyped with the same metal, care must be taken to prevent adhesion between the two surfaces, the process being, in fact, the exact reverse to that of covering one metal with a firm coating of another. A film of foreign matter must be interposed, sufficient to prevent adhesion without so far impairing the electrical conductivity of the surface that it refuses to receive any deposit. The surest method of accomplishing this is to impart the thinnest possible wash of silver to the surface by a momentary immersion in a silver-bath, which may be done without sensibly affecting even the finest lines of the engraving; then by pouring over the silvered surface a small quantity of water containing sufficient tincture of iodine to give to it a pale sherry colour, and rubbing lightly with a cloth, or with cotton wool, a scarcely visible film of silver iodide is formed upon the surface, which will guarantee an easy parting of the plates. When the plate to be copied is not particularly valuable, and the silver treatment is deemed unnecessary, it should be rubbed gently with turpentine containing a small quantity of bees'-wax in solution; on the spontaneous evaporation of the liquid, a mere trace of residual wax will be distributed evenly over the whole plate, and will usually suffice to prevent adhesion: but if the engraved lines are numerous and fine, the former method is to be preferred.

Having prepared the plate to receive the deposit upon its face, the back must be well painted with a water-resisting and non-conductive varnish, so that there shall be no tendency for copper to deposit where it is not required; common copal-varnish answers the purpose very well, and is readily removed at any time by means of turpentine. The plate is now ready to be suspended in the bath, and must be gripped by some metal support which will serve to suspend it from the cross-rods of the bath, and thus to open electrical communication with the battery. Rings or bent wires may be soldered lightly to the back (before coating it with the insulating varnish); or, when there is danger of buckling the plate by the heat necessary to this treatment, the plate may be rested on two or more S-shaped copper strips, bent upwards into hook-shape at the bottom to receive it, and downwards at the top that it may be hung upon the cathode-rods. Or a sliding-frame may be made to grip the plate, such as that sketched in fig. 85. This consists of an upright of varnished wood, with a fixed cross-piece of the same material near the lower end; the cross-piece is made with a longitudinal groove to support the plate; and along the bottom of the groove is a strip of clean copper, to which are soldered at the ends two copper wires insulated, except at the joint with the strip, by means of an india-rubber sheath, and these passing out of the solution above are unsheathed and bent into hook-form, thus serving at once to support the whole arrangement, and to make connection between the plate and the battery. Sliding on the upright rod is a similar grooved cross-bar above the former, but with groove reversed; this may be fixed in any position by a screw at the back, and is merely intended to clamp the plate firmly in position, and to press it into contact with the copper strip in the lower groove; the insulated wires pass through the upper support, and thus act as guides. In using this apparatus no wires need be soldered to the plate; the upper cross-bar is made to

Fig. 85.
Electrotype-plate supporter.

slide upwards, the plate is placed with its lower edge upon the copper strip, and the upper groove is brought down upon its upper edge, and is then clamped in position. The whole frame is now suspended in the bath from the cathode-rods above; connection is thus made with the battery, and deposition at once commences.

All pairs of electrodes placed in the bath in parallel are must be approximately equi-distant, for reasons which will shortly be explained.

Arrangement of Baths and Plates.—With metallic plates of known area, most of which are probably of the same size, the plates in any bath, or the baths themselves, may be arranged either parallel or in series, according to the strength of the current and at the will of the operator. If the dynamo be evolving, let us say, 5 volts pressure (the electro-motive force of batteries may be altered by adjusting the connections), then as each pair of electrodes requires only from 0·7 to 1·0 volt, either five baths, each with all its plates parallel, may be placed in series; or five pairs of plates in each bath may be arranged series-fashion, all the baths being now in parallel are; if the electro-motive force at the brushes of the dynamo be 2 volts, two baths would be placed in series, and so on. Having determined the number of couples which should be connected in series

11

to suit the available *current-pressure*, the distribution of plates
in parallel arc will depend upon the relation of the aggregate
surface of such plates to the *volume* of the current. It has been
seen that a current of from 1·7 to 2 ampères per square decimetre
(or from 0·11 to 0·13 per square inch) is most suitable for copper-
deposition in the acid bath (when the liquid is kept in motion
and of correct density). If there be, let us say, 20 equal pairs of
plates with a total cathode area of 400 square inches, and all the
couples are placed parallel, the best current-strength ranges
from 400 × 0·11 = 44, to 400 × 0·13 = 52 ampères ; but if the
E.M.F. be 2 volts, so that the plates are divided into two groups,
each consisting of ten pairs of electrodes, and these two groups
are placed in series, a current of only 22 to 26 ampères is needed,
because only half the total surface is arranged in parallel arc ;
while if the voltage be 4, the twenty pairs of plates are sub-
divided into four groups in series with five parallel couples in

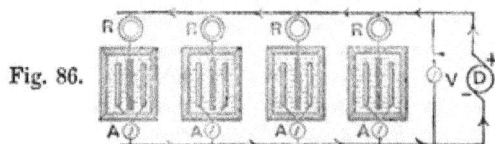

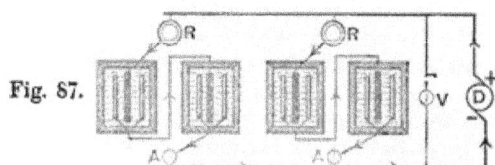

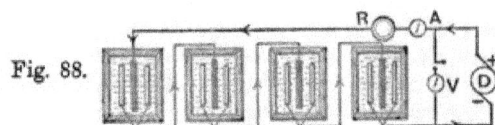

Fig. 86.

Fig. 87.

Fig. 88.

Figs. 86 to 88.—Modes of arranging plates.

each ; and the current required will be only 11 to 13 ampères,
and so forth. All this follows, of course, from the law explained
on p. 98, that a given current of sufficient potential deposits
the same weight of metal in every cell placed in series. It will
be observed that the product of ampères multiplied by volts is
in all these cases the same, showing that the total number of

watts evolved; in other words, the work done is practically identical under all conditions, and the different dispositions are only to be made in order to suit the current to the work.

To make this quite clear, the preceding diagrams are constructed to illustrate the dispositions to be observed in the cases of four cathode-plates, each measuring 40 square inches, this being equivalent to four plates each 8 inches by 5.

In fig. 86, the electro-motive force at the dynamo brushes is 1 volt; all four plates are, therefore, arranged parallel, presenting a total of $4 \times 40 = 160$ square inches of surface, which will require (say) 20 ampères (at 0.0125 ampère per square inch).

In fig. 87, the difference of potential is supposed to be 2 volts. The plates are now placed in two series of two parallel pairs each: the total surface in each group is now $2 \times 40 = 80$ square inches, and the current required is 10 ampères.

In fig. 88, the initial pressure is 4 volts, so that all plates are disposed in series, each one presenting an area of 40 square inches, the total current demanded is thus only 5 ampères.

The arrangement of ammeters, resistance frames, and voltmeters, marked respectively A, R, and V, is shown for each of the above dispositions of baths.

When batteries are used, it is best to arrange them in parallel arc, which will also require a parallel grouping of the plates, because of the low electro-motive force of a single voltaic cell; nevertheless with Bunsen- or bichromate-cells, each of which gives an E.M.F. of nearly two volts, the battery-elements should be placed as before in parallel arc, but the electrodes in two large-series groups. It is impossible to lay down any fixed rule for the number of cells of any battery required to accomplish a given work, because the volume of current depends upon the condition of the battery, upon the temperature, and upon the internal resistance of the battery itself; and the latter varies greatly according to the strength of the exciting liquid (and of the depolariser, when one is used), as well as on the external resistance in the circuit, due to that of the copper-baths and connecting wires. By the use of an ammeter all doubts upon questions of current-strength may be solved, and the operator will know precisely the value of the current with which he is working. When an ammeter is not available, recourse should be had to the table given on p. 347; from this it will be found that 1 ampère should deposit 18·26 grains of copper in an hour; a current of 0·12 ampère per square inch of cathode-surface should therefore deposit about 2·2 grains of copper per hour on every square inch of plate, or 35 grains upon a plate about 4 inches

long by 4 inches wide.   To ascertain the average current-strength
by this means, a deposited electrotype-plate should be weighed
carefully after drying; the weight, expressed in grains, divided
by the number of hours which were required for its deposition,
gives, of course, the weight deposited per hour; and this number
divided by the superficial area of the plate in square inches,
gives the weight of copper deposited per square inch per hour.
Finally, this last number, divided by 18·26 (the weight of copper
per ampère per hour) is the average strength of the current per
square inch expressed in ampères.

To consider an example : *a plate 5 inches by 4 (= 20 square
inches area) is deposited in 10 hours, and then weighs 500 grains ;
find the average current volume per square inch of surface.*

The weight divided by the number of hours

$$= \frac{500}{10} = 50.$$

This number divided by the area in square inches

$$= \frac{50}{20} = 2·5.$$

So that 2·5 grains were deposited on each square inch every
hour.

This number divided by the copper equivalent of 1 ampère in grains

$$= \frac{2·5}{18·26} = 0·137.$$

Therefore, the current-volume = 0·137 ampère per square inch.

From a simple calculation of this description, at least a rough
estimate of the relation of current-volume to cathode-area may
be made, and the disposition of the apparatus governed accordingly.

Character of Copper Deposits—Adjustment of Current.—Further
indications are made to an experienced observer by the character
and rapidity of the deposit.   A film of copper should be observed
to cover the whole surface of the metal plate immediately it is
placed in the bath, and this copper should have a rich reddish-
yellow colour ; a dark-brown deposit usually indicates a more or
less spongy copper and too strong a current ; while any indication of gas-evolution at the cathode is a sign of the current being
far too powerful, and will certainly be accompanied by a pulverulent coating ; on the other hand, a tardy formation, and a
copper possessing an extremely pale shade of colour, shows that
the current is too weak.   With measuring apparatus at command,

it is only necessary to divide the indicated current-volume by the total area of the cathode-plates placed in parallel arc, and to compare the quotient with the prescribed number of ampères per unit of surface, to know in which direction and to what extent there is error in the battery-power applied.

To regulate the current is then a simple matter ; if the current is too strong, greater electrical resistance must be added to the circuit, either by increasing the distance between each pair of electrodes, or by introducing one or more wires of the resistance coil. The latter is the safer method ; if the former, however, be adopted, all the anodes must be shifted equally, otherwise some of the cathode-plates will be nearer to their respective anodes than others, with the result that the local resistance is decreased, and the proportion of current passing through them is increased, and with it the amount of deposit ; some plates may thus be receiving an excessive current, while others are insufficiently served. Should the current be too weak, either the anodes must be brought nearer to the cathodes, if this be consistent with safety ; or a plate may be removed from the bath, thus distributing a slightly diminished current over a much smaller area, and so giving greater current-volume per square inch ; or the battery-power must be increased, unless there were at first added resistance in the circuit from the wire coils, when it will, of course, suffice to remove or reduce this extra resistance.   In causing the plates to approach, care must be taken to prevent their forming a short circuit by coming into contact.

The current having been adjusted, the electrolytic action must be allowed to proceed steadily, with frequent agitation of the liquid in the bath, until a sufficient thickness of deposit has been attained.   This may be ascertained by removing the plate from the bath and rapidly turning up one extreme corner very slightly, by carefully inserting the edge of a thin-bladed penknife between the two surfaces.   The actual thickness required depends upon the treatment which the plate is to receive subsequently ; if it is to be backed or strengthened by another metal the thickness of stout writing-paper will suffice ; but more often the plate will be required to depend upon its own strength alone, and then the depositing should be continued until it is from $\frac{1}{16}$ to $\frac{1}{10}$ of an inch thick, which may require from ten days to a fortnight to accomplish.

The precipitation completed, the plates must be separated by gently inserting the edge of a fine chisel between them on each side, and applying gentle pressure ; if they should not part readily, the chisels must be removed and re-inserted at fresh points—at the ends, for example—and this operation must be repeated until success is achieved. Often the rapid warming of one plate by the sudden application of hot water, followed by an immediate application of the chisels, will effect the separation of a refractory pair of plates. Force must on no account be used; will only result in bending and spoiling one or both of the plates. But usually there should be little difficulty encountered in removing the deposit from the original surface.

The plate has now only to be cut square and finished by mechanical means; it may also be coated with a thin film of hard iron, nickel, or brass by processes to be described in the Chapters dealing with these metals, in order to impart to it higher resisting qualities than the softer copper alone possesses.

Typographical Matter.

When "formes" of type set up for printing are sent to be electrotyped, they must first be carefully examined to see that they are properly prepared for moulding; for success in this branch of work is only secured by patient attention to details.

Preparation of Forme.—In the preparation of the forme by the compositor, due regard must be had to the requirements of the process. In the first place, only strong wrought-iron chases must be used for holding the type in place, thin, cast-iron, or wooden chases being liable to give way under the intense pressure applied in moulding; screw chases are to be preferred if available. The type and rules should be made with a fair amount of bevel, so that a better impression is given to the wax. All the spaces between the letters, and all furniture should be high; low spaces cause much trouble, because the wax is forced deeply into them and, on withdrawal from the mould, stands unprotected, and because the cavities formed by the letters are so deep in the wax matrix, that it is only with difficulty that they receive a good coating of copper; and this is a worse evil than the former, inasmuch as it causes weakness of deposit at the very point where strength is most needed. And, lastly, the leads between the lines should be square, not bevelled, as the wax becomes forced into the fine space made by the bevel, and cannot possibly be removed cleanly, which demands extra labour in the subsequent distribution of the type. If wood-blocks are present in the page they must be exactly type high; if less than this they must be brought up to the required level by packing slips of paper beneath them ; it is preferable that they should be, if anything, a trifle higher than the rest of the page, so that they may give a sharper impression in the printing press.

When the type has been set up in weak chases, they must be carefully removed by "dropping" upon an imposing-surface, and others must be substituted. If the spaces are low, the forme must be planed (made perfectly level) and *floated* by pouring over it a thick cream of plaster of Paris, which must be well rubbed into the spaces by hand, and the excess removed by means of a hard brush applied while it is still pasty ; then after thoroughly drying, it is again brushed to remove every trace of plaster from the angles of the letters. The forme must now be firmly locked ; and at this stage a proof should be taken, that accidental slips or errors may be finally detected and rectified. The surface of the type is then thoroughly cleansed from dirt and stale ink by brushing with potash solution (followed by water), or with benzene, while wood-blocks are similarly cleansed with benzene or turpentine alone. The forme is now dried and brushed over with fine plumbago applied with a soft brush ; the whole surface should thus be perfectly glazed, but all excess must be carefully removed from the finer lines and from the beards of the letters. The plumbago coat not only assists the separation of the type from the wax-mould, but by leaving a trace of the black-lead on the latter, helps to ensure it being thoroughly coated before it is placed in the copper-vat.

Preparation of the Wax-Surface.—Meanwhile the wax should have been prepared to receive the impression. A cast-iron moulding-box (fig. 89) is gently heated, which in reality is only a tray of sufficiently great external dimensions with shallow sides of about a quarter of an inch in depth all round, and with a continuation of the bottom-plate on one of the shorter sides for about three inches beyond the box, to allow of its being supported by hooks from the conducting rods of the bath. It is then filled with the wax-and-turpentine composition formulated on p. 153; this should be quietly melted in a steam-heated vessel, and poured into the box placed upon a level surface until the latter is filled to the brim. Air-bells form-ing on the surface during cooling are removed at once by a touch with a hot iron rod. If the con-traction which occurs during solidification reduces the surface-level, more wax must be added before the former is quite set, as it is better that the composition should overflow than that insufficient

Fig. 89.
Moulding-
box.

should be used, while, if the contraction cause the material to crack away from the side of the box at any point, the fault may be made good by the careful application of a warm iron rod.   In short, it is necessary that a perfectly smooth, level, and unbroken surface of wax shall be presented after solidification.   This surface is then dusted over with the finest plumbago, which is lightly brushed in with a soft brush; the excess being removed, the box is ready for taking the impression of the type.   In place of these iron boxes, brass cases of the same size and shape are coming into use; these have a special "electric connection gripper" which serves to suspend the mould in the solution, and at the same time to make connection with the surface of the mould only, and not with the metal box containing the wax, the back of which, therefore, does not require the protection of stopping-out varnish, as does the iron tray above described, to prevent the deposition of copper where it is not required.   Steam-heat should be applied to melt the wax, as it affords less danger of overheating and burning the composition; this is readily done by using two concentric pots joined air-tight at the rim, the inner one containing the wax, and the space between the two being utilised for the steam; two pipes, therefore, communicate with the outer jacket, one to introduce the steam, the other to carry it off.   Waste steam may often be utilised thus.

Taking the Wax Impression.—The next process consists in carefully placing the forme of type, face downwards, upon the centre of the wax surface and submitting it to intense pressure. For this purpose, either a hydraulic or a toggle-press is usually employed; the action of the former is too well known to need further description, that of the latter is indicated in the sectional diagram (fig. 90).   The counterbalanced swing-cover, C, closed in the drawing, is indicated by dotted lines in its open position for the purpose of introducing the mould and type between it and the rising pressure-plate, P; these having been brought into place, the cover, C, is brought down into a hori-

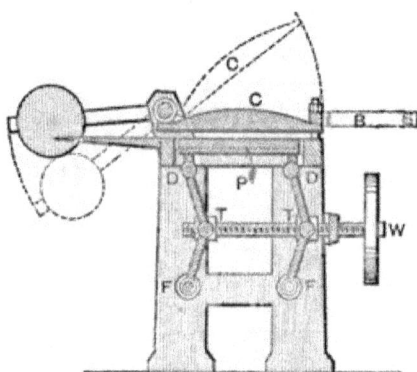

Fig. 90.—Toggle-press.

zontal position and retained there by the movable bar, B. On turning the hand wheel, W, pressure is applied to the two >-shaped
toggle-joints, T T, which are pivoted at their centre and held by the
framework beneath at F F, and attached to the plate at D D, the
result of the pressure being that the >-shaped jointed bars tend
to straighten, and being constrained from moving at their lower
end, thrust the plate carrying the mould with great force
upwards against the cover, and in so doing, press the face of the

Fig. 91.—Hoe's toggle-press.

type into the wax. The pressure must be gradually, steadily,
and evenly applied, until, for large surfaces, it may amount in
the aggregate to several tons. Excessive pressure renders it
almost impossible to separate the forme from the wax without
damaging the latter, whilst an insufficient pressure does not give
the depth necessary for printing, to prevent the inking of the
paper in the spaces between the letters; the mean between the
two pressures is soon learnt by experience, and may be judged
with tolerable accuracy by the force applied on the hand-wheel
or pump of the press.

When the pressure is released and the press opened, the surfaces of mould and type must be separated by inserting bent screw-drivers gently between them at either end, and applying slight leverage until they are almost disengaged; after similar assistance, applied if necessary at the sides also, and when the forme is quite free of the matrix, the latter is removed by lifting it vertically upwards; it must be inspected to ensure that it is perfectly sound in every part.

Trimming the Wax Impression.—All the wax which has been forced up around the sides or into deep spaces must now be carefully pared away with a sharp knife, and other spaces, the levels of which are so dangerously near to that of the type face that there is danger of their printing off

Fig. 92.—Building-knife.

with the type, when they have been electrotyped, are filled up by means of a heated knife (fig. 92) or a *building-tool* (fig. 93). These tools are heated to a temperature a little above the boiling point of water, so that a fragment of wax placed against them melts, and passing down to the point of the tool may be run on to any desired point. This is a critical operation, which requires much care and no little skill to accomplish satisfactorily.

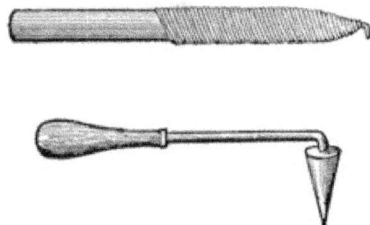

Fig. 93.—Building-tools.

Black-leading the Impression.—The surface must now be most completely black-leaded. This is done by sprinkling a quantity of the finest plumbago over the whole, and then gently stippling and beating it into the wax by means of goats'-hair brushes. The great volume of black dust produced is very unpleasant, and many operators use black-leading machines, which not only prevent the scattering of dust, and consequent loss of valuable material, but effect a more certain covering of the matrix, which will be recognised as a matter of vital importance when it is remembered that a small area of surface insufficiently coated will give rise to a flaw in the electrotype plate, while even a speck will produce a pin-hole. The black-leading machine (fig. 94) is a large rectangular frame with a box beneath, a cover over the whole, and a trellis-table to support the wax matrix; a reciprocating motion is imparted to the table by a hand-wheel placed outside the case, but connected with the necessary mechanism within; and this serves also to produce a vertical reciprocating or dabbing motion to a brush which extends across the whole table at its centre.   On sprinkling the mould with a fair supply of plumbago, placing it face upwards on the table, and actuating the hand-wheel, the wax cast will be drawn to and fro beneath the moving brush, which will force the plumbago into every interstice; the excess of black-lead falls into the box, while the cover prevents the escape of dust into the air.

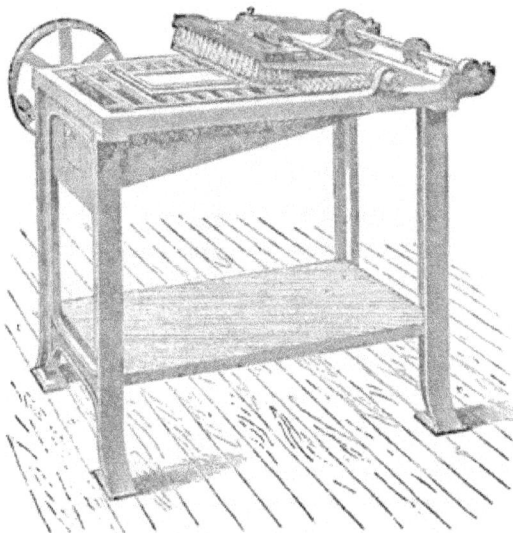

Fig. 94.—Hoe's black-leading machine.

The cover may be made of glass, so that the progress of the operation may be watched, as far as the dust will permit; occasionally even the hand-brushing is conducted under a shade, but this is not altogether satisfactory, as the operator is somewhat cramped in his position, and, therefore, less able to ensure thorough work.

The silvered or gilt plumbago, the tin-powder or the precipitated copper process for metallising the mould, may be used here if desired. Any system which necessitates the use of phosphorus is to be avoided, because this element may render the copper superficially brittle, and this defect is fatal to work that has to withstand the wear of the printing-press.

When the mould is thus rendered superficially a conductor of electricity, every portion which is not required to receive a coating of copper—for example, the back of the frame and the edges of the wax—is painted over with melted wax by means of a soft brush.

Making Electrical Contact.—It is now ready for the electro-depositing process. Electrical connection must first be arranged for by embedding a frame-work of warm copper wire around the wax edge of the mould, then black-leading the surface of the wire to ensure good contact with the plumbago coating upon the wax; and attaching the end of the wire itself to the cathode-rod of the bath. Sometimes the end of the wire only is embedded, but then the depositing action of the current has to spread itself over the whole of the plumbago surface from one point; whereas, by using the frame, it starts simultaneously from the whole circumference and gradually covers the surface towards the centre. When the electric contact gripper is used no further trouble need be taken.

The smaller cavities in the wax would remain filled with air if plunged at once into the copper-vat, especially as the plumbago has a somewhat repellent action upon water until it is once wetted; and as any air space of this kind prevents deposition locally by destroying contact between the wax and the solution, the mould is finally prepared for the bath by placing it in a tray and flowing an ounce or two of spirits of wine over it, to facilitate the wetting of the plumbago, then filling the tray to a depth of about 3 inches with water, and directing a high-pressure jet of water upon the surface from a *rose* held at a height of a few inches above the surface. After washing in this way for a minute or two, the surface should be inspected n various lights while still under water; any air-bells yet adhering to the surface of the wax are thus at once seen, and the washing must be continued until they have disappeared.

Depositing the Copper.—The tray of wax is at once transferred to the acid copper-bath in which it may be suspended after the manner recommended for steel-engravings. It is advisable to increase the electro-motive force of the current employed beyond the normal at first, in order to force the copper deposit over the comparatively weak-conducting surface afforded by the plumbago. Copper should be deposited immediately on the exposed metallic surfaces of the conducting wire, and should gradually spread from this until the whole surface is covered, when the potential of the current should again be reduced; the metal should then continue to precipitate evenly over the entire plate until it has attained to a thickness of the $\frac{1}{100}$ up to the $\frac{1}{35}$ of an inch, which is generally adequate.   Progress is tested, as in depositing upon metallic plates, by gently lifting one corner with a pen-knife; from four to fifteen hours usually suffices.   When sufficient copper has been deposited, the frame is removed, rinsed with water, rested upon a level or slightly-sloping board, and suddenly flooded with hot water on the back of the newly deposited metal; this immediately releases the latter from the wax so that it may be detached at once (every precaution being taken against bending it) and examined by holding it up to the light.   If many holes be visible the plate should be discarded, if only a few, and these small ones, they may be made good in the next process.   The wax matrix can rarely, if ever, be safely used a second time. These electrotype-plates will be subsequently strengthened by "*backing metal*," hence they need not be so strong intrinsically as those which have to bear the strain of the press unsupported; and, therefore, a more intense current may be used than is permissible—for example, in the reproduction of engraved plates; but on no account must the volume be so intense that hydrogen is deposited with the copper, for not only is the metal itself weak (even if it be coherent at all) under these circumstances, but the clinging of the bubbles of gas to the work is certain to produce pin-holes in the plate.   With a 20 per cent. solution of copper sulphate, slightly acidified and constantly agitated, a current of 0.2 to 0.225 ampère per square inch (3 or 3.5 ampères per square decimetre) is quite the maximum that should be adopted.   For work which will be carefully used a thin deposit may suffice, such as might be deposited by 0.13 to 0.16 ampère per square inch (2 or 2.5 ampères per square decimetre) in three or four hours; but for plates which may have to withstand rough treatment or long wear, fifteen or twenty hours may have to be given.

Backing the Copper-Sheet.—After examination, the trace of wax which adheres to the copper is removed by a rinsing with caustic potash solution, followed by a thorough washing with water.   The next operation is to protect the thin shell with a strengthening metal.   The back of the copper sheet is painted over with a solution of zinc chloride containing a little sal-ammoniac (ammonium chloride), or borax; it is then rested on an iron tray which is suspended in contact with the surface of melted backing-metal.   Granulated tin-lead alloy or foil containing 50 per cent. of each of these metals is then placed upon the copper sheet, and the heat is continued until the white metal has just melted—not higher, lest the copper become

oxidised. The tray is now removed to a level place and a small
ladle full of backing-metal, which has been well skimmed, is
slowly poured over the surface, commencing at one corner, until
a depth of about one-eighth of an inch is attained. It is then
allowed to cool. The backing-metal is an alloy of 91 per cent.
of lead, 5 of antimony, and 4 of tin (by weight); it should not
be overheated—a temperature of about 600° F. is suitable. A
rough practical test is to dip a scrap of white paper into the
molten bath, when it should become only just discoloured, any
stronger signs of scorching showing that the temperature is too
high. The object of the preliminary coating with tin and lead
is to ensure a sound union between the copper and the backing-
metal, such as could not otherwise be guaranteed.

After subdividing with a circular saw, if necessary by reason
of the treatment of separate blocks or pages upon the same plate,
the copper is examined with a steel straight-edge, and if not truly
level, the positions of defective portions are marked on the back;
it is then straightened by gentle blows with a polished hammer,
taking every care that the face be not damaged. After obtaining a plane surface, the excess of backing-metal is shaved
off in a specially constructed lathe or hand shaving-machine; it
is then trimmed and again tested with the straight-edge; irregularities are again rectified, and it is finally reduced to exactly
the required thickness (usually that of a small pica) by a
hand planing-machine; after finally bevelling at the edges it is
mounted on wood, type high.

Any backing-metal which has found its way to the surface
through pin-holes in the copper may generally be removed,
unless the soldering-fluid has also penetrated, and so caused the
two clean metallic surfaces to unite. Unevennesses and defects
of this character must be set right by a competent workman
with a knowledge of engraving, to whom the final examination
of the finished plate should be entrusted.

Gutta-percha composition-moulds from type-formes are treated
in the same manner as wax-impressions.

WOOD-BLOCKS.

Wood-blocks, like typographical matter, may be copied by
wax; but since they are liable to be damaged by extreme
pressure, it is safer to mould them in gutta-percha rendered
plastic by heat, or by pouring the melted gutta-percha and lard
mixture over them as described on pp. 151, 152; then, after rendering them conductive, they are electrotyped, trimmed, backed,

planed, and mounted in the same way as those produced from wax-matrices.

Art Electrotyping.

In this group may be arranged the reproduction of medals, medallions, busts, and statues, or objects of vegetable and animal origin and the like.   Among the principal points to be observed are the choice of a suitable moulding-material, and the carrying out of the casting on the one hand, and the arrangement of the anodes in the bath on the other.

Moulding.

Medals.—Medals, coins, (or medallions if metallic) may be coated directly with copper after the manner of copying engraved steel-plates, the copper-matrix then being used to electrotype upon, provided that they are not in any degree undercut.   Only one side can be treated at a time, and the back must be protected by a stopping-off varnish, while the face is brushed over with the solution of wax in turpentine, already described, to prevent adhesion.   If possible, the connecting wire should be soldered lightly to the rim of the medal; but as this is rarely allowable the object may be slung in a wire loop, or it may be placed in a copper tray as described by Urquhart, and depicted in fig. 95; these trays are made of thin sheet copper painted on the outside with Japan black to prevent local deposition, but left bright inside to make connection with the medal; they are supported in the bath by the hook shown above, the medal being merely fitted into the bottom of the tray.   They may be made of various sizes to take any coin or medal of which a copy may be required.   When the object is slung from a wire loop, its position must be shifted from time to time to prevent the formation of

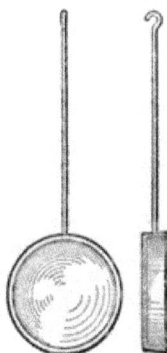

Fig. 95.—Copper trays.

wire-marks.   Medallions made of a non-conducting material must be made conductive by plumbago or thin metal "leaf;" a process which is rarely admissible.

It is usually safer, in any case, to prepare a mould from the medal in preference to taking an electrotype-matrix.   To this end the medal is rubbed lightly over with plumbago by means of a brush to which a circular movement is given; it is then placed upon a flat surface, preferably protected on the under side by a disc of chamois leather, if there be designs on both sides and the "relief" of the lower one is nearly as high as its surrounding rim ; it is then moulded with plastic gutta-percha or with the fluid composition as explained in the section dealing with moulding materials. Either of these methods can be recommended, but any of the other materials described in the beginning of this chapter may be employed as there directed. The mould is then rendered conductive with plumbago or metal and is ready to be electrotyped, the methods of doing which have been already dealt with in full.

Busts and Statues.—The moulding may be first effected in the elastic composition (see p. 156) by placing the object, slightly oiled on the surface, if permissible, within a box with tapering sides slightly greased, and gently pouring in the warm liquefied mixture, until the object is covered and the box completely filled, taking care that no air-bells form upon the surface of the former during the operation. The whole is now allowed to stand in a cool place until solidification is complete, when the box is removed, and the composition is cut through to the statue or bust from top to bottom, with the aid of a sharp knife, along a line previously determined by the shape of the object to be the most convenient. Being elastic the mould may now be opened out and withdrawn from the object, even if it be somewhat sharply undercut ; once removed it returns to its original shape by virtue of its elasticity. The interior cavity, which of course has taken the form of the moulded object, may be hardened and waterproofed as above described, and then coated with a conductive film and subjected to electrolysis ; but owing to the difficulty in rendering it completely water-resisting, and to the fact that wherever liquid may penetrate the mould will swell out of shape, it is not advisable to adopt this plan. It is better to prepare a special wax-composition by melting together a mixture of bees'-wax, rosin, and Russian tallow, in the proportion of 50 : 40 : 10 respectively, and pouring this, just at the moment before it sets, into the hollow space within the elastic mould, which should be re-enclosed for the time in its original containing-box ; the wax-mixture must not be too hot, or the two compositions will unite and the whole operation will be ruined. When the wax has solidified throughout, the mould is stripped from it, and an exact wax-reproduction of the original bust should result. This in turn is placed in a suitable box, and the space around is filled with a cream of plaster of Paris, which must be allowed to harden ; it is then removed from the box, dried and heated over a trough, in an oven or stove, to a temperature sufficient to melt the composition from the interior ; the side of the plaster block, which had been in contact with the bottom of the box, and upon the surface of which the base of the wax-object is visible, is, of course, placed downwards, so that as the wax melts it runs into the tray prepared for its reception. A small proportion of the wax is absorbed by the plaster, and thus renders it non-absorbent. The interior of the cavity in the plaster is now rendered conductive by black-leading or metallisation, and is ready for the electrolytic process. When, however, the whole plaster-mould is to be immersed in the vat, the outside must also be waterproofed by painting it with wax and subsequently applying heat ; and it is desirable also in this case to cut a small aperture through the plaster at the highest part of the object (the top of the head in the case of a bust) so that a constant circulation of the electrolyte may be effected during the time of deposition ; this channel also must be made impervious to water.

A similar but shortened process is applicable to the copying of models moulded in wax, if the original may be destroyed—the operation is taken up at the second stage of the above cycle, the plaster being poured around the object at once, leaving only an opening at a convenient point, through which the composition may be melted out, and the copper solution and anode introduced. When the original wax-model may not be sacrificed, the longer process of taking a first impression in elastic composition must be resorted to, but the fracture of projecting portions of the brittle wax must be carefully guarded against.

Other systems of moulding are also in vogue. Lenoir's method for reproducing statues in a manner approaches in principle to that of the foundry. He moulds the figure in gutta-percha in a sufficient number of different parts, the sections being so disposed and marked that when united together they form a complete mould of the object ; the different internal surfaces are black-leaded and then fitted around a skeleton-anode of wire ; a convenient number of apertures are made above and below to afford communication between the exterior and interior of the mould, for connecting the anodes with the battery and for the circulation of the solution ; and the arrangement is ready for electrolysis. A knowledge of the moulder's art is very valuable, if not indispensable, in determining the most suitable method of dividing up the surface of the statue into sections.

Large statues moulded in plaster have been copied by rendering them impermeable by liquid, coating the whole exterior with plumbago, and then immersing them as cathodes in an acid copper sulphate bath, until a thickness of about one-sixteenth of an inch of copper has been deposited. They are then cut through at suitable points, where the marks of the joins will be least conspicuous on the finished reproduction, and the plaster being completely removed, the outside is joined to connecting wires and covered with stopping-out varnish, and the inside is rendered dirty by the turpentine solution of wax, or by painting with dilute ammonium sulphide, which gives a superficial tarnish of copper sulphide; the excess of the ammonium sulphide must be washed away, and copper is then deposited upon the different sections of the copper matrix individually; when a thickness of copper of at least one-sixth of an inch, but preferably a quarter or even a third of an inch, has been acquired, the thin copper mould is stripped away, and the separate portions of the electrotyped statue are mechanically finished off and fitted together to form the complete figure.

The mould having been prepared and rendered conductive by any of these processes, it is finally arranged to receive the deposit of copper; the main point now to be observed is that the anode shall be as nearly as possible equidistant from every part of the cathode-surface. If this be not attended to—for example, in electrotyping statue-moulds—the chief recesses in the mould will receive the thinnest deposit of metal, whereas they will afterwards be subjected to the greatest wear, being the most prominent portions of the finished surface, and should, therefore, by preference have increased rather than diminished in thickness. This matter needs careful consideration, and the ingenuity of the workman may often be taxed to find the best possible arrangement. For shallow-cut medals or plane surfaces with no design in high relief, a flat anode placed at some little distance may suffice; for reasons fully given on p. 100, the electrodes must not be allowed to approximate too closely. But for surfaces which are raised at any point, and which, therefore, produce deeply-cut moulds, the anode-surface should be dished out into an approximate representation of the mean lines of the original object. It may then be placed nearer to the cathode, and thus impose less resistance in the circuit. But in dealing with statue-moulds, the problem is more difficult. Lenoir meets it by using an anode of thin platinum wire, bent backwards and forwards into a framework or skeleton of the figure, of course of smaller size, so that there shall be no danger of contact with the plumbagoed mould. The mould is then built up around this (vide supra), and the whole of the cavity is filled with the copper-solution; the metal is deposited, and the wire-skeleton is finally removed by withdrawal through one of the cavities in the plaster. Planté used a similar skeleton composed of perforated lead sheet, fashioned roughly into the required shape, and this being comparatively inexpensive was left within the statue when the process was finished. When the statue is moulded in sections, there is, of course, less difficulty in adapting a suitable anode.

The lead and platinum anodes do not dissolve in the solution; the strength of the latter must be kept up by adding crystals of copper sulphate from time to time, as the copper which it contains initially becomes exhausted. It is mainly for this reason that apertures must be provided in the mould-walls for the circulation of the liquid, which may be maintained by placing a muslin bag or copper wire-box, containing crystals of copper sulphate, above the head-aperture, as this produces a gradual downward flow of heavy liquid containing fresh supplies of copper salt. Another result of using insoluble anodes is that a current with higher electro-motive force is necessary to effect the deposition (see p. 29); and, again, when platinum is used, oxygen gas is evolved, and for this reason the head-aperture must be at the very highest point to allow the gas to escape, otherwise an accumulation of gas forms at the summit of the figure so that the mould will not be in contact with the solution, and from that time can receive no further deposit of metal. The lead anode, especially at first, combines with and thus absorbs a large proportion of the oxygen, to form lead peroxide; and in proportion as this is formed less electro-motive force is required because the heat of the lead undergoing oxidation is a substitute for that of the copper dissolving at the anode in ordinary electrotypy. When copper anodes are used it is more than ever of importance that they should be of the purest electrotype-copper, to prevent the formation of insoluble mud, which would deposit upon the interior surfaces of statue-moulds, and give rise to much inconvenience.

Care is needed to ensure that the anodes at no time short-circuit the current, and stop the process by coming in contact with the cathodes. This is especially liable to happen in electro typing statues or busts, because the slightest movement may alter the relative positions of the surfaces inside the moulds, and it is impossible to watch the progress of the deposition. Lenoir has suggested that the outside wires of his platinum-skeleton should be encircled by an extended spiral of india-rubber filament, which is an insulator; but although for a time this would be successful, it is probable that by the gradual growth of the deposit the precipitated copper might creep up to the anode and effect contact at some point, at which it happened originally to approach the cathode too nearly. Such short-circuiting would, of course, be fatal to the deposition upon all the moulds which might happen to be in the same circuit; Lenoir, therefore, introduced into the circuit of each individual mould, a short length of thin iron wire sufficient to carry the comparatively small current required for the electrolysis, but which would heat, and almost immediately fuse, by reason of its high electrical resistance if subjected to the much stronger current entailed by a short circuit. In this way the iron acts as a safety-valve, automatically breaking the circuit in the branch in which the accident has occurred, and restoring it to the remaining electrotypes in the bath. It is in fact what is known as a fusible *cut-out*, such as are now usually made of lead foil for electric-lighting circuits. The more modern form would answer the same purpose, being made to melt and break the circuit as soon as it is subjected to an undue intensity of current, the thickness of the lead being determined by the strength of the current which is to call it into use.

For this class of work the ammeter should always be employed; it would not only indicate short-circuiting as soon as it occurred by registering the greatly-increased current flowing in the circuit, but it would also show whether the process was taking its normal course. Any undue approach of the electrodes would diminish the resistance and give rise to an increased current-volume, while any break or defect in the wires would be shown by the diminished ampèreage recorded by the instrument. It, alone, affords an opportunity of judging of the progress of the work within a closed mould.

Having decided upon the best arrangement of anodes, the method of ensuring the most rapid covering of the mould demands attention. When the matrix or mould is of metal, no difficulty arises, because of its high conductivity; it is only necessary to connect any part of the mould with the generator to ensure an immediate deposition over the whole surface exposed. But when connection is made between the battery and cathode at only one point in a large non-conductive mould, a long time must elapse before the deposit will spread to the more distant portions of the plumbagoed surfaces; but the deposit is more uniform, and the result, therefore, more satisfactory, in proportion to the rapidity with which the mould is initially covered with copper. In order to convey the current to several parts of the mould at once, light guiding-wires may sometimes be temporarily arranged so that their points rest lightly on the plumbagoed surface; these act as so many nuclei or starting-points for the deposit, and, as soon as the copper has spread from them and covered the intervening surfaces, they are no longer required and may be removed. These guiding-wires are specially useful in carrying the deposit into the deeper or under-cut portions of the mould, into which it is often difficult to drive it at the outset, but which continue to receive a deposit when once they have been covered by a better conducting surface than the plumbago. The wires cannot be well arranged in the internal cavities required for reproducing busts and statues, as they are liable to make contact with, and to disturb the position of the anodes. They may, however, be sometimes passed per-manently through the walls of the mould itself by adjusting them within the casting-box, so that their points rest very lightly upon raised portions of the wax-model (preferably at such points that they may not mar a flat surface on the finished figure by any mark indicative of their position). The several wires are collected into a bundle together outside the mould, and are then connected with the negative wire from the battery. The tip of the wire should be flush with the internal surface of the plaster or composition in the finished matrix, and being black-leaded will not adhere to the deposited metal. The deposit may often be coaxed into a refractory corner by using a supplementary anode of stout copper wire or thin sheet, which, being connected with the battery (positive pole), is held tempo-rarily with its surface very close to, but not touching the part to be covered; thus the local resistance of the solution is much diminished and the deposit is readily started, and, when once formed, will continue to increase without difficulty.

In all art-work of the description to which we have been latterly referring, the conductive film must be of the finest quality in order to transmit the current with the utmost rapidity from the points of original contact to the remainder of the surface. The silvered plumbago offers great advantages in this connection; and if ordinary plumbago be employed, only the best description is permissible.

The finished electrotype generally presents a dirty appearance, owing to the black-leaded surface with which it has been in contact; it may, however, be cleaned by rubbing with turpentine or benzene, sometimes after a preliminary plunge into boiling oil.

Reproducing Natural Objects.—Animal or vegetable objects are often simply coated with a thin film of copper, and used in this condition for ornamental purposes, an additional deposit of gold or silver being added to that of the copper. To effect this, a conductive surface is first formed upon the object to be coated; warm spirits of wine are shaken with crystals of silver nitrate until no more of the solid is dissolved. The object is then painted superficially with this solution, and placed under a glass bell-jar, or clock-shade, together with a saucer containing a few drops of a solution, made by dissolving a small fragment of vitreous phosphorus in an ounce of carbon bisulphide. The vapour of phosphorus evolved reduces the silver nitrate to the metallic state, and thus covers the whole surface of the object with a thin but continuous conductive film of silver, and enables it to receive an electro-deposit of copper of any desired thickness by merely suspending it as the cathode in an acid copper-bath. The greatest care is required in using this solution of phosphorus; it is liable to produce painful sores if it fall upon the skin of the operator and be not immediately washed off, and to cause spontaneous ignition, even after the lapse of a considerable time, if it come in contact with organic fabrics.

Instead of merely covering the object with a film of copper, it may be reproduced by taking a mould in suitable material, black-leading its internal surface, and electrotyping it in the manner of statues or busts. The copying of any insect or leaf becomes thus a question of moulding.

Many other applications of electro-metallurgy in connection with copper are used, especially in connection with the multiplication of drawings and designs: among such processes may be mentioned—

Glyphography, which requires a flat copper plate to be either coated with two layers of composition, one black, the other white or, preferably, to be itself rendered black by exposure to ammonium sulphide solution; it is then coated with a white material. The required design is scratched through the wax until the black surface of the copper is visible; when the drawing is complete—which is readily seen, because the lines appear black upon a white ground—the whole surface is coated with plumbago and electrotyped to a thickness of about the $\frac{1}{32}$ of an inch. This is supported with backing-material, like an ordinary electrotype-block, and is ready for printing in the typographical printing-press, the lines of the drawing being, of course, in relief upon a flat surface in the finished plate.

Stilography is somewhat similar as to the mechanical part of the process. The copper plate being covered with a mixture of 67 per cent. of shellac and 33 of stearine, with sufficient lamp-black to render it black, is varnished and sprinkled lightly with silver dust.   The latter is then removed along the lines of the intended design until the black composition is seen beneath : the whole is then plumbagoed and electrolytically coated with copper.   The lines are not sufficiently raised for ordinary type-printing, a second plate must, therefore, be taken from the first, reproducing the etched lines of the original, and this is used for printing as from an engraved surface.

Galvanography consists in building up a picture in coloured varnish, the gradation of light and shade being given by varying the thickness of this film.   After black-leading, the surface is coppered, and, being cleaned with oil of turpentine, is used like the last as an engraved plate.

It is obvious that any of the photo-mechanical printing-pro-cesses of the present day, in which the printing-surfaces in relief are obtained from photographic reproductions of any drawing or suitable object, may also be aided by the sister art of electrotypy : the electrolytic part of the process is practically the same in all, and differs not from those already instanced, so that it is unnecessary to describe any of them in further detail.

Electrolytic Etching.—By the reverse of these processes the same result is attained.   The design is traced on the waxed surface of a copper plate, taking care that the etching-tools lay bare the metal in all the lines.   The plate is now introduced into the copper-vat in connection with the anode wire instead of the cathode, and the copper dissolves at all places where it is exposed to the action of the solution ; but since the whole plate is in-sulated with the exception of the lines of the etching, it follows that along these only the copper is attacked.   The lines may thus be bitten-in to any required depth ; the depth may be determined at the will of the operator by adjusting the distance between the electrodes, the points of nearest approach being those which receive the deepest cut.   The resulting plate is used precisely as an ordinary etched copper plate ; and, indeed, it is such, the processes employed to produce them being identical except in the method of biting-in the lines.   To obtain any-thing but crude results, however, by these processes demands much experience and attention, as it is frequently necessary to stop-out some of the finer lines to prevent further action at different periods of the process, and practice and artistic skill alone can guide the operator in this matter.

An ingenious process for obtaining *nature-prints* of leaves and similar bodies is sometimes used.   The leaf is placed between two plates, one of polished steel, the other of soft lead, and is then passed between rollers which exert a considerable pressure.

The leaf thus imparts an exact impression of itself, and of all
its veins and markings, to the surface of the lead ; and this
impression may be electrotyped and the produced copper plate
used for printing in the ordinary way.

The subject of reducing copper from its ores and refining the
crude metal will be dealt with in a separate chapter later in the
Work.

CHAPTER IX.

THE ELECTRO-DEPOSITION OF SILVER.

THE electro-plating of articles with silver was one of the earliest applications of electrolysis, because it produced a material analogous to, but cheaper than, the older "silver-plate," in which the base metal was covered *mechanically* with a layer of silver; and even at the present day, when electrolysis is used to obtain coatings of so many different metals for such varied purposes, the deposition of silver must, perhaps, take the foremost place, both in respect of universality of practice and value of results.

Deposition by Simple Immersion, or Whitening.

Silver, as compared with most metals, is very electro-negative, and hence all the base metals are capable of exchanging places with it when dipped into a solution of one of its salts. This process is, however, used only to impart the thinnest possible wash of silver, more especially to small articles such as nails and hooks; so thin, indeed, is the film that the name *whitening* is thoroughly descriptive of the process. It is clear that the coat of silver can be but of the thinnest, because, as soon as the metal is covered with the slightest covering of silver, it becomes protected, partially at least, from further contact with the solution.

There are two principal methods of silvering by simple immersion—firstly, by dipping the article into a solution of silver, either hot or cold; secondly, by rubbing a semi-solid paste of a silver compound over the surface of the object. For both processes the objects must be clean, and must present bright metallic surfaces to the action of the depositing compound. Formulæ for making-up such silver mixtures are numerous; those principally used are included in the following table, in which they are arranged under the respective class-headings of solutions and pastes :—

TABLE XIII.—SHOWING THE COMPOSITION OF SILVER SIMPLE-IMMERSION MIXTURES, RECOMMENDED BY VARIOUS AUTHORITIES.

No.	Authority	Special Application of Mixture	Most Suitable Temperature.	1 Silver Chloride.	2 Silver Nitrate.	3 Potassium Cyanide.	4 Caustic Potash.	5 Potas. Carbonate.	6 Potas. Bicarbonate.	7 Potassium Binoxalate.	8 Potassium Bitartrate.	9 Potassium Ferrocyanide.	10 Alum.	11 Sodium Carbonate.	12 Sodium Chloride.	13 Sodium Hyposulphite.	14 Ammonia.	15 Ammonium Chloride.	16 Levigated Chalk.	17 Water.	Special Method of Preparation.
		SOLUTIONS.																			
1	Elsner	…	Boiling	7·5	…	57·5	…	…	…	…	…	…	…	27·5	15	…	60	…	…	1000	Dissolve 1 in 14 and add to rest. Filter.
2	Gore	…	"	…	10	60	…	…	…	…	…	…	…	…	…	…	…	…	…	10·00	"
3	Roseleur	…	Hot	1	15	50	…	…	…	…	64	…	…	…	64	…	…	…	…	q. s. 1000	Pour 2, in 100 of 17, into 3 in 300 of 17.
4	Wahl	Buttons, &c.	Boiling	80	…	26	…	…	…	…	…	…	…	…	80	…	…	…	…	1000	Dissolve 5 and 9 in 17, boil, and add 1.
5	"	Copper & Iron	Hot	2	4	12	32	…	20	…	…	…	…	…	…	…	…	…	…	10·00	
6	Watt	…	…	…	…	…	…	80	…	…	80	120	…	…	80	…	…	…	…	1000	
7	Watt	…	…	…	…	…	…	…	…	…	…	…	…	…	…	…	…	…	…		
		PASTES.																			
8	Gore	…	…	1	…	…	…	…	…	…	8	…	2	…	8	…	…	…	…	q. s.	Add enough 17 to make paste.
9	Kühn	…	…	…	40	…	…	…	…	…	…	…	…	…	…	100	…	20	…	q. s.	Prepare as required, and apply with rag.
10	Roseleur	Lamp Reflectors	…	15	…	…	…	…	…	…	…	…	…	…	…	80	…	800	80	1000	Mix 1, 3, and 17, add enough 16 to make paste.
11	"		…	2000	…	300	…	…	…	3000	3000	…	…	…	4200	…	…	…	…	1000	
12	"		…	50	…	…	…	…	…	…	…	…	…	…	…	…	…	…	q. s.	1000	Add enough 17 to make paste.
13	Stein	Brass & Cop., small goods	…	1	1	3	…	…	…	…	…	…	…	…	…	…	…	…	…	q. s.	"
14	Watt		…	4	…	…	…	…	…	…	5	…	…	…	…	…	…	…	…	q. s.	"

NOTE TO TABLE.—The black figures in the last column refer to the numbers of the vertical columns denoting the various ingredients.

Immersion Solutions.—It is obvious that since the deposition of the silver is due (and is also proportional) to the amount of the base-metal which dissolves from the object under treatment, the solution gradually becomes exhausted of the former and contaminated with the latter; and if the articles are of copper or brass, as they most frequently are, the fact of the contamination, and in some degree its extent, are rendered apparent by the blue colour imparted to the solution by the dissolved copper. Most of the liquids are used hot, and a momentary dip suffices to effect the required purpose. One of the best is No. 3 (Roseleur's), prepared by making up into a paste 1 ounce of silver chloride with 4 pounds each of powdered potassium bitartrate (cream of tartar) and sodium chloride (common salt), then adding a proportion of this to boiling-water, contained in a copper vessel, immediately before it is required for use. The articles, held in a copper sieve or porcelain colander, are plunged into the solution, where they become coated instantaneously; but for the sake of security they should be stirred around with a piece of wood or with a porcelain or glass rod; they may then be removed, thoroughly washed by rinsing in two or three vats of water, and dried in hot boxwood sawdust. As this bath works best when old, and consequently highly charged with copper, care must be taken that no pieces of iron or zinc or other very electro-positive metal be clinging to the goods, or a certain proportion of copper will be deposited with the silver, which will in consequence acquire a pinkish colouration.

Cyanide solutions may be made to give a good whitening-effect, as indeed may any of those specified in the above table. An interesting process of Roseleur's is not included in this table; the liquid is prepared by slowly adding a solution of silver nitrate to one of sodium bisulphite, until the precipitate, which forms upon admixture, begins to dissolve but slowly in the solution on shaking. The copper or brass objects are dipped into the bath, cold, and immediately become covered with silver by simple exchange; but after this, unlike the behaviour of other solutions, the film continues to increase in thickness, not, however, on account of any further solution of the base-metal, but owing to a chemical action inherent in the bath itself, which causes the deposition of the silver, not only on the metallic objects immersed, but even on the walls of the bath, on glass, or on any substance introduced. This is due to the ready decomposability of the silver salt employed, and to the tendency of sulphurous acid to absorb oxygen, which it does at the expense of a portion of the silver oxide, depositing an amount of silver corresponding to that of the oxygen used up. This reaction occurs but slowly in the cold, so that there is time for a gradual building up of the silver into a coherent and adhesive deposit. If the liquid be heated, the action becomes too rapid and the quality of the coat suffers accordingly.

Pastes.—The use of pastes is especially applicable to the wash-silvering of comparatively large and flat surfaces, such as the dials of barometers, and for the application of local deposits, or even of preliminary protective films to bodies which are subsequently to be plated with an electro-negative metal. The simplest paste is that made by rubbing together 1 part of silver chloride with 2 or 3 parts of potassium bitartrate (cream of tartar) until they are in a condition of the finest powder, and then working the mixture into a creamy paste by the addition of water. Many operators vary these proportions, or add other ingredients, but the mixture, as it stands, will be found to give excellent results. Roseleur's paste for silvering lamp-reflectors (No. 12 on above list) is rubbed on to the surface with a wad of soft rag, allowed to dry *in situ*, and is then rapidly removed with a fresh piece of soft linen.

In applying the pastes generally, a piece of soft cork or a pad of wash-leather may conveniently be employed. Many of the so-called "plate-restoring powders" used for restoring a white colour to worn electro-plate, which shows the brass foundation in places, consist of one or other of these mixtures or of modifications of them. Occasionally plate-powders containing mercury are sold; they are, however, fraudulent, for they purport to give a film of silver to the discoloured objects, but instead impart one of the less expensive mercury, which is in every way to be condemned, for not only is the mercury itself objectionable, but it is gradually absorbed by the base-metal, leaving the surface dull, while repeated applications cause the object to become brittle and useless.

The thickness of silver on whitened goods is usually so infinitesimal that they will not bear scratch-brushing, or any of the ordinary methods of polishing; but friction by contact with dry sawdust in a rotating barrel may be satisfactorily substituted.

SINGLE-CELL PROCESS.

This process is not largely used for silver-deposition, and is quite unsuitable to establishments where there is much work in hand. It may, however, be effected by using an ordinary

cyanide plating-solution, containing a porous cell with a zinc rod
or plate immersed in potassium cyanide solution, with the usual
connections between the zinc and the objects which are being
coated in the outer cell. Steele prepared a solution for single-
cell work by converting 1 part of silver into silver chloride,
washing and dissolving it in 60 parts of water, in which was
also placed the mass resulting from the fusion of 6 parts of
potassium ferro-cyanide with 3 of potassium carbonate. No
porous cell was used, the object to be plated was simply connected with a plate of zinc, and both together were plunged into
the prepared solution. In a similar manner articles immersed
in hot silver-baths have been sometimes treated by simply
binding zinc wire around them, so that a greater thickness of
deposit would be given than that impartable by simple immersion. The separate-current process may be said to be universally
applied to electro-silvering, as the plant may be made of any
size, and the process is under perfect control.

THE SEPARATE-CURRENT PROCESS.

In working silver solutions with a battery or dynamo-electric
machine, the solutions must be well watched, and the current
prevented from becoming excessive, as a good fine-grained
minutely-crystalline deposit can never be yielded by a large
current-volume. Resistance-coils should, therefore, be at hand,
or some other suitable means of regulating the current under all
conditions of the bath, and under all dispositions of the
electrodes within it.

The Battery.—The Smee- or Daniell-cells are, perhaps, to be
most recommended ; the former being arranged in groups of two
in series, when more than one cell is employed, so that the
electro-motive force may be twice that given by a single pair of
the plates. A single Daniell-element gives an electro-motive
force very suitable to the work (1 volt), and if several cells are
used they should be placed in parallel arc. Some operators
prefer the original copper-zinc cell, probably because its prime
cost is less than that of Smee's, owing to the absence of
platinised silver. It is, however, less effective, and becomes very
badly polarised as soon as its action commences, but by using
a number of couples and plates of large size, it is quite possible
to obtain excellent results with it. If a dynamo be used, it
should have a very low electro-motive force, because it is less
convenient to arrange the silver-baths or the individual electrodes in each series-fashion, than it is in the electrotype-copper

TABLE XIV.—Showing the Composition of Silver-Baths for Separate-Current Process, recommended by various Authorities.

No.	Authority	Special Application of Bath.	Silver Chloride	Silver Cyanide	Silver Iodide	Silver Nitrate	Silver Oxide	Silver Carbonate	Potassium Cyanide (96%)	Potassium Iodide	Sodium Carbonate	Sodium Chloride	Ammonia	Water	Special Method of Preparation, &c.
			1	2	3	4	5	6	7	8	9	10	11	12	
1	Böttger	Cast-iron	…	…	…	15·6	…	…	31·2	…	…	15·6	…	1000	Dissolve 4 in 250 of 12; add 7, and then 10 in 750 of 12.
2	Elsner	…	…	…	…	18	…	…	30	…	…	…	…	1000	Dissolve 4 in part of 12, 7 in rest of 12; mix.
3	Gore	To "whiten"	…	8·7	…	…	…	…	14	…	…	…	…	1000	(Use 3 to 10 Smee-cells in series.)
4	Gore	To finish	…	34	…	…	…	…	20·5	…	…	…	…	1000	(„ 3 to 10 „ parallel.)
5	Parkes	…	…	10·2	…	…	3·7	…	30·6	…	…	…	…	1000	Dissolve 4 in 5·0 of 12; and 7 in rest; mix. Filter, if necessary.
6	Pfanhauser	…	…	…	…	13	3·7	…	50	…	55	…	…	1000	
7	Roseleur	…	…	12·5	…	…	…	…	18	…	…	…	…	1000	Dissolve 7 in 12, and 2 in mixture.
8	Volkmer	…	25	25	…	…	…	…	28	…	…	…	…	1000	
9	Wahl	Steel plates	…	31	…	…	…	…	150	…	…	…	…	1000	
10	Wahl	„	…	4·2	…	…	…	…	35	…	…	…	…	1000	
11	„	Striking } Plating }	32·3	…	…	…	…	…	37·5 / 37·5	…	…	…	…	1000 / 1000	(Rogers Plating Co., U.S.A.)
12	Watt	Striking } Plating }	…	3·3–4·2	…	…	…	…	75–100 / 75	…	…	…	…	1000 / 1000	(Meriten-Plating Co., U.S.A.)
13	Watt	…	…	25·4	…	…	…	9	q.s. + x.s.	…	…	…	…	1000	Mix 6 in 12; add just sufficient 7 to dissolve 6, and then slight excess. (Requires no quicking.)
14	„	German silver	…	…	15·2	…	…	…	q.s. + x.s.	…	…	…	…	1000	
15	„	To obtain very white deposits }	9·8	…	…	…	…	…	q.s. + x.s.	…	…	…	…	1000	
16	Weiss	…	9·8	…	…	…	…	…	35	…	35	11	x.s.	1000	Dissolve 1 (freshly precipitated from 7 Ag.) in 11; add rest of ingredients.
17	Zinin	Iron and steel	11·1	…	…	67	…	…	44	…	34	…	…	1000	Dissolve 4; add 8: use weak current.
18	Zinin	…	…	…	…	…	…	…	…	500	…	…	…	1000	

NOTE TO TABLE.—The black figures in the last column refer to the numbers of the vertical columns denoting the various ingredients. x.s.=excess.

vats; the current, moreover, must be under absolute control by the use of measuring-apparatus and resistance-coils.

The Solution.—Most of the solutions used largely in practice have the double cyanide of silver and potassium for their basis; and doubtless the solution of this body in a liquid containing an excess of potassium cyanide, constitutes the simplest and best plating-bath for general work. The composition of the principal mixtures suggested is embodied in the foregoing table.

Silver-Baths.—The silver-baths are generally prepared, as required, by dissolving metallic silver in nitric acid, precipitating it with potassium cyanide, washing thoroughly, and dissolving it in excess of the potassium salt. Ten parts of pure silver yield 12·4 parts of pure silver cyanide. The water and all the chemicals used in preparing the solutions must be pure, as the presence of much foreign matter acts injuriously upon the deposit; the potassium cyanide especially should be examined, as it frequently contains only 50 or 60 per cent. of the pure salt (see p. 338). For a like reason it is better to prepare the silver cyanide separately, and to wash it thoroughly, before mixing it with the remaining ingredients of the solution; by simply adding potassium cyanide in excess to the nitrate or chloride of silver, a clear bath is prepared, but it contains, in addition to the silver cyanide, a quantity of the potassium salt corresponding to the silver compound used, which is generally objectionable. Thus, for example, on adding potassium cyanide to silver chloride, the silver cyanide is formed which is required for plating, but with it is an equivalent of potassium chloride produced by exchange.

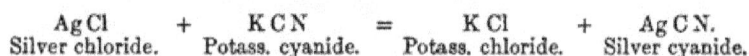

$$ AgCl \quad + \quad KCN \quad = \quad KCl \quad + \quad AgCN. $$

Silver chloride.　　Potass. cyanide.　　Potass. chloride.　　Silver cyanide.

Again, the proportion of potassium cyanide to silver in the bath, although variable between wide limits, is by no means an indeterminate quantity. Having produced the insoluble silver cyanide, as in the above equation, by the use of one equivalent of potassium cyanide, a second equivalent of the latter is necessary to form the double cyanide of silver and potassium, which alone is soluble in the bath; and in addition to this an extra proportion of the potassium salt (*free cyanide*) must be employed to ensure the perfect solution of the anodes, for a reason which may be stated as follows :—In passing the electric current through a solution of silver-potassium cyanide, silver is deposited on the cathode, cyanogen (CN) on the anode; then if the anode be made of silver, it is attacked by the cyanogen and converted into silver cyanide; this body, as we have seen, requires an extra equivalent of potassium cyanide to render it soluble, and although the exact amount of this salt is set free in the liquid by the decomposition of the double compound, it is not all in direct contact with the anode for a sufficient space of time to permit the double salt to re-form and dissolve; it diffuses into the solution, and leaves a portion of the insoluble simple silver cyanide clinging to the surface of the anode in the form of an incrustation. It is to avoid this that an excess of free cyanide is introduced into the bath, which then allows the silver-plate to remain bright throughout. The following two equations show the requirement of the minimum two equivalents of potassium cyanide :—

1. $AgNO_3$ + KCN = $AgCN$ + KNO_3 (washed away).
2. $AgCN$ + KCN = $AgK(CN)_2$.

Thus 108 parts of silver, or $108 + 14 + 48$ ($AgNO_3$) = 170 of silver nitrate, requires $2(39 + 12 + 14) = 130$ parts of potassium cyanide (two equivalents = $2KCN$); while $108 + 12 + 14 = 134$ parts of the pure silver cyanide ($AgCN$) require one equivalent, or $39 + 12 + 14 = 65$ parts of the potassium salt, to form the double compound. Thus 108 parts of silver require a minimum of 130 parts of potassium cyanide, and should have, in addition, at least 50 to 75 per cent. extra cyanide, supposing the latter to be pure; when the commercial salt, containing (say) from 50 to 70 of pure KCN, is employed, the minimum would range from 180 to 250 parts, and the added quantity from 70 to 150. So also 134 parts of silver cyanide call for 65 of the pure potassium cyanide as the minimum allowance.

But, although a certain amount of free cyanide is necessary, a great excess must be avoided, because it would dissolve the anode too freely and increase the strength of the bath, and, worse than that, would tend to produce a somewhat scaly and non-adhesive deposit upon the cathode.

The condition of the bath in respect of free cyanide may be readily tested by withdrawing a little of the liquid in a glass vessel and adding to it a few drops of silver nitrate solution; a precipitate is thus produced which should at once redissolve in the liquid. If it dissolve but slowly even on stirring, it is an indication of a deficiency of cyanide, and the time that elapses before it vanishes completely affords a rough gauge of the amount of free cyanide present. In practice the appearance of the anodes is itself indicative of the condition of the bath; if they are covered with a black deposit during the passage of the current, there is insufficient cyanide present, while if they are quite bright and white, the cyanide is in excess. The best results are obtained when the anodes present a greyish appearance while the current passes, but immediately become white and brilliant when it ceases, showing a complete solution of the thin film upon the surface. A further precaution which may be taken as a check upon the working of the bath is to occasionally weigh the electrodes separately both before and after the process; the loss of weight shown by the anode at its second weighing should be just balanced by the gain upon the plated objects or cathodes. Any departure from this equilibrium indicates either an incorrect ratio of anode to cathode surface, or a wrong proportion of cyanide in the bath; or, thirdly, an unsatisfactory adjustment of the one to the other. But if the surfaces of the electrodes are well arranged (see further on, under *anodes*), a loss of anode-weight unaccounted for by the increase in cathode-weight, clearly points to an excess of cyanide in the bath, or *vice versâ*. Such an abnormal action brings about an alteration in the character of the bath, and must be rectified by the addition of cyanide of silver or potassium, as the case may be. Another very useful test may be made by dipping a strip of bright copper into the bath; if it become coated with silver by simple immersion, the liquid contains too much cyanide, and the deposit yielded by it will be bad, especially upon copper articles, or upon those which have been coppered.

Baths are also liable to alteration by exposure to the air, gradually absorbing carbonic acid, which takes the place of an equivalent of hydrocyanic acid in the bath, so that a certain proportion of the cyanide is removed, and the electrical resistance of the solution is increased. Fresh cyanide must, therefore, be added from time to time, and these frequent additions, coupled with the gradual accumulation of other substances dissolved from impure anodes, or from the cathodes, give rise to a corresponding increase in the density of the solution. Accompanying the increased density is a greater sluggishness of the solution, and hence a greater tendency to separate into layers during electrolysis, the heavy silver-laden liquid from the anode sinking to the bottom of the vat and accumulating there, while the lighter potassium cyanide finds its way to the top. Thus the electrolytic action becomes irregular, a greater quantity of silver is deposited upon the lower portions of the cathodes, while the coating upon the upper part is not increased, but may even be dissolved, after the manner described on p. 101. A similar action is observed in working concentrated, and, therefore, sluggish baths, with an exceedingly weak current from a battery which has "run down." Under these circumstances the anode

13

is immersed in a liquid saturated with silver, the cathode in one containing little silver but much free cyanide, and an opposing electro-motive force is thus set up which tends to re-dissolve the deposit. Nevertheless, a moderately (one or two years') old solution will be generally found to give a better deposit than one newly made-up, provided that the ratio of free cyanide to silver be rightly maintained; hence a certain proportion of an old plating-liquid is commonly used in making-up a new bath. When this is impracticable, the effect of age may be imitated by boiling the liquid for two or three hours, or by the addition of a few drops of ammonia solution. The evils attending excessive concentration may be remedied by appropriate dilution, except when the extreme density is due to the accumulation of foreign matter; or, within certain limits, by maintaining the cathode-objects in gentle motion, which exposes changing surfaces to the liquid, and also prevents the separation of the latter into layers; or, thirdly, by stirring the liquid well every night after work is over. Even a concentrated solution, however, may conduct well, and may be made to give a good deposit; but a dilute bath has a lower conductivity—and is, therefore, more tardy in action—while the metal will have a characteristic dead-white lustre. The specific gravity of the solution should lie between 1·05 and 1·10, pure water being regarded as unity; Gore lays down the limits between which a good deposit is attainable as 1·036 and 1·116. But the dissolved bodies are not the only impurities which find their way into the solution; insoluble matter from the anodes and dust from the air gradually collect, and when the liquid is kept well stirred, remain in suspension in it, and, becoming entangled in the precipitating silver, produce an uneven deposit. It is, therefore, advisable to filter the solution through blotting-paper from time to time as required.

The worst enemy to the plating-solution is organic matter: little by little it accumulates, and although exerting no prejudicial influence in small quantities, it is fatal to successful work when a certain limit is reached. The cyanide solution, most unfortunately, is capable of readily dissolving many organic bodies, and the most jealous examination must be made of all objects which are to be introduced into it. Guttapercha in any form is especially to be avoided; moulds or stopping-out varnishes containing this body should, therefore, be excluded when silver-plating is to be effected in the cyanide bath.

The cyanide solution is generally used cold; for coating small articles, however, or objects made of iron, tin, zinc, or lead, upon which a film of copper has been first deposited, it is occasionally heated.

The bath may be prepared electrolytically, although it is rarely so treated on a large scale, by dissolving 1½ to 2 ounces of pure potassium cyanide in a gallon of distilled or rain-water, and passing a current through it from a large weighed silver anode to a small silver or platinum cathode, the weight of which also should be known, until the excess of silver dissolved into the bath from the former over that deposited upon the latter, shows the bath to be sufficiently charged with the precious metal. For this purpose the electrodes are removed from time to time, rinsed, dried, and weighed, until at last the desired strength of solution is reached. Here the large anode is used with a small cathode that the action of the current upon them respectively may be disproportionate; it is required to add to the weight of silver in solution, so that the case is different to that of an electro-plating bath, in which the solution is to be maintained of uniform density, and which, therefore, demands approximate equality of electrode-surface.

There is no doubt that the cyanide solution possesses many advantages over other possible baths, and with ordinary care will give but little trouble. The main objection to its use is its highly poisonous character, which always involves risk to the operator, and which renders the atmosphere unwholesome, even in fairly-ventilated rooms. Many attempts have been made to substitute safer solutions, but in no case yet with sufficient success to proclaim the introduction of a serious rival to the cyanide bath. One inventor has used a silver salt dissolved in sodium hyposulphite solution; but this bath, although it is said to give good results, gradually decomposes, especially on exposure to light, and, becoming brown at first, gradually deposits its silver in the form of a black sulphide. Zinin has more recently found that the solution of silver iodide in potassium iodide (No. 18 in the table of solutions) could give very good results with a small current of low electro-motive force. He recommends it especially for the production of thick deposits, as in electro-typy. One practical objection to its use is, of course, the high price of the iodides. This is not a fatal objection, if the process be a good one, but is certainly adverse to its general adoption.

The metal obtained from the solutions named has a frosted appearance, due to its being built up of an incalculable number of minute crystals, the facets of which disperse the rays of light falling upon them, instead of reflecting them uniformly; but the slightest friction upon the surface suffices to unite the crystals into an even plane, which reflects the light perfectly, and has the lustre of polished silver.

It was early found, however, that the presence of a minute quantity of carbon bisulphide in the plating-vat caused the precipitation of the metal in the bright condition, and although the use of the *bright plating-solution* entails greater difficulties than are met with in the ordinary processes, yet it is largely used for certain classes of work ; for example, in those to which it is not easy to apply friction, such as those carrying remote or sharp angles or interior surfaces. In no case, however, is the whole plating-process conducted in the brightening-vat, but the bright deposit is given finally, when an almost sufficient weight of silver has been deposited in the usual way.

Of all the substances which have been recommended as brightening agents for the silver-bath (and these include, *inter alia*, silver sulphide, collodion, a solution of iodine and gutta-percha in chloroform, chloride of carbon, and chloride of sulphur), the only reagent practically employed is carbon bisulphide. A mere trace of this suffices to effect its object, while a slight excess produces spotted deposits and brown stains, probably of silver sulphide, and a large excess overshoots the mark altogether, and often gives a dead-white film.

To prepare the bright-bath, place a quart of an old plating-solution in a large bottle (a Winchester quart bottle, for instance) and add to it 3 ounces of carbon bisulphide with, or without, 1 or 2 ounces of ether, shake vigorously for a minute, and add $1\frac{1}{2}$ pints more of the old solution. Again agitate thoroughly, and allow it to stand for two or three days ; there is, at most, $\frac{1}{25}$ of an ounce of carbon bisulphide in each ounce of this liquid. A separate plating-bath must be used for brightening ; then every night, after the day's work is done, one ounce of the mixture just described is added to every 10 gallons of an ordinary plating-solution, specially devoted to this class of work and contained in the special vat. An ounce of old plating-liquid may be added to the mixture in the Winchester bottle in place of that removed. This quantity (1 ounce per 10 gallons) is the maximum amount of brightening-mixture permissible. If the required effect can be produced with less, so much the better ; but on no account should a larger proportion be used, as the risk of spoiling the whole bath would amount almost to a certainty.

The bright-bath requires a stronger current than the ordinary solution, and it deposits a harder metal, the bright film beginning to show itself at the bottom and gradually extending upwards. Any disturbance of the solution during electrolysis may cause it to yield a dull deposit; a whole batch of objects should, therefore, be prepared for immersion simultaneously, and introduced consecutively with the utmost rapidity possible; then, when a sufficient deposit has been given, which usually requires from ten to twenty minutes, the current is stopped, and the pieces are removed and are at once well washed in clean water.

These liquids have a great tendency to deposit the black sulphide of silver, to check which the cyanide solution is added, as described, to replace the portion added to the bath; the brown stains above alluded to are doubtless traceable to the same cause. If the pieces are not thoroughly washed immediately upon removal from the brightening-bath, they will become rapidly tarnished. Gore has shown, too, that the deposited metal contains appreciable quantities of sulphur, which may, in part, account for the variation in the physical characteristics of the metal.

The Anodes.—The silver anodes must be of the purest silver obtainable; the ordinary standard silver for coinage contains 7·5 per cent. of copper (most foreign currency has even a larger proportion of base metal); it should not, therefore, be used as such, but if it be the only form readily available at any time, fine silver should be prepared by the method described on p. 340. The anode may be of cast or rolled metal, but if the latter be selected, it should be annealed by heating to a dull-red heat, and subsequently cooling it before immersion in the vat, so that it may be softer and more readily soluble. As a general rule, the anode should present an area of surface equal to that of the cathode; but, as will now be readily understood, a somewhat smaller anode surface must be employed when the bath contains excess of cyanide, and a greater area when the cyanide is deficient; the object being always to equalise the action at the electrodes, in so far as it is represented by solution or deposition of metal. The anode should never be suspended in the solution by means of copper-wires (unless they can be so arranged that the copper never comes into contact with the bath) because they dissolve under the action of the current, and passing into the solution render it impure. Silver wire has not the same objection; but as it dissolves rapidly, especially at the surface of the liquid, it gradually becomes weakened until it is no longer capable of supporting the weight of the anode. Platinum wire, being quite insoluble, is free from both these objections, and is the best material to use. If copper or even silver supports are adopted, they should be kept from contact with the liquid in the manner explained in Chapter V. (p. 108).

The Vat.—Any of the ordinary vats described in Chapter V. may be employed, except those lined with gutta-percha mixtures, which are more or less soluble in the cyanide liquid. Enamelled-iron lined with thin wood is, perhaps, mostly to be preferred, necessarily so if the solution be heated. The disposition of conducting wires and the manner of imparting a reciprocating motion to a frame, from which the various objects are suspended, so that they may be kept in constant motion, has also been explained in Chapter V. The vats should be considerably larger than the objects to be plated, and may indeed be made of any reasonable size, remembering that a large bulk of solution generally gives a better deposit than a small one, especially in the case of bright-plating baths. When several vats are to be worked from the same battery or dynamo, they should be coupled in parallel arc because, as a rule, the work to be silvered is very irregular in shape and size, and under these conditions the series arrangement is less satisfactory. A well-fitting cover may be made for the vat, to preserve it from atmospheric dust when it is not in actual use.

The Character of the Metal Deposited.—Like most other metals, silver, which is deposited by a current strong enough to evolve hydrogen simultaneously, is dark in colour, powdery, and non-adherent; it is in the spongy condition, and is useless as a coating. A weak current, on the contrary, gives a strong, malleable metal, adherent and coherent, and minutely crystalline. Some operators consider the commonly-employed current-strength of 0·032 ampère per square inch (0·5 ampère per square decimetre) too high, and prefer to reduce it to 0·013 ampère (0·2 ampère per square decimetre); but for all ordinary work the larger current-volume will be found satisfactory, and will, of course, deposit a given weight of metal in a shorter period of time.

The metal should have a pure white colour, any departure from this indicating the presence of impurities. A pinkish shade probably points to the existence of copper in the precipitate. A yellowish shade or tarnish, which is apt to appear upon surfaces that have been for some time exposed after removal from the vat, is probably due to a small percentage of a sub-cyanide of silver deposited with the metal, which gradually changes colour on exposure to light. It is found that a dip into potassium-cyanide solution, or even a stay of two or three minutes in the plating-bath after the current is cut off, suffices to prevent this, doubtless by dissolving the objectionable sub-salt. It has already been stated that the silver deposited by the carbon bisulphide brightening-solution contains a small proportion of sulphur, which

is possibly accountable for the alteration of structure indicated
by the different nature of the deposit.

The thickness of a coating of silver may vary from an almost
imperceptible film to a depth of $\frac{1}{64}$ of an inch on electro-plate,
or of $\frac{1}{10}$ of an inch on silver electrotypes.

Owing to its open crystalline nature, the deposited silver, if
peeled from the surface on which it is precipitated, lacks the
metallic ring emitted by the rolled metal when struck.

The Process of Electro-Silvering.

Brass, copper, bronze, German silver, and similar alloys are
best adapted to the electro-silvering treatment; the softer metals—
lead, tin, Britannia metal and pewter—though sometimes plated,
are less well suited because they are not structurally so capable
of resisting the final mechanical treatment of polishing and
burnishing; iron and steel, zinc and other metals may also be
silvered. But whenever the metal is attacked by the cyanide
bath, so that silver is deposited by simple exchange and without
the aid of the current, it should receive a thin coating of copper
or be subjected to the process of quicking, this coating with
mercury is, however, often used even when the metal has no
action on the bath to render adhesion doubly sure. The explanation of the process is given on p. 124.

Organic matter must as far as possible be eliminated for
reasons already given; hollow sheet-metal objects, therefore
(brass candlesticks, for example), which are often filled up with
pitch-composition, as a support to the thin metal of which they
are made, must be gently heated to effect its thorough removal
prior to electro-plating; this operation must be conducted with
care because cheap articles are frequently made in several pieces,
which are held in place by the composition, and, therefore,
become separated when it is removed. All non-metallic handles
or appurtenances should be, if possible, detached from objects
before plating, because there is not only a risk of their being
damaged by the solution, but liquid is sure to penetrate into the
sockets and interstices, from which it can afterwards be removed
only at the expense of much trouble.

The processes preliminary to the actual electro-deposition are—

(a.) Stripping, or removal of an old coat of silver, if any exist.

(b.) Polishing, if necessary.

(c.) Cleansing, consisting of—1, boiling in caustic potash to
remove grease; 2, dipping in sulphuric acid to remove oxide;

and 3, scouring with sand or a dip into mixture containing nitric acid according to the nature of the metal. (See Chapter VI.).

(d.) Preliminary coating with copper, if necessary.

(e.) Quicking, if required.

Stripping.—When old goods are to be re-plated, every trace of the original coating must be removed, in order that the new deposit shall be regular, and uniformly adhesive. In the choice of a stripping solution, the operator must be guided by the character of the basis metal, from the surface of which the silver is to be dissolved, as it is essential that the liquid should not be able to attack this to any serious extent, when it is laid bare by the removal of the precious metal.

For brass, copper, or German silver, a mixture of concentrated sulphuric and nitric acids is generally employed. The most rapid method consists in heating a sufficiently large quantity of strong sulphuric acid in a stoneware vessel, and adding to it, immediately before use, a small quantity of potassium nitrate (saltpetre) or sodium nitrate (Chili saltpetre); this is at once decomposed, a small proportion of the sulphuric acid being neutralised and a corresponding quantity of nitric acid being liberated in the liquid. Such a mixture when used hot is capable of dissolving the silver from the articles, which should be suspended in it by copper hooks or preferably by copper tongs, but should show no very great corrosive effect on the copper or basis-metal. Nevertheless it is not entirely without action, and the process must, therefore, be watched most carefully, especially towards the end, when most of the silver has been dissolved, so that on the disappearance of the last trace of covering metal, the article may be removed and plunged into a large volume of water without loss of time. With this object in view the pieces under treatment should be frequently removed from the liquid for inspection. The extremities of long articles which have not been properly reversed during their first electro-silvering, and the more prominent portions of every object having a thicker coat than the remainder, are denuded last; and it is often advisable to so place the goods towards the end of the stripping-process, that only these portions are immersed in the liquid. It is essential that the concentration of the bath be well maintained; any dilution increases its tendency to attack the base-metal, a comparatively small addition of water sufficing to render the action even violent. For this reason the mixture, which absorbs water vapour from the air with great avidity, must be stored in tightly-closed vessels when not actually in use; and the objects to be stripped must be dry when placed in the vat; indeed the introduction of moisture in any way into the hot sulphuric acid would cause a sudden generation of steam, almost explosive in its violence, so that it is alike dangerous to the operator and destructive to the bath. In course of time the accumulation of potassium bisulphate in the bath (from the decomposition of the saltpetre added each time before use) is rendered evident by the deposition of crystals. When this is observed it will generally be found that so much acid is neutralised that the liquid is no longer serviceable; a fresh quantity of sulphuric acid should then be prepared, the old bath being reserved to recover from it the silver which it contains.

On account of the great care necessary in conducting this process, only one object should be treated at a time, if at least it be of moderate size, for the operation proceeds with great rapidity. When this is inconvenient, by reason of the number of pieces to be treated, extra precautions must be taken to guard against too prolonged action in any individual case.

With a cold solution the action is slower, and consequently under better control; hence a mixture of 10 parts of concentrated sulphuric acid (specific gravity 1·84) and 1 part of strong nitric acid (specific gravity = 1·39) is often preferred, this being applied at the ordinary temperature of the room. Except that it is not heated, and that a greater number of pieces may be treated with safety, the method of use is the same as that just described; and water and moisture must be equally rigorously excluded.

A different stripping-process must be employed for zinc, iron, lead, tin, Britannia metal or pewter, or for any alloys of these, which would be vigorously attacked by the acid mixture suitable for copper.

This process consists in suspending the articles as the *anodes* in a strong solution (say 10 per cent.) of potassium cyanide, opposite a plate of platinum, copper, or brass, and connecting the former with the positive or copper pole of the battery, so that the process of electro-plating is reversed, and the current flows in the electrolyte from, instead of to, the pieces. Thus, the goods being the anode, the silver is dissolved from them and deposited upon the platinum cathode after a time, at a rate depending upon the volume of the current which is being applied. An old silver-bath may be utilised for this purpose, the metal deposited upon the cathode plate being, of course, recoverable. In any case the same solution may be used repeatedly; and the current may be stronger than that permissible for plating, because the object is no longer to produce a good deposit, but to dissolve an old one with the utmost rapidity. But when a strong current is employed, the silver may be deposited in the pulverulent condition, so that particles frequently become detached and fall into the liquid, to prevent which the cathode plate may be enveloped in a case of parchment-paper or even of fine muslin. Silver baths in current use must never be employed as stripping-solutions, because they would gradually dissolve small quantities of the base metals from which the silver had been removed, and would thus become too impure to yield a good deposit; only disused baths are permissible. As soon as the pieces are completely stripped, they are removed from the vat, plunged into water and well washed.

The process is, of course, equally applicable to copper and those alloys which are often treated by the more rapid acid method.

Polishing, Washing, and Copper Coating.—After the original silver case has been removed, it is often necessary to pass the goods to the polishers to buff and finish, prior to the cleansing and quicking operations, after which it is transferred to the plating-vat without loss of time. Iron and steel cannot be quicked, because this metal is one of the few which refuse to amalgamate, or alloy with mercury; Britannia metal also is not usually quicked; but copper, brass, or nickel-silver are fitted in this way to receive an adherent deposit. Zinc should receive a preliminary wash of copper in the alkaline copper-bath, and it was at one time customary to submit the tin-lead alloys (pewter, Britannia metal, and the like) to the same treatment, but there is no great difficulty in directly silvering them with good results. Steel, which is, in some hands, still coppered before silvering, may also take a perfectly sound and adhesive deposit by dipping the cleaned articles at once into a striking-solution.

Suspension of Objects in the Bath.—The suspension of objects in the silver-bath is effected by thin copper wires (commonly of about No. 20 of the Birmingham wire-gauge). New wire should be used each time, because the deposit upon an old wire is apt to be loosened by re-bending, and to crumble off in the bath. The manner of attaching the wire depends upon the nature and shape of the goods. Some articles afford natural places of vantage from which they can be slung, such as cream ewers, cups, and the like, which carry metallic handles, or perforated objects, or those which, being unfinished, have rivet holes, through which the wire may be threaded. Spoons and forks are best supported in slings made by forming the wire into a loop around the shank, or by bending it into three quarters of a circle at the end, and at right angles to the wire itself, leaving a horizontal space (the

remaining quarter of the circle), through which the shank may
be slipped, but which is not wide enough to allow the handle or
bowl to pass (fig. 96). Plates, salvers, and the like, should be
hung by wires bent lightly around them, these having their
ends joined by twisting together. Obviously the methods of
attaching wires are innumerable, and must be determined by
the circumstances of the case; the guiding rule in making the
connection is that the wire shall be so arranged that the object
cannot escape from its hold; yet on the other hand, it should
be so loosely fixed that the relative position of wire and object
may be shifted at any moment without difficulty, so that fresh
surfaces are brought in contact, and wire-marks are not formed
on the deposited metal.

The wiring is best done before the final potash and acid
cleansing-processes. Many firms place a piece of glass-tubing
over that portion of the wire which is in contact with
the solution between the cross cathode-rod of the vat
and the suspended object, so that silver may not be
uselessly deposited upon it; others use gutta-percha or
india-rubber as an isolating medium, but as these are
slowly attacked by the cyanide liquor, glass is preferable, and there is no difficulty in adapting it. Having
ascertained the length necessary to protect the portion
of wire which is to be immersed, this distance is
measured off on a piece of narrow glass-tubing; a fairly
deep mark is then made at the desired point with a
triangular file; now placing a hand on either side of the
nick, with the thumbs immediately beneath it on the
other side of the tube, a steady bending pressure is so
applied that the file mark is on the outside, the thumbs
on the inside of the bend; almost immediately the
tube should break with a clean even fracture at the
place of the file-mark. In experienced hands accidents
are not likely to happen, but in early attempts at
breaking tube in this way it is perhaps safer, though even then
scarcely necessary, to envelope the hands with a thick cloth.

Fig. 96.
Sling for
spoons.

Arrangement of Objects.—In arranging the objects in the bath
they must be introduced gently, that any sediment may remain
undisturbed, and should be placed alternately with anode plates,
so that every piece is equidistantly between two anodes, and will
receive equal weights of deposit on the two sides; thus, whatever
the size of the bath, the number will be always one in excess of
that of the rows of cathodes, and all will be in parallel circuit. If
there be only one row of cathodes there will be two anodes; if two

rows, then three anodes, and so on. Several objects may, of course, be suspended side by side from the same cathode-rods, and will be influenced by the same pair of anode-plates—provided that free space is left between adjacent objects. The anode- and cathode-rods should be parallel to one another, in order that the spaces of conducting liquid between the different pairs of electrodes may be equal. It is preferable also that, as far as practicable, only goods of the same shape, or, at least, of the same diameter, should be suspended from the same rod. On account of the great irregularities in form of objects to be plated, a considerable distance must be left between the electrodes, so that the portions nearest to the anode may not be so near, as compared with those more remote, that they receive an undue share of the deposited metal. Allowance must also be made for the motion imparted to the objects in the bath, so that the opposing electrodes may not make contact at the end of each swing. Again, two different kinds of metal should not be suspended from the same rod (for example, copper and Britannia metal), as the local current set up between them, both being immersed in the same exciting liquid, and being in metallic connection through the suspending-rod, will tend to cause the gradual solution of the more electro-positive metal, and will diminish the deposition of silver upon it, until it is quite protected by a perfect layer of the precious metal. The solution is thus injured by the introduction of foreign matter. If the electro-chemical difference between the two metals be so great as to cause an electro-motive force greater than that of the depositing current (which could rarely, if ever, happen were ordinary care bestowed on the process), no deposit would occur on the more positive metal, which would rapidly dissolve and cause a double thickness of coating to be given to the object made of the more negative metal. Otherwise the local back electro-motive force simply retards the action of the current in regard to the positive metal, until it is sufficiently covered to prevent further action; and unless this retardation be very protracted, the difference between the weights of metal deposited on the plates and that which it was desired to precipitate, will not be very appreciable, so that the chief injury is done to the bath.

Use of the Striking-Bath.—An additional reason for guarding against this contingency is that lead and its alloys conduct electricity less satisfactorily than brass or nickel-silver, and far less so than copper, and hence they need a somewhat greater length of time to acquire the thin wash of silver which suffices to protect them.

Many operators, therefore, prefer as a preliminary step to dip the articles, immediately after quicking, into a silver-bath worked by a stronger current until, almost immediately, a thin film of this metal has been imparted, when they may be transferred to the ordinary vat, in which the remainder of the deposit is to be built up. The first, or *striking-bath*, may contain less silver than the usual solutions (half-an-ounce to the gallon commonly suffices), but the proportion of free cyanide is often greater. Large silver anodes are used; and, indeed, everything must be done which tends to reduce the resistance and increase the rapidity of deposit, in order that the action may be almost instantaneous, and that a momentary dip into the vat may be sufficient to give the required deposit. But, on the other hand, it need scarcely be remarked that the volume of current must not be so great that a pulverulent or spongy deposit results. The electrical connections of the striking-vat may be similar to those recommended for the plating-bath; but a smaller bath with two large anodes, one at either end, and with a single cathode-rod, to which the negative battery-wire is attached, and which is lowered into the bath by hand, is really all that is required.

How to Ensure Uniformity of Coating.—After the transference of the goods to the plating-vats, they may be left with less constant attention until a sufficient thickness of film has been obtained, provided that the current is constant and that the arrangement for imparting motion to them during the action is working satisfactorily. All that is necessary is to slightly shift the position of each piece relatively to its supporting-wires from time to time, to ensure uniformity of deposit at these points, with an occasional momentary removal from the bath for an examination as to the regularity of the action. Should spots appear upon the surface, the article must be removed from the bath, rinsed, scratch-brushed, and then cleansed by a dip into hot potassium cyanide or caustic potash solution. Finally, after rinsing once more, they are re-quicked and introduced again into the bath. Since, in spite of the gentle motion imparted to the objects, the solution is certain to vary in density, and to produce a thicker deposit upon the lower portions of articles, long objects, such as spoons and forks, suspended upright in the bath should be reversed at intervals of (say) half-an-hour, so that if the bowls were downwards at first, the handles would be so after the first shift.

In immersing the quicked and struck articles into an empty vat, that is, into one which contains only anodes suspended within it, those first immersed would receive too strong a current, unless their superficial area were very considerable, and would be covered with a spongy silver precipitate, until, at last, the cathode-surface had been increased by the introduction of more objects, sufficiently to produce the right proportion of current-strength per unit of area. Meanwhile, however, the original pieces would have suffered serious injury. To obviate this, either the current-volume may be reduced at first by the inter-position of wire-resistances, which are gradually lessened as fresh objects are introduced, or one or more of the anode-plates (according to the size of the vat) are hung upon the cathode-rods at the outset, and are transferred to their proper places, one by one, as each batch of objects is immersed which presents a total area equal to that of one plate. In this way a large proportion of the current is at first occupied in transferring silver from one anode-plate to another; and the bath from the very beginning is under the same conditions as it is when filled with goods undergoing the silvering-process. Thus even a small object intro-duced alone should receive a normal current throughout.

Some electro-platers, in order to secure a more perfect coat, are in the habit of removing the articles after a certain amount of silver has been deposited and submitting them to a preliminary scratch-brushing, after which they are well rinsed and cleansed and again returned to the bath; but it is very doubtful whether any real advantage accrues from this practice.

When a sufficient thickness of metal has been deposited, which may be known, as explained in Chapter V., by ascertaining the mean strength of current, the total cathode-area, and the time occupied, or by the use of the plating-balance, the pieces are removed from the vat, transferred (if necessary) to the brighten-ing-bath, where they are left undisturbed for a few minutes, and are then plunged into a slightly warm solution of potassium cyanide to remove any silver subcyanide left in the pores of the metal, and thoroughly washed in several waters held in succes-sive tubs (*vide* instructions for washing coppered goods on p. 145); they are next dipped momentarily into a vat containing water mixed with 2 or 3 per cent. of sulphuric acid, and are again rinsed in water, and taken to the scratch-brush for preliminary polishing, and to the burnishers for the final treatment. The potassium cyanide dip may be dispensed with, if the objects are left in the plating-solution for a few minutes after disconnecting the current.

How to Thicken the Coat Locally.—An extra thick coating of silver may sometimes be imparted to those portions of goods which will have to stand the chief amount of wear in use. This may be effected in many ways according to the appliances and ingenuity of the plater. The application of stopping-out varnish after a certain time to the parts which are to receive a thinner coat is rarely admissible, because, although it would have the desired effect, it gives too defined an outline of the thicker deposit, and this has to be obliterated mechanically, an imperceptible gradation being generally required. This method would be suitable if the whole of one side of the object had to be thickened, but an equally good result would be attainable by the use of only one anode (adjacent to this side). Another plan is to introduce, towards the end of the operation, a subsidiary anode, corresponding in shape to the part which is to receive the greater deposit, and which is placed in greater proximity to it, in proportion to the increase of substance to be acquired. Yet another system may be adopted in plating the bowls of spoons and similar objects, which are worn most largely in use at the most prominent parts of the curve. After plating in the usual way the spoons are so placed in a shallow-bath that only the convexity of the bowl is immersed, and receives a coating; the depth of immersion must be frequently altered in a slight degree to ensure that no distinct boundary marks are produced on the surface. It is true that these marks may be mechanically removed, but there is no reason why they should occur at all.

Thickness of Deposit.—The thickness of the deposit and, consequently, the duration of the process is, of course, governed solely by the class of work under treatment. Many common goods, especially those made of white metal, receive a film so thin as to be beyond the range of practical measurement. This is naturally useless to resist even the slightest wear; it gives simply an ornamental covering for a short time. As a general rule for ordinary electro-plating, a deposit of from 1 to 2 ounces per square foot of surface-area may be deemed a good well-wearing coating; a single page of this book represents approximately the thickness of a film equal to 1 ounce per square foot. This will occupy from three to nine hours in coating according to the strength of the current. A thinner deposit than that of 1 ounce per square foot is not to be recommended, as even two or three years of ordinary wear would suffice to lay bare the base metal at the edges, and in all the more prominent parts. In working, however, for the trade, the craftsman is rarely allowed to decide what, in his judgment, is best fitted for the work, but must do as he is ordered by his customer, and will be paid at the rate of so much per unit weight of silver deposited.

Silver Electrotyping.

Silver is occasionally used in special cases for copying works
of art or even valuable engraved steel-plates. Ordinary wax
and gutta-percha moulds, such as are used for copper electrotyping, are not admissible for silvering, because they are to some
extent attacked by the cyanide solutions. The simplest method
of obtaining replicas of works of art in silver is to obtain first a
thin electrotype-shell of copper from the intaglio-mould, and
then to deposit silver upon this in the cyanide-bath. The copper
protecting-film may be of the thinnest, so that it shall not
destroy the sharpness of the lines, but it must, of course, be
subsequently removed, after the required thickness of silver has
been deposited, and the whole electro separated from the mould.
This solution of the copper may be effected by treatment with
warm hydrochloric acid or (better) with a warm solution of iron
perchloride, either of which will attack the copper but leave the
silver untouched. On the removal of the copper, the pure
silver surface has the required form in practically undiminished
sharpness and brilliancy. The silver may be built up to a
thickness of one-eighth of an inch or more. It is rarely,
however, that this process is required; and practically the sole
application of electro-silvering is to be found in the coating of
other metals to endow them with properties which they do not
of themselves possess.

Ornamenting Silver Surfaces.

There are many ways of altering the appearance of electro-
silvered goods; but to give a description of these, many of
which are purely mechanical, is beyond the scope of this work.
Let it suffice then to say that—

A dead lustre may be obtained by depositing upon the silver
a thin film of copper, which has a slightly roughened surface of
excessively fine grain, and then again upon this a thin layer of
silver.

Oxidised silver, which is an entirely misleading term, inasmuch
as oxygen plays no part in its formation, is made by dipping the
object into, or painting it with, either a solution of platinum,
which covers the whole surface with a thin layer of that metal
by "simple immersion," or one containing sulphides which
imparts to the silver a superficial film of black silver sulphide.
This latter solution is made up by dissolving three-quarters of

an ounce of potassium polysulphide ("liver of sulphur"), or of ammonium sulphide, in each gallon of water, and applying it to the silver at a temperature of 150° F. The potassium compound is to be preferred; some operators add to it about twice its weight of ammonium carbonate. A few seconds' immersion in either of these liquids usually suffices; the articles are then rinsed in water and dried.

Antique silver is produced by rubbing into, and leaving upon, the parts of an object which are not in relief, a thin layer of black-lead, finely crushed and stirred into spirit of turpentine; some prefer to add a little ochre to the mixture in order to produce a warmer ground tone of colour.

Niello-work is prepared by tracing a pattern upon bright silver with silver sulphide or with mixtures of lead, copper, and silver sulphides, prepared artificially; when placed in position the object is heated to their fusing point in order to ensure adhesion. It is, in fact, a process of enamelling. Clearly, however, it is quite inapplicable to many classes of electro-plate, while to any it must be applied with the utmost care, in order to avoid the stripping or buckling of the coated object or the fusing of the basis metal.

Satin finish may be produced, according to Wahl, by the application of fine sand propelled forcibly upon the surface of an object with the aid of an air-blast—a process analogous to that largely used at present for decorating glass. Any process which will destroy the polish upon the silver with equal fineness and regularity would, of course, answer the same purpose; but the sand-blast is probably the simplest and most economical extant.

Obviously the enumeration of the methods of decorating silver surfaces is by no means exhausted in the few words given above; they are innumerable and capable of infinite variation according to the taste and skill of the artificer.

14

CHAPTER X.

THE ELECTRO-DEPOSITION OF GOLD.

Advantages of Gold-Plating.—Owing to its high power of resisting atmospheric influences, combined with the richness of its colour, and the brilliancy of the polish which it is capable of receiving, and to the fact that all these properties are manifested even by the thinnest imaginable film of the metal, gold is very frequently deposited; none the less so, perhaps, because being a costly material, gilt objects of low value may pass for articles of much higher worth.

. Gold is a very electro-negative element, so that any of the common metals is capable of replacing it in any of its compounds. It may, therefore, be readily deposited by simple immersion, although the electrolytic process is more satisfactory.

DEPOSITION BY SIMPLE IMMERSION.

Solutions.—Many baths have been used at various times, the principal of which are included in the following table.

Roseleur's Process.—Of all these, Roseleur's solution (No. 4) is the best for treating small articles—of jewellery, for example —made of copper, bronze, or brass. The gold chloride crystals should be dissolved in a small proportion of the water, and added to the solution of sodium pyrophosphate in the remainder, and the mixture warmed until the yellow colour of the liquid has disappeared. The solution thus made up, however, is too readily decomposable, as, indeed, is indicated by the gradual change of colour to a dark red-purple which it undergoes on standing; hence the hydrocyanic (prussic) acid is added as a check upon the rapidity of the spontaneous reduction. It is omitted by some gilders, but the bath is under better control when it is used; when working too slowly, more gold chloride is added, or when it becomes deep purple in colour, fresh hydrocyanic acid is introduced.

In using this bath, the object must present a clean bright surface such as may be imparted to it by pickling, scratch-brushing, and cleaning; it is then quicked by a momentary

TABLE XV.—SHOWING THE COMPOSITION OF GOLD SIMPLE-IMMERSION MIXTURES RECOMMENDED BY VARIOUS AUTHORITIES.

No.	Authority.	Special Application of Mixture.	Most Suitable Temperature.	PARTS BY WEIGHT OF INGREDIENTS.									Special Method of Preparation.
				1 Gold, as Gold Chloride.	2 Gold Sulphide.	3 Hydrocyanic Acid.	4 Pota-alum Cyanide.	5 Caustic Potash.	6 Potassium Bicarbonate.	7 Sodium Pyrophosphate.	8 Ammonium Sulphide.	9 Water.	
1	Braun	Zinc goods	...	...	4	...	...	...	...	...	q. s.	1000	Dissolve 2 in excess of 8 and 9.
2	Elkington	...	...	5	...	...	...	...	305	...	...	1000	Mix 1 with 155 of 6 in part of 9; dissolve 150 of 6 in rest of 9; mix.
3	Gore	...	Warm	8·6	...	...	...	...	500	...	...	1000	Boil for two hours.
4	Roseleur	Copper, Bronze, and Brass	Hot	10	...	0·8	...	...	...	80	...	1000	"Quick" articles first.
5	„	Large Bronzes before Electro-gilding	Hot	1	...	...	9	108	20	...	...	1000	No quicking needed.
6	Wahl	...	...	7·5	...	...	...	...	562	...	...	1000	

NOTE TO TABLE.—The black figures in the last column refer to the numbers of the vertical columns denoting the various ingredients.

plunge into a dilute solution of mercuric nitrate, rinsed in water, and immediately transferred to the gold-bath, which should be nearly boiling. It is more economical and satisfactory to use three gold dips in succession, each solution being richer in gold than that previously applied; this is readily effected by using old baths for the first two operations, and by arranging a system in which, as soon as the final vat ceases to yield a good deposit, it is made the second instead of the third bath; that which had been the second being now the first, and the old first, now practically exhausted, being discarded. Thus the pieces are most thoroughly washed before they enter the last bath; and no gold is lost, as the small quantity left in the third solution, when it is no longer serviceable, is used up during the time that it is acting as the second and the first. An immersion of a few seconds in each liquid should suffice; and the resulting deposit, which is, of course, very thin, should have a good yellow colour and require only slightly scratch-brushing or burnishing to impart a final polish, rinsing, and lastly, drying in hot white wood sawdust; for this resinous woods, oak, or walnut, which tend to discolour the work are to be avoided. It may sometimes be necessary to improve the appearance of the gold by colouring methods, which will be described at the end of this chapter.

If it be desired to obtain a thicker coating by this method, it is only necessary to repeat the process several times, re-quicking each time before passing the articles through the gold-baths. The deposit is thus gradually built up, because at each quicking-stage a small proportion of mercury deposits upon the surface, and then exchanges for an equivalent of gold, when placed in the gilding solution, the latter gradually accumulating mercury in place of the more precious metal. A really thick coating, however, cannot well be built up by this tedious process, and the electrolytic process is more convenient and more expeditious.

Elkington's Process.—Elkington's process of *water-gilding* (No. 2) employed potassium bicarbonate in place of sodium phosphate; but its use is more troublesome, and it permits only a semi-exhaustion of the bath, leaving the remainder of the gold to be recovered from the residual liquid by chemical means.

Roseleur's Process for Large Objects.—Roseleur's bath (No. 5) rapidly precipitates gold upon articles which have not been previously quicked; the deposit is not of high quality, but the process is well adapted and largely used for coating large objects with a wash of gold, prior to submitting them to the electrolytic process.

Other Solutions.—Of other solutions for simple immersion gilding, perhaps the most interesting are:—that of gold chloride in ether, which is applied to gilding iron and steel goods; and that of Braun's, a solution of gold sulphide in ammonium sulphide, which is adapted to the direct gilding of zinc, because the latter metal would dissolve but slowly in such a liquid, and the coating is, therefore, the more likely to be adherent.   This sulphide solution is quickly oxidised, and should be preserved from unnecessary exposure to the air.

DEPOSITION BY THE SINGLE-CELL PROCESS.

The Elkington bicarbonate process above alluded to, when used to deposit upon silver or German silver, demanded that a piece of zinc or copper should be attached to the objects, and thus became practically a single-cell process. Steele also, in the specification of a patent granted to him, claimed the use of a cyanide bath in which the object to be coated was immersed in contact with a piece of zinc; but, inasmuch as a considerable proportion of the gold was found to deposit upon the zinc itself, owing to the wide difference between the electro-chemical relations of the two metals, zinc and gold, the method is not practically used.

DEPOSITION BY THE SEPARATE-CURRENT PROCESS.

The Battery.—Almost any of the ordinary battery-cells may be used.   A current of fairly high potential is required, but no great volume is essential.   The Bunsen-cell is well adapted for the work, but the resistance in the circuit should be sufficient to reduce the current-intensity to 0·006 ampère per square inch (0·1 ampère per square decimetre).

The Solution.—As with silver, the double cyanide solution will generally be found to give the best results, provided that due care is paid to all the details of the process. The number of other solutions prepared with the object of supplanting the poisonous cyanide compounds, and of the modifications of the cyanide-bath itself are, as usual, innumerable. The chief of them are included in the following table.

For ordinary use, a bath containing three-quarters of a troy ounce of gold, dissolved and converted into cyanide, together with about 7 ounces of good potassium cyanide in every gallon of solution, should give a good deposit at a temperature of 120° to 140° F. ; it should be boiled prior to use. The proportions,

TABLE XVI.—Showing Composition of Gold-Baths for Separate-Current Process, Recom. by Various Authorities.

No.	Authority	Special Application of Bath	Most Suitable Temperature (F.)	Gold in Form of				Potassium Cyanide	Potassium Sulphocyanide	Potassium Ferrocyanide	Potash	Potassium Carbonate	Sodium Bisulphite	Sodium Hyposulphite	Sodium Phosphate	Ammonium Chloride	Ammonium Sulphate	Water	Special Method of Preparation
				Chloride	Cyanide	Oxide	Ammonium. ret	5	6	7	8	9	10	11	12	13	14	15	
1	Bacquerel	...	Hot or Cold	6·5		8·5				100	25							1000	Add 3 (neutral) to 7 and 8 in 15.
2	de Briant	...								100	25	9·4						1000	Dissolve 1 in 15; add 9 till just cloudy.
3	Fizeau	...		4														1000	
4	Gore	...		6·8				100										1000	Dissolve 1 in 15; add 6 till ppt. just re-dissolves. Just re-acidify with hydrochloric acid.
5	"	...		11				80										1000	
6	"	...		274				200										1000	
7	Kirk	...		10–15				50										1000	
8	Levol	Silver		q. s.				q. s.										1000	
9	Lerebour	...	Hot	0·6		6·2		125						4				1000	Boil together for half-an-hour.
10	M.J.L. (Gore)	Copper, Brass, Silver		1·6						25								1000	Boil 1 and 7 with 250 of 15, filter, and add rest of 15.
11	"							4·3								3		1000	
12	Pfanhauser	...	Warm	0·35				20							60			1000	Dissolve 1 in 260 of 15, and add to 5 in 800 of 15. Boil half-an-hour.
13	Roseleur	...	Cold	10														1000	
14	"	Silver, Copper, Ger. Silver	122°–176°	1			4	1				10	12·5					1000	Dissolve 12 in 500 of 15; cool; add 1 in 100 of 15 slowly; dissolve 5 and 10 in rest of 15. Mix.
15	"	Iron & Steel	122°–176°	1		7·2		0·5				12·5			50			1000	
16	"			15				12				15						1000	
17	"	Watch Movements					4	100		20								1000	
18	de Ruolz		60°–77°			7·2		10										1000	Dissolve 4 in 5 and 15. Boil for fifteen minutes; cool.
19	Wagner		130°	0·7		2·1		z. s. Little										1000	
20	Watt							z. s.										1000	
21	"		100°–150°	4			2·1	z. s.										1000	
22	"			1			2·1	z. s.										1000	
23	"		135°	1			0·31	3		60					10		q. s.	1000	Dissolve 7 and 13 in some of 15, add 1 a little of 5 and rest of 15. Precipitate 1 with 14, filter; dissolve ppt. in 5 and 15.
24	Weiss		90°					3										1000	
25	Wood			6·8				27·4										1000	

NOTES TO TABLE.—The black figures in the last column refer to the numbers of the vertical columns, denoting the various ingredients. The weights of gold, in columns 1 to 4, represent the amounts of actual metal to be converted into the compounds specified. Ppt. = precipitate.

however, may be widely altered without greatly prejudicing the character of the work; and it is well to vary the composition of the bath with any change in the conditions of depositing. Indeed, it is best not to adhere slavishly to any given formula, but rather to modify it by dilution or strengthening, by adding cyanide or gold, according to the work in hand and the methods of the operator. Formulæ given in books should be regarded only as guides, which, in describing the experiences of others, seek to add to those of the reader, or to afford him a basis from which to start, although it may frequently, perhaps generally, happen that the solutions described will give good results in his hands even at the first trial. A solution to be used for cold gilding should contain, at least, two or three times as much gold, and proportionately cyanide, as one which is to be employed when heated. A weak solution, however, will usually give a better deposit than a stronger one at the same temperature.

The cyanide gilding-solution is comparable with the corresponding silver-bath; a certain amount of free cyanide is required to promote the solution of the anodes, a deficiency of this salt being indicated by the formation of a slimy deposit upon the anode-surface. Too great an excess of cyanide, however, causes the anode to dissolve too rapidly, and thus to yield too strong a bath; but it also tends to attack gold without the aid of the current, with the result that the deposit is produced excessively slowly, and may not even be formed at all in the deeper recesses of an irregularly-surfaced article, the simple solvent action of the bath neutralising the depositing power of a weak current at these more distant points; moreover, when each side of an article is coated singly, the film of gold upon the side more remote from the anode may be redissolved during the covering of the second surface. When such actions as these are perceived they must be neutralised by the addition of a further supply of gold cyanide.

The gradual accumulation of dirt and various impurities, soluble and insoluble, renders the bath unfit for use after a time. The use of the liquid for depositing gold upon articles made of base metals, causes a slow absorption of these metals into it by simple exchange in the few seconds during which they are immersed before a protective cover of gold is imparted to them; and as soon as the proportion of silver or copper thus introduced becomes appreciable, the colour of the gold precipitated by the bath becomes influenced, as will be explained hereafter. But these impure baths may, of course, be successfully applied to the production of a deposit, when the particular shade of colour which they yield is sought, or they may be used for deeper colours by adding to them a further quantity of one of these colouring metals as may be desired. The presence of much organic matter gives a dark colour to the solution, and causes it to yield a brown deposit, which can never be converted into a good coat; such a bath should, therefore, be discarded.

The cyanide-bath is often made up by simply adding the chloride or some other salt of gold to the solution of potassium cyanide, but this is an objectionable practice, because it needlessly introduces impurities into the solution (*vide* p. 191). Gold cyanide should, therefore, be prepared and well washed so that only the pure salt is introduced into the bath. The bath may also be prepared electrolytically by passing a fairly strong current from a large pure gold-anode to a small gold or platinum-cathode through a 3 or 4 per cent. solution of potassium cyanide heated to a temperature of 120° to 140° F., until, as in the case of silver-baths similarly made up, the difference between the loss of weight of the one electrode and the gain of the other, indicates that sufficient precious metal has passed into the solution.

Heated solutions suffer a gradual loss of water by evaporation, which must be made good from time to time, preferably every evening after the day's work is over.

In order to obtain certain shades of colour upon the gold deposit, solutions of certain metals are added to the gold-bath, which by being precipitated simultaneously with the more precious metal influence its tint. The effect of varying the strength of either current or solution and of modifying the working of the bath will be discussed in the section relating to the character of the deposit; but it should be noted at this point that a red colour, or a greenish shade merging almost into white, may be produced by adding to the gold-bath (preferably a cold one) a sufficient quantity of a copper cyanide solution on the one hand, or of silver upon the other. The proportion cannot, of course be rigorously prescribed, because they will vary with the shade of colour to be produced, a few trial-experiments sufficing to indicate the amounts suitable for any given tint. In preparing such baths the added metal must be introduced very gradually, so that an excess may be avoided, remembering that it is generally more convenient to add a little more copper or silver if required, than to reduce the relative proportion by the introduction of a further proportion of gold. It has been noted previously that an old gold-bath, which is beginning to yield a coloured deposit, may with advantage be used as the basis for a solution intended to produce the same shade intensified.

The Anode.—The anode should be made of the purest gold obtainable; the presence of silver and copper which, either singly or together, are nearly always alloyed with gold in the arts in order to render it harder and more durable, is fatal, because these also dissolve under the combined influence of the electrolyte and the current, and produce a bath which deposits a coloured gold.

When pure gold is not to be had, and the means for preparing it from the alloys of commerce are not available, it is better to substitute a platinum sheet as anode, which will not be attacked by the solution. In this case the bath rapidly decreases in strength, and must be replenished from time to time by the addition of gold cyanide, until so great a quantity of foreign matter has accumulated that a good deposit is no longer produced. The gold anode, however, is to be preferred, as it maintains an even constitution of solution.

For large articles which require a thick covering the surfaces of the two electrodes should be approximately equal in area; the anode should be completely immersed in the liquid by means of platinum suspending wires, to obviate unequal corrosion of the plate. For small objects, which need but a few minutes' exposure to impart a sufficient film, a smaller anode is often used; but the bath should be examined at intervals, when it is much worked, and, if necessary, a further amount of gold cyanide must be added. It is convenient to have ready means for altering the position of the anode in the liquid, so that a greater or less surface may be immersed at will, and hence also for regulating the strength of current and with this the rate and character of deposit. To impart the almost imaginary film of gold which is to be found on cheap jewellery, Roseleur recommends the use of a platinum anode, maintaining the strength of the bath by adding to it crystals of gold chloride.

The Vat.—Earthenware or porcelain vats are best suited to cold solutions; and a deep porcelain evaporating-basin, obtainable from any chemical-apparatus dealer, and from many druggists, is the best containing-vessels for hot plating-baths, provided that only small objects are to be treated. For larger work enamelled iron should be employed, but it is more important than ever that the enamel should be perfectly sound. A cover should be made to exclude dust, or the liquid may be stored in stoppered bottles when not in use. Generally speaking the duration of the gilding operation is so short, and the objects are often so small, that it is unnecessary to arrange the conducting-wires around the vats, they may simply be attached to the anode-plate and the pieces respectively.

Character of the Metal Deposited.—The deposit of gold is, perhaps, more susceptible of change by varying external conditions than that of any other metal. A large proportion of the value of gilding depends upon the colour of the metal precipitated, and this is most readily affected not only by the presence of foreign matter, as recently explained, but by changes in the strength of the current or of the bath.

A current which is too strong will, of course, deposit the gold as a black powder; but within the limits between which coherent and adhesive deposits are yielded, a stronger current produces a deeper coloured coating than a weak current. Hence, speaking generally, any influence which tends to increase the current-volume gives rise to a metal possessing a warmer hue. A feeble current, a small anode, and a cold solution, alike give pale yellow deposits; but a stronger battery, an increase of anode-surface (and hence less resistance), or warming the liquid, increase the current-strength, and a deeper yellow tone prevails. Motion imparted to the pieces also causes the production of a lighter colour.

The best colour for the pieces to present on removal from the bath is a very deep yellow, inclining to brown; a pure gold colour will become too pale when the lightening and polishing action of scratch-brushing or burnishing has done its work. A dark, almost black, colour produced by an excess of gold or too powerful a current, will never yield a good tint subsequently, nor indeed will even a brown deposit do so.

The Electro-Plating Process for Gold.—*Stripping.*—As with silver, so with gold, old coatings should be entirely removed before attempting to deposit a fresh layer of the precious metal. For a simple stripping-bath, some workers recommend a mixture of nitric acid with a little common salt, which produces by chemical reaction nitro-hydrochloric acid or aqua regia; but if this mixture be used, the utmost care must be taken to stop the action immediately the gold is removed, as the basis metal would be most powerfully attacked by the mixture. Others use strong sulphuric acid, to which about one-tenth of its volume of strong nitric acid and one-fifth of strong hydrochloric acid have been added. Equally with the other, however, great care is necessary to prevent the attack upon the basis metal by this solution, and it must be most scrupulously preserved from contact with water, for very moderate dilution would render its action upon the basis metals intensely vigorous (see p. 200). Large articles may be stripped, according to Wahl, by making them anodes in a bath of the strongest sulphuric acid, the cathodes being of copper. But the simplest plan is to make the old plated goods the anodes in a 10 per cent. solution of potassium cyanide; such a solution *per se* may even suffice to dissolve a mere wash of gold, but for any appreciable depth of deposit the solvent action should be aided by attaching the pieces to the positive wire of a battery; the action must not be unnecessarily prolonged as many of the basis metals are dissolved in this way—silver especially so.

The Process.—The articles, whether new or old (but in the latter case, after stripping), are well polished and cleansed by potash and acid dips, and after a thorough rinsing are placed in the gold-bath. Many platers quick the objects previously to this by passing them through a weak solution of mercurous nitrate; but if they are clean and the baths are well prepared, they should take a good coat of gold in the vat without this preliminary process, and as gold is prone to amalgamate or alloy with mercury, and thus to become discoloured, it is well to guard as far as possible against damage to finished goods by avoiding the use of mercury in all operations connected with gilding.

Small wares, when only a few pieces are to be dipped at a time, may be slung upon a copper wire attached to the negative pole of the battery, and are thus plunged into the gold-bath (usually hot), the wire being still held by the right hand so that they may be gently moved about in the solution; the left hand meanwhile grasps the wire attached to the anode, and is thus able to regulate the proportion of its surface which is immersed in the liquid, and hence also to alter the volume of current at will. The deposit should take place immediately, and as soon as the articles are completely covered, the anode may be partially withdrawn from the liquid to reduce the current-strength, but not sufficiently to give rise to a yellow coating of gold. This method of working has the merit of affording control over the colour of the deposit; if too pale it may be brought to the slightly brownish yellow which ultimately gives the richest tone, by lowering the anode and thus increasing its surface; or if too dark it may be correspondingly improved by decreasing the area of anode exposed.

After a few seconds' immersion, the pieces should be lifted from the solution and examined. If the coating be sound and of good colour, the position of the wire-support should be shifted

slightly by a gentle shake, and the goods returned to the bath.
They should be re-examined from time to time, and finally
removed, washed, and scratch-brushed lightly or burnished.
The washing should be effected in several waters as explained in
reference to copper (p. 145). If at the first inspection the
colour is found to be wrong, the fault must be remedied by
altering the position of the anode to a proportionate extent;
if the object is imperfectly coated owing to the presence of
grease-marks (which is scarcely likely to be the case owing to
the comparatively ready solubility of grease in the hot cyanide
of the bath itself), it must be removed, rinsed, and re-dipped in
potash, again rinsed and returned to the plating-vat. Bad solutions sometimes cause discoloured patches on the surface of the
article, and these should be obliterated by scratch-brushing before
continuing the deposition. The beer or other organic matter
from the liquid used in scratch-brushing must, of course, be most
carefully washed away before replacing the goods in the gold-
bath.

The continuance of a black or dark-brown deposit must not
be permitted, as it is impossible to produce a good coloured
lustre by polishing such a coating. If the colour be due to
excess of gold in the solution, or to too strong a current, the
remedy is obvious; but if it be due to organic matter contained
in an old bath, the use of the latter must be discontinued. For
some classes of work, however, such as the coating of interior
surfaces—of tankards and the like—a slightly dark colour is
often preferable, and an old bath which is not absolutely past
use may find an application here. The duration of the gilding-
process rarely exceeds a few minutes, as a comparatively thin
film of the metal suffices for most purposes. When thick
deposits are required, the pieces may be removed from the bath
two or three times during the process, and be scratch-brushed,
well washed, and returned.

Large objects cannot, of course, be treated in this manner, but
should be suspended in the bath as in ordinary plating-vats, but,
if possible, they should be gently moved from time to time;
very large pieces are more often treated in cold baths because of
their greater convenience of application. Small perforated
articles are best hung upon a wire on which they are separated
from one another by small glass beads; but the wire should now
and again be sharply shaken, to prevent the formation of
permanent marks upon the goods at the points of contact; so
also in slinging chains in the gold-bath, the relative positions of
the links should be shifted from time to time with the same

object in view.   To gild the interior of a vessel, it must be well
cleansed and prepared to receive the deposit and then very
thoroughly dried on the exterior, especially around the rim.
Having been rested upon a level surface, and attached by a wire
to the negative pole of the battery, it is carefully filled to the
brim with gold solution at a temperature of 120° to 140° F., so
that none of the liquid splashes or creeps over the margin (hence
the necessity for absolute dryness outside).   A gold anode
attached to the positive pole of the battery is dipped to a
sufficient distance in the liquid and gently stirred round within
it ; in a few seconds the interior surface will be completely
covered with gold, and in four or five minutes a sufficient
deposit will have been effected.   The anode is then removed,
the liquid returned to the heated gold-plating-vat, and the
article thoroughly washed, polished, and dried.   Attention is
chiefly to be bestowed in securing an even margin at the edge of
the vessel ; if it be not sufficiently dried initially, the gilding-
solution will gradually creep up the damp portions ; and
wherever the liquid has penetrated, gold will be deposited, and
thus a wavy line instead of a straight one marks the junction
between the outside of the vessel and the lining ; splashings
which are in liquid connection with the main portion of the
solution will, of course, bring about a similar result.

It often happens that the vessels which are to be gilt
interiorly, have an irregular outline at the top, so that some
parts of the surface which should receive a coating are above
the surface of the liquid, as, for example, in the case of ewers
and other lipped wares.   All these portions may, however, be
covered by means of a *doctor ;* this consists of a piece of soft
rag folded several times around a thin strip of gold connected as
an anode.   On saturating the rag with gilding-solution and
applying it as a brush to the parts which are to be treated, the
object being, of course, connected up as a cathode, the current
passes and electrolyses the liquid in the rag, depositing the gold
upon the metallic surface and dissolving it from the anode.   As,
however, there is not free motion in such an absorbed solution, it
is advisable to re-moisten the rag in the gold-bath from time to
time.   The doctor is best applied during the time of gilding the
rest of the interior, as there is then less likelihood of a line
forming, which would indicate the level of the liquid in the vessel.
Such a line, if at all pronounced, is not easy to obliterate without
risk to the gilding around.   Nevertheless, with care, the rag-
gilding may be applied either before or after the other process
without evil consequences.   The more elaborate device proposed

by Wagener and Netto for other purposes (p. 115) could be applied to this work if desired.

The time required for gilding is far less than that demanded for silvering, because, as a rule, the coated articles are not required to withstand such severe wear, and a much thinner deposit is sufficient; but thimbles, pencil- or watch-cases, or any object which will be put to the test of rougher usage must receive a fair proportion of gold. For other goods which are to receive but a thin film, the actual thickness must be regulated by the colour, which depends largely in the early stages of gilding, upon the colour of the basis metal; thus brass, copper, or bronze, which are themselves yellow or reddish metals, become thoroughly gold-like after an immersion of a few seconds, while silver, which is a white metal, pales the colour of the gold precipitated upon it until sufficient has been deposited to form a film completely opaque. Occasionally silver is covered with a preliminary wash of copper, in order to enhance the colour of a mere wash of deposited gold. It must be borne in mind, however, that a mere film of gold will not entirely protect the silver or other basis metal from the action of the atmosphere; so that such surfaces are very liable to tarnish when brought into large towns, or wherever the air is at all polluted with hydrogen sulphide.

Dead-Gilding.—It is sometimes desired to produce a surface with that dead lustre which in richness of effect is so charming to the eye. This may always be accomplished by ensuring that the surface is dead before gilding; if it be not, it must be rendered so, either mechanically, by rubbing with a fine powder, such as that of Bath-brick, which will impart the necessary degree of roughness without causing deep marks or scratches; or chemically, as in the case of copper, by dipping momentarily into a strong acid; or electrolytically, by imparting a preliminary frosted film to the article before gilding. Of these processes the first is self-explanatory. The second is convenient for small wares, being especially adapted to copper, brass, or bronze goods, and depends for success upon the microscopically uneven etching of the surface of a minutely crystalline, or not absolutely homogeneous, material. It is effected by plunging the goods, suspended from a wire or contained in a perforated porcelain or platinum-wire basket, into a bath containing 100 parts of nitric acid (specific gravity = 1·33), a like quantity of sulphuric acid (specific gravity = 1·84), and 1 part of common salt. They are almost immediately removed and plunged into a large volume of water, so that the acid clinging to them is at once washed away (an insufficient volume of wash-water would, for the first moment, only dilute the acid, and cause it to attack the metal too violently). If sufficiently frosted, they are very thoroughly washed, and, with or without quicking, according to the practice of the establishment, are passed on to the gilding-vat. If still too bright, the acid dip is again and again repeated, until the requisite degree of dulness has been imparted. It need hardly be observed that previous to this process, the goods must have been thoroughly cleansed by the usual dips. The third method is equally convenient and simple, and is especially suitable to silver articles. Advantage is taken of the beautiful dead lustre of electro-deposited copper; a thin film of this metal is deposited upon the cleansed silver surface, and without scratch-brushing or burnishing, but after washing well, is at once protected by the gold deposit. Occasionally the copper is made still more dead by very rapidly passing the pieces through the acid dip, immediately before gilding; but in this case great care must be taken that the copper is not entirely removed at any point, because at such places an irregularity of gilt surface would be apparent. The addition of aurate of ammonia to the gold-bath also tends in the direction of giving a good dead lustre.

Dead-gilded work must never be exposed to friction, either in manufacture or in use—or it will rapidly become brightened; this class of gilding is, therefore, not applicable to general work, but is well suited to surfaces which will be kept under glass, such as the dials of clocks or philosophical instruments, or for the ground work of a raised design which will protect it from being rubbed, and which, being itself burnished, may be made to produce a very fine effect by the skilful contrasting of the two styles of gilding. When the mechanical method of producing the dead surface before gilding is resorted to, as is often done with sword mountings and the like, or where the two kinds of gilding are blended on the same object, the Bath-brick treatment should be confined, as far as possible, to those portions which are to be dead-surfaced, as the labour of brightening the others afterwards would be greatly increased. So, too, in imparting the final polish to the bright portions, the tools must not be allowed to touch the other surfaces, or the latter will at once lose their characteristic appearance.

Some difficulty is at times experienced in giving the first coating of gold to dead surfaces, and the progress of the deposition must be carefully watched. Should such difficulties arise, the current-strength and that of the solution may be increased; indeed with filigree work, in which several depths of surface are presented, there may be difficulty in forcing the deposit into the deeper interstices, unless this treatment is resorted to. But as these alterations of conditions both tend in the direction of giving a brown deposit, which it is quite impossible to remedy if the surface is to remain dull, and almost so if it should be brightened (because of the difficulty in causing the polishing tool to penetrate to all parts), the countervailing precautions of using a new bath and keeping the pieces in active motion in the liquid must be taken, because *cæteris paribus* either of these measures tends to lighten the colour of the deposit. When the surfaces are once covered with gold, the progress should be steady.

Gilding the more Electro-positive Metals.—Any metal which will itself deposit gold from the cyanide solution demands especial care in its treatment. When it is possible to effect the gilding without preliminary protection by copper, as, for example, in the case of nickel-silver, iron, or steel, the rapidity of deposition must be checked by using a weaker solution, a less intense current, and a lower temperature during the operation; unless these points are attended to, the deposit will probably not be sufficiently adhesive to withstand the subsequent treatment of polishing. Very commonly these metals are first protected by a coat of copper or brass, imparted by the alkaline (cyanide) bath; lead, Britannia metal, zinc, and similar metals are always so treated. A thin film of copper is deposited upon the well-cleaned surface in the cyanide-vat (p. 137); then, if wished, this may be slightly thickened in the acid copper-bath, and having been scratch-brushed (and quicked if desired) it is ready for gilding.

The dead-gilding of large surfaces of zinc is frequently demanded by the requirements of art, and is readily accomplished by a modification of the last process. The metal being well prepared is coppered thinly in the cyanide-vat, washed, scratch-brushed to ensure an even and adherent deposit, and a slight increase of copper is given in the same bath; then, after washing, a thicker frosted deposit is made in the acid copper-bath, and now the surface is ready for the gold-vat, which should be used almost at the boiling point. A sufficient amount of gold having been thrown down, the plate is thoroughly washed and dried, preferably in a drying-oven. It is necessary to avoid handling the dead-coated portions at any stage of the process, as the slightest mark becomes visible upon the finished surface. Many operators, according to Roseleur, give the coppered objects a thin coating of silver by simple immersion before gilding, and after drying, and burnishing those portions which are to be brightened, impart a second thin coat of gold, wash, and again dry it.

When silver objects have been joined with tinman's (soft) solder, which is an alloy of tin and lead, the gold frequently refuses to deposit upon it, especially if the bath is in bad order or the current weak. Watt obviates this difficulty by painting the line of the solder with a solution containing 5 per cent. of copper sulphate crystals and a like proportion of sulphuric acid, and lightly resting a piece of iron in the liquid in contact with the solder; galvanic action is immediately set up, iron dissolves, and an equivalent of copper is precipitated over the whole surface covered by the solution. Thus, a copper surface is substituted for one of solder, and the entire object may be gilt in the usual way. When only a wash of gold is to be given, the coppered line should preferably be silvered before gilding, to ensure uniformity of colour.

Ornamentation and Treatment of Gilt Surfaces.—It is sufficiently obvious that, from the number of varied tints which may be imparted to electro-deposited gold, as well as from the different lustres obtainable, the means of ornamentation depending upon the combination of these alone, or of part-silvering and part-gilding, are almost infinite in number and variation of effect. Thus, a silver (or silvered) article may be in part gilded, by painting with a stopping-off varnish (see p. 344) those portions of the design that are to remain as silver; then, after drying and hardening the varnish by heating gently in a stove, the remainder may be caused to receive a deposit of gold of any desired colour. Or a pattern may be left blank upon the varnished surface, to receive a coat of red gold; the varnish may then be washed off the remaining portion by the application of turpentine and subsequently spirits of wine, and the gilt pattern being in turn protected, the ground work may be covered with a green or yellow gold. But the various combinations of flat and relief surfaces, of red, green, and yellow, gilding with silver, and with gold of bright or dead lustre, would be with greater fitness discussed in a manual of decorative art than in these pages, wherein it must suffice to point out the means by which the ideas of the artist may be carried into effect.

It occasionally happens that the gold deposit has not a good colour, so that a special colouring process must be resorted to. Usually the electro-deposited metal, being under better control at the time of precipitation, does not require this treatment, but the gilding produced by simple immersion may do so. The required shade may, of course, be imparted by electro-gilding; but when the layer of gold is not too thin, the older jeweller's method may be adopted, by which the subjection of the object

15

to the action of oxidising agents in a state of fusion, or in hot concentrated solution, effects the oxidation and removal of the alloyed metal upon the surface, leaving the gold unaltered. Such a mixture is made by fusing together in an earthen pipkin equal parts of alum, nitre, ferrous sulphate, and zinc sulphate; when thoroughly melted it is mixed by stirring, and is painted over the surfaces of the objects to be coloured. These are then suspended in the central space of a specially-constructed circular furnace, which has an inner concentric lining of vertical fire bars (fig. 97 shows this furnace in vertical cross-section), the annular space between the two being filled with incandescent fuel. Here they should remain until a moistened surface, caused to touch any of the pieces, produces a slight hissing sound, by which time the colouring mixture will have fused; they are then removed, and at once pickled in dilute sulphuric acid (water containing 2 or 3 per cent. of the acid), until the solid crust and the oxide of copper, produced by the oxygen of the nitre, have dissolved and left a clear gold surface. Any copper uncovered, or any base metal insufficiently protected, will, of course, be corroded and the piece spoiled; copper surfaces will have become covered with red cuprous oxide, which is insoluble in the weak acid. If results of this nature have been obtained, the only remedy is to strip off any gold which may remain, and re-cleanse and re-gild the objects with a thicker coating.

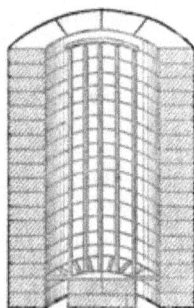

Fig. 97.
Colouring-furnace.

A mixture made from 2 parts of potassium nitrate and 1 part each of alum, sodium chloride, and zinc sulphate by rubbing them into a thick paste may be applied in the same way; it is painted over the surfaces to be covered, and the pieces are heated on a clean iron plate over a charcoal fire until the mixture has darkened in colour, when it is removed by dipping the articles into dilute sulphuric acid. A third mixture, which is often similarly employed, consists of 6 parts of potassium nitrate, with 2 of ferrous sulphate, and 1 of zinc sulphate. But however useful these mixtures may be in imparting a good colour to inferior but solid alloys of gold with copper, they are certainly not generally suitable to the treatment of the thin films of gold imparted by electrolysis to articles of base metal; and the electro-colouring process, by judiciously combining gold and copper or gold and silver solutions, or by merely adjusting the strength of the plating-current is far more reliable and satisfactory.

Gilding of Watch-Mechanisms.—The peculiar semi-dead lustre noticeable in the internal movements of watches is produced by a preliminary mechanical process known as *graining*, which has been fully described by Roseleur. It carries sufficient intrinsic interest to warrant the insertion of an outline sketch of the process at this point.

The various parts are carefully polished to obliterate completely all file- and tool-marks; they are next strung upon a brass wire and boiled in a 10 per cent. solution of caustic potash or soda, to remove grease; and, being then well rinsed with water, should show no marks of imperfect wetting, which would indicate the presence of unremoved greasy matter. (This would point to the necessity of further treatment with potash.) All iron or steel portions are now protected by covering them with a stopping-off varnish, applied in the melted condition by means of a thin warm glass rod, and consisting of—

Clear rosin,	10 parts.	Best red sealing-wax,	4 parts.
Yellow bees'-wax,	6 ,,	Finest polishing-rouge,	3 ,,

The rosin and sealing-wax are melted together, the bees'-wax is added, and, finally, the rouge is stirred in and thoroughly incorporated. The pieces may be momentarily immersed in an acid dip, but this stage is frequently omitted; in either case they are now fastened by flat-headed brass pins to a level surface of cork, cavities being made when necessary, to allow for spindles or projections upon the articles. Held thus in position, they are rubbed well by a rotary motion, with a brush dipped in powdered pumice, and wetted with water; after perfect rinsing they are (with the cork-support) passed rapidly through a weak quicking-solution containing 1 part of nitrate of mercury and 2 of sulphuric acid in 5,000 parts of water, and are ready for the process of graining. Impalpable silver-powder must be procured for this purpose; it may be prepared either by grinding the finest leaf silver (thin silver-foil) with honey upon a ground-glass slab with the aid of an artist's muller, and then washing away the honey with boiling-water, the mixture being placed upon a good blotting-paper filter (p. 54); or by placing strips of clean copper in a very dilute solution of silver nitrate, collecting the spongy silver precipitated by "simple immersion," washing it free from adhering copper solution and drying it. The silver-powder is then most intimately mixed with finely-powdered and -sieved tartar (potassium bitartrate) and common salt, in the proportions of 3 of silver with 10 to 30 of tartar and 40 to 100 of salt; the ingredients should of preference be dried individually at a temperature slightly above 212° F., and mixed warm. The mixture is now made into a thin paste with water and spread evenly over the surfaces of the pieces, and is rubbed persistently over the whole by an oval brush with stout hard bristles; a circular motion should meanwhile be imparted both to the brush and to the cork, but in opposite directions. The use of a large quantity of paste or of much salt produces a large grain, while less paste and more tartar yield a smaller grain; the desired effect of roundness is imparted to the grain in direct proportion to the extent of the circular motion applied to the work. After this operation and subsequent washing, the surfaces are scratch-brushed with a straight brush of very thin brass wires more or less annealed. It is recommended to keep three brushes : one which has been heated to dull redness and cooled, and is very soft in consequence ; one somewhat less heated, and so only moderately hard ; and one but slightly annealed, and, therefore, much harder. The scratch-brushing must also be effected under the influence of a double rotary motion, applied to cork and brush, and is aided by a decoction of liquorice or soapwort as a lubricant. The grain should finally be perfectly uniform, even when viewed under a magnifying lens.

The pieces are now removed from the cork, and being attached singly to suitable holders, are suspended in the gold-bath, for which Roseleur recommends the use of the solution No. 17 (quoted on p. 214), containing gold fulminate. The gold should be slowly deposited by a moderate current and with platinum-anodes. The pieces are finally re-scratch-brushed with either of the lubricants above-named ; and the varnish is removed from the steel portions by the application of warm oil, benzene, or turpentine, followed by immersion in an alkaline solution almost boiling.

Platinum- or gold-powders may be substituted for the silver in the production of grain, but as their use presents no advantage, and they are far more costly, they are rarely so employed.

CHAPTER XI.

THE ELECTRO-DEPOSITION OF NICKEL AND COBALT.

Advantages of Nickel.—The introduction of nickel-plating is one of the more recent applications of electrolysis. Until about the year 1870 or later, the high cost of metallic nickel, combined with the impurity and unsuitability of the metal which was then available, rendered nugatory all attempts at electro-nickeling on a large scale. But with the improvements in the metallurgy and manufacture of nickel, which has placed comparatively pure anodes on the market at a reasonable rate, arose the new industry of plating with nickel, which has perhaps advanced with more rapid strides than any of its numerous rivals in the same field. The extreme hardness of deposited nickel, which enables even a thin coating to resist so well the wear and tear of hard use ; the brilliant polish which from its hardness it is capable of taking ; combined with its pure white colour, almost rivalling that of silver, and its non-liability to tarnish under ordinary atmospheric conditions—all tend to popularise the use of the metal, not only as a protective coating for more oxidisable metals, but as an ornamental addition to those less attractive in appearance. In the first-named capacity it is applied to exposed portions of machinery which do not actually present working surfaces that are liable to friction, especially in domestic appliances such as the sewing machine, or in bicycles or gas-engines, &c., in which it plays the dual part of protecting and decorating. In the second capacity it is applied to small articles of brass or zinc, such as pencil-cases and umbrella-fittings, as well as to larger surfaces, such as restaurant coffee-urns and the like. A thousand other applications might be named, but the above sufficiently represent the classes of work to which electro-nickeling is daily applied, and which are constantly extending in all directions.

The nickel-plater, therefore, has mainly three classes of work to treat—iron or steel, brass, and zinc ; none of them are likely to give trouble provided attention to detail is rigorously observed. More than any other metal, perhaps, nickel requires the most punctilious care in order to obtain an adhesive deposit; no metal is so sensitive to any undue change in the manner of treatment, but, on the contrary, none repays the operator so well for his attention to the preliminaries and requirements of working. But care is most largely needed in the re-nickeling of old work, for it is found that the metal cannot be induced to give a deposit in any degree adhesive upon an old surface of nickel.

Nickel well deposited is extremely hard, so hard that it cannot be burnished, and is somewhat brittle. Thick coatings are especially liable to flake off in use, unless exceptionally well deposited, and even the thinnest films will part from surfaces which are not chemically clean. Fortunately, however, thick deposits are rarely required, on account of the high resistance of the metal to wear, which enables even a thin film to rival in durability a thick coat of any other metal. Nevertheless, the opposite extreme to which manufacturers tend, both by reason of the greater difficulty experienced in depositing thick coats, and of the less cost of producing the thinnest possible wash, that to outward appearances is indistinguishable from a good deposit, is greatly to be deprecated, because a non-durable film brings discredit upon the process and upon the manufacturers who use it. It must be remembered that the cost of a slight increase of thickness is by no means proportional to the extra metal and to the battery-power used; inasmuch as one of the largest items of expenditure is to be found in the preparation of the object to receive the deposit,—and this is constant, whatever the thickness of the metal precipitated.

The nickel precipitated by too strong a current is grey and pulverulent, and is said to be "burnt;" but, on the other hand, a feeble current produces a hard but very brittle deposit, which will probably become separated from the plate during the final process of polishing. Schaschl, using Pfanhauser's citrated bath, has found that a film 0·2 millimetre thick deposited on thin sheet-iron by a current of about 1·0 ampère per square decimetre, became torn along the line of the bend when the plate was sharply bent over upon itself; but that by first annealing the plate at a dull red heat, the nickel was so far softened that the plate could be hammered down to a quarter of its original thickness without incurring the slightest injury. Copper and brass, however, on which nickel had been deposited by the normal current, withstood this treatment without any preliminary heating.

Nickel is practically never deposited by simple immersion or by the single-cell process, the battery deposition is, therefore, the only method which demands attention.

Nickel-Plating by the Separate-Current Process.

The Battery.—The Bunsen- or bichromate-cells are well suited for this work, the electro-motive force best adapted for nickeling being higher than that required for most metals. Two or even three Bunsen-cells coupled in series may be employed, and these should be found to last for one day's work, being renewed every morning. The chief objection to this battery is to be found in the red fumes of nitrogen peroxide evolved during use; but this is of small consequence if the cells be kept, as recommended, in a separate and well-ventilated chamber. The current should be somewhat stronger at first, until the whole surface is just flashed over with nickel, when it should be reduced to the normal strength. Thus at the outset it may conveniently be 0·1 ampère per square inch, or 1·5 ampères per square decimetre at 5 volts' pressure, and should be subsequently reduced 0·02 ampère per square inch, or 0·3 per square decimetre, at 2 volts' pressure.

The Solution.—The number of solutions which have been successfully employed is very great, and any of them may be made to give good results; that being so, the simplest is probably the best. The following list includes some of the principal formulæ recommended.

For general work, the solution made by dissolving 8 pounds of the nickel-ammonium sulphate in each gallon of water, with the addition of just so much ammonia if it be acid, or of citric or sulphuric acid if it be alkaline, as will suffice to render it exactly neutral. Nickel-platers should always be supplied with blue and red litmus, which by turning red or blue respectively on immersion in the fluid, indicate acidity or alkalinity; when neutral, the solution should not change the colour of either paper. Although, theoretically, the bath should be neutral, in practice it is found better to maintain it in the faintest degree acid; because secondary reactions, which constantly take place during electrolysis, tend to liberate ammonia, and thus render it alkaline, and alkaline-baths give trouble by depositing a basic compound of nickel in the form of a greenish powder. On the other hand, excessive acidity must be avoided as it may entirely prevent the deposition of nickel upon the cathode. The cause of the increasing alkalinity is, no doubt, to be ascribed to the simultaneous decomposition of the nickel sulphate with a small

TABLE XVII. Showing the Composition of Nickel-Baths for

No.	Authority.	Special Application of Bath.	Nickel Acetate.	Nickel Carbonate.	Nickel Chloride.	Nickel Citrate.	Nickel Nitrate.	Nickel Phosphate.	Nickel Sulphate.	Nickel-Ammonium Sulphate.	Cobalt-Ammonium Sulphate.	Potassium Citrate.	Sodium Bicarbonate.	Sodium Bisulphite.	Sodium...
			1	2	3	4	5	6	7	8	9	10	11	12	13
1	Adams, . .	...	...	...	...	...	...	...	...	50-80	...	...	...	...	..
2	Boden, . .	...	...	...	...	...	26·7	...	...	...	...	...	...	33	..
3	Desmur, . .	Small Goods	...	...	...	...	...	...	70	...	...	8	...	...	..
4	"Electricias,"	...	...	...	...	...	...	...	50	...	...	...	...	...	..
5	Hospitalier, .	...	...	...	...	...	...	...	100	...	...	...	...	...	..
6	Langbein, .	Printing surfaces	...	...	...	...	...	...	60-72	...	...	...	...	...	..
7	,, . .	...	...	4·5	...	...	...	...	50-60	...	...	...	...	...	..
8	Nägel, . .	...	...	...	...	...	...	...	54	...	...	...	...	...	..
9	Pfanhauser,	...	...	...	...	. .	...	...	50	...	...	...	50	...	..
10	,,	...	...	...	...	...	...	...	50	...	...	...	...	...	..
11	Potts, . .	...	27·5	...	...	...	...	...	...	...	...	...	...	...	..
12	Powell, . .	...	...	...	...	...	...	...	...	50	...	...	...	...	..
13	,, . .	...	...	...	...	15	...	15	30	...	...	...	...	...	..
14	,, . .	...	...	...	...	15	...	25	...	...	...	...	...	3	2
15	Roseleur, .	...	...	...	...	...	...	...	...	40	...	...	...	...	..
16	Volkmer, .	...	...	...	...	...	...	...	111	...	...	...	...	...	..
17	Watt, . .	Tin, Britannia Metal, &c.	...	...	...	...	...	...	...	33·3	...	...	...	...	..
18	,, . .	Iron	...	...	...	...	...	...	40	...	...	...	...	...	..
19	Weiss, . .	...	...	...	...	...	...	...	50	...	...	...	...	...	..
20	,, . .	..	...	...	...	...	...	...	50	...	...	...	...	...	..
21	,, . .	Iron and Steel	...	...	...	...	...	...	50	...	...	...	...	...	..
22	,, . .	Zinc	...	...	...	...	...	...	42	...	...	17	...	...	..
23	,, . .	...	...	...	...	42	...	...	...	...	...	...	...	...	..
24	,, . .	Hard deposit	...	...	...	...	...	...	40	10	...	...	...	..	
25	Weston, . .	...	...	...	...	...	...	...	50-67	...	...	...	...	...	..

NOTES TO TABLE.—The black figures in the last column refer to the numbers of the vertical columns representing the ordinary temperature.

SEPARATE-CURRENT PROCESS, AS RECOMMENDED BY VARIOUS AUTHORITIES.

14	15	16	17	18	19	20	21	22	23	24	25	
Ammonia.	Ammonium Carbonate.	Ammonium Chloride.	Ammonium Sulphate.	Ammonium Tartrate.	Calcium Acetate.	Acetic Acid.	Benzoic Acid.	Boric Acid.	Citric Acid.	Tannic Acid.	Water.	Special Method of Preparation.
...	...	...	...	...	...	...	...	...	...	...	1000	Neutralise, if necessary, with ammonia.
26·7	...	...	...	...	...	...	...	...	...	...	1000	Dissolve 12 in 25; then add rest.
...	...	...	...	...	...	...	...	...	...	...	1000	Warm sol. of 8 in 25; add 11 slowly.
...	...	...	...	36·5	...	...	...	...	...	0·25	1000	Stir all with 150 of 25; then add rest of 25.
...	...	...	...	...	...	...	...	...	...	...	1000	Boil, cool, and filter.
...	...	...	19·22	...	...	...	...	...	4·5	...	1000	
...	...	...	...	...	...	...	...	25-30	...	...	1000	
190	...	...	...	...	...	...	...	...	...	...	1000	
...	...	...	...	...	...	...	...	...	...	...	1000	
...	...	50	...	...	...	...	...	...	...	...	1000	
...	...	...	...	...	25	q.s.	...	...	...	...	1000	
...	q.s.	...	20	...	...	...	...	...	5	...	1000	Add 15 last, till just neutral.
...	...	...	...	...	...	...	7·5	...	...	...	1000	
37	...	...	...	...	...	...	...	7·5	...	...	1000	
...	q.s.	...	...	...	...	...	...	...	...	...	1000	Pour sol. of 15 into sol. of 8 till just neutral; avoid alkalinity.
...	...	22·33	...	...	...	...	...	...	...	...	1000	
...	...	...	6·6	...	...	...	...	...	...	...	1000	
...	...	...	6	...	...	...	...	...	...	...	1000	
...	q.s.	...	50	...	...	...	...	...	q.s.	...	1000	Boil 7 and 17 with 25, add 15 till neutral, then 23 till just acid.
...	...	...	15	...	...	...	...	...	q.s.	...	1000	
...	...	...	25	...	...	...	...	...	5	...	1000	
...	...	25	...	...	...	...	...	...	...	...	1000	
...	...	42	...	...	...	...	...	...	...	...	1000	
...	...	...	17	...	...	...	...	...	...	...	1000	(Mixed Cobalt-nickel precipitate.)
...	...	...	...	...	...	...	...	15-30	...	...	1000	

the various reagents. All the solutions except No. 3, which may be used nearly boiling, are generally employed at proportion of the ammonium sulphate as would be predicted (see p. 33). The electrolysis of the nickel salt alone deposits upon the cathode a weight of metal equal to that dissolved from the anode, and is thus without influence upon the composition of the bath. But the ammonium salt under electrolysis deposits sulphuric acid upon the anode, which, by combining with it, adds an equivalent weight of nickel to the bath, while it throws down the elements of the hypothetical body ammonium ($N H_4$) upon the cathode. Thus the sulphuric acid is neutralised by the nickel, and the bath is enriched to that extent by the fresh nickel introduced, while the ammonium is broken up into ammonia ($N H_3$) and hydrogen (H), the former dissolving in the solution and tending to alkalise it, the latter escaping as a gas from the surface of the object being plated. In depositing nickel, hydrogen is usually deposited to some extent with the metal (in part due to the reaction above considered); but this must be minimised as far as possible by regulating the strength of the current and of the solution, remembering that a strong current and a weak solution alike favour the evolution of hydrogen. Whenever much hydrogen is given off, the metal becomes grey and pulverulent, and the absorption of hydrogen by the nickel undoubtedly tends to increase its brittleness.

The reactions in the nickel-bath may, therefore, be thus expressed :—

	At cathode.	At anode.	Forming with Ni anode.
Main reaction,	Ni	$S O_4$	$Ni S O_4$
Subsidiary reaction,	$N H_3 + H$	$S O_4$	$Ni S O_4$

While the bath is in use it should be frequently tested with litmus-paper, and rendered faintly acid, if necessary, with sulphuric acid. It is advisable to agitate the bath at intervals in order to maintain an equality of density throughout; but this is not so necessary as in the case of copper deposition, because the duration of the nickeling process is so much less than that of the other, rarely requiring an exposure for more than two or three hours.

Some operators have endeavoured, with considerable success, to prevent the formation of the basic salt of nickel by the addition of a small proportion of an organic acid (citric or tannic), or of a feeble inorganic acid such as boric acid. This last-named solution, as formulated by Weston (No. 25), certainly gives admirable results, and although the simple solution previously recommended may be made to yield a most excellent deposit without difficulty by careful attention to the working of the vat, Weston's bath allows more latitude in working, and may perhaps be preferred by many. It is made by dissolving from 8 to 10 ounces of the double sulphate of nickel and ammonia in each gallon of water, and adding to it from $2\frac{1}{2}$ to 5 ounces of boric acid.

Various other nickel salts have been substituted for the double sulphate, among them, the chloride and acetate; but in nearly every case a double rather than a single salt is preferred, as yielding better results. When the single salt is used in making up the bath, a quantity of the corresponding compound of ammonia is added to the solution at the same time. In one bath (No. 24) a small proportion of cobalt is added, which causes a joint precipitate of the two metals that is said to possess greater hardness than nickel alone.

The Anodes.—The nickel anodes must be as pure as it is possible to obtain them. They are to be had either cast or rolled, of almost any shape. The cast plates are less dense and, as a rule, are more readily soluble than the others; either kind may be used, but the latter may be procured thinner than the former, and the prime cost of the nickeling plant is thus reduced, moreover, they are more uniform and reliable in composition, and are less liable to become spongy during treatment owing to unequal solution. They should present a total area in excess of that of the cathodes, in order that the bath may be kept saturated with nickel, which otherwise would not be possible, owing to the inferior solubility of the metal.

The anodes should be supported by nickel hooks, and may with advantage be made with lugs at the upper corners, as indicated in fig. 61, so that the hooks do not enter the solution; even, however, with these anodes, the use of brass or copper supports is to be avoided, for, becoming splashed with the solution, they would in time become corroded, and the copper passing into the vat would seriously damage the nickel-bath.

The Vats.—Any glass, earthenware, enamelled iron, or lined wood tank may be used. If the solution is to be heated, the enamelled iron is, of course, preferable. The vat should always exceed in size that of the work to be treated by 15 or 20 per cent.

The Process of Electro-Nickeling—*Stripping.*—It is even more important in nickeling than in silvering or gilding that an existing film of nickel shall be entirely removed, or the new deposit will most certainly lack adhesive properties.

Small articles may be treated by persistently rubbing with fine emery-cloth until the desired end is accomplished. More often a chemical method is employed, which consists in dipping the articles for a brief period into an acid bath that will readily attack nickel. Such a bath may be prepared as recommended by Watt, by gradually and carefully adding one volume of nitric acid and two of sulphuric acid to one volume of water, constantly stirring meanwhile with a porcelain or wooden rod to prevent the action from becoming too violent. The liquid is allowed to cool, and should be transferred for use to a glazed earthenware pan or dish sufficiently large to hold any object which it will be required to treat in it. It is employed cold or only very slightly warm, and should be placed outside the operating-room in a well ventilated place, for example, in the battery-cupboard, because unwholesome and irritating acid fumes are evolved during the process.

The plated article is suspended from a copper wire and plunged beneath the liquid in the bath, where, however, it should not be permitted to remain for more than a few seconds at a time, for a thin coating is almost instantaneously dissolved, and the acid is then free to attack the basis metal beneath. It is, therefore, frequently removed from the vat and closely examined, so that the action may be stayed at the moment when the last trace of nickel has disappeared; it is then transferred to a large volume of cold water, and after washing twice or thrice in fresh water, is ready for the subsequent stages of the process.

Some operators prefer to strip by electrolysis, by making the object the anode in an old nickel-bath. Attention is equally necessary in conducting this process to guard against any attack upon the basis metal; but since it is impossible to entirely prevent all action, no bath which is to be afterwards employed for depositing the metal should be used for this purpose, as it will become gradually charged with impurities. A 10 per cent. solution of sulphuric acid in water may be equally readily adapted to the electrolytic stripping.

Iron or steel articles are best treated by either of the first two processes, brass or copper by either of the two last-named, yet with due care any of the three methods may be applied to all classes of work.

Preliminary Preparation for the Bath.—It has already been pointed out that the nickel deposit cannot be burnished, by reason of its extreme hardness. Ordinary methods for imparting the final polish to electro-plated goods are not, therefore, applicable to nickel-coated wares. It is thus essential that the highest possible polish should be given to the objects prior to immersion in the plating-vat, remembering that as they are when placed in the bath, so they will be when finished. Even traces of pre-existing scratches, or tool-marks, cannot be obliterated, except with the greatest difficulty, when once nickel has been precipitated upon the surface. If, therefore, the goods passing into the hands of the plater are in any degree rough or unfinished, they must be most carefully polished until every scratch has ceased to show ; extra care should be taken with large blank areas of surface, unbroken by the lines of a design or by a change of shape in the article itself, because even the slightest flaw becomes more visible on such surfaces than upon smaller articles.

More than usual care must also be bestowed upon the cleansing operations. In silver- and gold-plating, especially with warm solutions, the cyanide liquor compensates for any slight inadequacy of cleansing, by its power of dissolving the offending grease, so that the surface is finally cleansed by the electrolytic bath itself (which, however, is gradually spoiled by the absorption of organic matter), but the nickel-bath has no such solvent action ; so that it cannot be too strongly impressed upon beginners that the success of their work is dependent upon the absolute chemical cleanliness of the pieces to be plated. After polishing, every trace of grease is first removed in the potash-vat, and of tarnish in the acid dip (for iron goods) or the cyanide-bath (for brass, copper, or zinc). Then after a thorough rinsing in water, the goods are transferred without loss of time to the plating-vat. After passing the potash-bath the surface of the article should be handled with brass tongs or with clean rags, and must on no account be touched with the hands.

Copper and brass articles always, wrought-iron and steel generally, are at once nickeled without further preliminary treatment ; zinc is also occasionally treated in the same way, but inasmuch as it is readily attacked by the nickel solution, and the latter is rendered worthless when contaminated with zinc, it is advisable to protect the objects with a covering of copper before immersion. Meidinger has suggested a covering of the zinc sheet with mercury, which would answer the same purpose, but care is necessary to guard against over amalgamation, which only renders the plate very brittle without affording any corresponding advantage. Cast-iron also should be covered with copper ; it is a common practice to first bestow upon the surface a wash of tin, then upon this one of copper, and, finally, the layer of nickel. The articles having been made perfectly bright and clean, and, if necessary, covered with copper, are ready for suspension in the bath.

Nickel-Depositing.—It has been pointed out that the current

should be somewhat stronger at first than subsequently; but it must not be so intense that the metal becomes burnt, as explained on p. 230, a fault which is by far the more serious of the two. Therefore, in introducing the cleaned objects, although the current must pass immediately they enter the bath, the objects first suspended must be protected from receiving an excessive current by the interposition of resistances, or by hanging one or more anodes from the cathode rods, as explained in speaking of electro-silvering on p. 206. The surface of the article should almost immediately be completely covered with a grey deposit of nickel. The goods are suspended by copper wires, which should be used only once, because, from the want of adhesion of nickel to old nickel surfaces, the metal deposited upon old wires is liable to strip off in flakes which, falling into the solution, may, in course of time, form a metallic bridge or connection between the electrodes, and thus produce a short circuit, or they may adhere to projecting portions of the cathode surface and interfere with the regularity of the coating.

The nickel solutions are usually inferior conductors of electricity, and there is in consequence a more marked difference than usual in the rate of deposition upon portions of a given object placed at different distances from the anode; and there is even less tendency for a current to pass through great lengths of solution when the basis metal is also a poor conductor of electricity; coating bad conductors with copper is, therefore, to be recommended as a distinct assistance in starting a deposit of nickel. Objects which are to be coated on all sides with nickel should, therefore, be quite surrounded with anodes, and should be placed as nearly as possible equidistant from them; and if they have an irregular form, they should be systematically inspected to ensure that all the deeper hollows are covered at once. While then, on the one hand, the pieces must be very carefully examined after they have been *struck* (*i.e.*, first completely covered with nickel), they must not, on the other hand, be kept too long out of the solution, so that they tend to become dry, because in that time they will have acquired an imperceptible film of oxide, which will effectually prevent the adhesion of the nickel afterwards deposited.

The thickness of the nickel need not, as a rule, be very great on account of its extreme hardness. Generally speaking, from half-an-hour to four hours will suffice for the deposition. Thick deposits are very liable to peel off, occasionally spontaneously in the bath, but more often during the period of administering the final polish; this is especially the case with iron and steel goods, which take a thick deposit less satisfactorily than those made of brass or copper. This peeling of the metal, whenever it happens, is annoying, because it necessitates stripping the remainder of the deposit with a recommencement of the process *de novo;* but if it occur in the bath, the separation of loose fragments may give trouble in a manner already described in this chapter.

When the thickness of coating is sufficient, the pieces are removed from the bath and thoroughly washed in cold water, then plunged into boiling water, so that evaporation may take place more rapidly, and dried completely—small objects in hot sawdust, large articles in a stove heated to the boiling point of water or very slightly above. They must then receive their final polish, and are ready for the market.

The nickeling of larger or irregular surfaces is conducted after the same manner as that of smaller objects: the conditions to be observed most particularly are—that the goods shall be thoroughly polished and absolutely clean ; that they shall be as far as possible surrounded by anodes, and equidistant from them ; that the whole surface is in fair conductive connection with the negative pole of the battery ; and that the solution is in good order, being neither alkaline nor more than feebly acid. To secure good connection it is often desirable to employ more than one wire, especially when considerable lengths, such as chains or rods of a feeble conductor, are under treatment ; these should be supported from the cathode-rods at intervals by copper hooks, so that several starting points are offered, instead of one, for the formation and spread of the deposit: there is thus a greater uniformity of coat. The hooks must be shifted from time to time, to avoid the formation of surface markings.

Small objects should not be coated in the perforated porcelain pans recommendable in plating with other metals, because of the difficulty in arranging the anodes. It is, indeed, possible to effect the nickeling in this manner, and the method is sometimes practically adopted ; but it is safer either to attach each article individually to a copper wire, or to rest several together on a very shallow and narrow metal tray which may be suspended between the two anodes. Small articles are especially liable to receive a burnt deposit when first placed in an empty vat— either a large number of articles should be introduced into the solution simultaneously, or one of the anodes should be made a cathode for the time being, as previously explained, or small pieces may be immersed while the current is already coating objects with larger surface.

Finishing.—When the goods are to be left as they come from the bath without further polishing, and, therefore, with a slightly deadened surface, they must not be touched with the hands upon any exposed surface, as the coating in this condition is peculiarly susceptible to grease-markings, and the stains will inevitably show after drying.

The pieces should be lifted from the vat by the suspending wires, plunged first into two or three cold wash-waters, and then into hot clean water. If the pieces are at all thick or substantial, the heat energy stored up in them, by a short immersion in the boiling water, will suffice on removal to evaporate the small proportion of liquid clinging to them ; but if the surface be large as compared with the mass, it may be necessary to finish the drying in a stove. To this end the objects are placed on a tray, the suspending wires unhooked, and the tray transferred to the oven, so that from first to last they are not touched with the fingers. So treated the dead surface presents an extremely attractive appearance. Distilled water should be used for the final washing-bath (heated) if possible, because it leaves no residue upon evaporation.

Britannia metal and zinc should be coppered before nickeling, and some prefer thus to treat even German silver ; this done, the objects are coated with nickel in the manner described.

Applications.—Among the many useful applications of electro-nickeling, that of coating the comparatively soft copper printing surfaces demands especial notice. A thin film of nickel, so thin that the size of the printing surface is not affected, will increase the hardness, and consequently the life of the plate enormously ; indeed it would appear that a nickel-coated copper plate will give about four times as many impressions as one coated even in the usual way with iron. A special advantage also attaches to its use—it enables copper type to be used with a red pigment (vermilion) which cannot be done without such protection, because the copper alone decomposes the mercury sulphide, which is the basis of the pigment, and thus destroys its colour, and at the same time tends to become brittle by the absorption of the reduced mercury.

ELECTRO-DEPOSITION OF COBALT.

The chemical properties of nickel and cobalt are so nearly allied that this chapter would appear to be the most appropriate place for introducing the subject of electro-plating with the latter metal. The deposit of cobalt is similar to that of nickel; it is

TABLE XVIII.—SHOWING THE COMPOSITIONS OF SOLUTIONS FOR THE ELECTRO-DEPOSITION OF COBALT, RECOMMENDED BY VARIOUS AUTHORITIES.

No.	Authority.	PARTS BY WEIGHT OF INGREDIENTS.								Special Method of Preparation.
		1 Cobalt Chloride.	2 Cobalt Nitrate.	3* Cobalt Sulphate.	4 Potassium Cyanide.	5 Ammonia.	6 Ammonium Chloride.	7* Magnesium Sulphate.	8 Water.	
1	Beardslee . . .	27-41	...	...	...	...	...	...	1000	Dissolve 1 in 8; add 5 till red litmus-paper just blued.
2	Becquerel . . .	34	...	...	...	q. s.	...	...	1000	
3	Böttger . . .	400	...	...	...	200	200	...	1000	Dissolve 1 and 5 in 8; and add 6.
4	Hartmann & Weiss	...	x	...	q. s.	...	...	...	1000	Add 4 to solution of 2, till precipitate just re-dissolves.
5	Thompson . . .	...	...	x	...	...	...	$20 x$	...	
6	Watt	27-41	...	...	...	...	...	...	1000	

NOTES TO TABLE.—The black figures in the last column refer to the numbers of the vertical columns representing the various reagents.

* Columns 3 and 7 (cobalt and magnesium sulphates) imply nearly saturated solutions of these bodies—not the solid substances.

16

equally brilliant, but is somewhat harder, and has been found by Professor Sylvanus Thompson to possess a higher resisting power for organic acids, which renders it more suitable for the internal coating of copper or other cooking utensils. It is only lately, however, that cobalt anodes have been procurable, and have thus enabled the process to enter practically upon the field of electro-metallurgical competition. Even now, its comparatively high price tends to check its adaptation to many purposes to which its application may be desirable.

The battery should consist of two Bunsen-cells, and it is well to place an ammeter in the circuit and to have resistance-coils at hand, because the deposit yielded by a powerful current is defective in adhesion; a fairly high electro-motive force but of low intensity of current is required. Many solutions have now been used of which some are indicated in the preceding table.

The double sulphate of cobalt and potash, corresponding to the nickel-potassium sulphate, above recommended, may also be used in 10 per cent. solution. Professor Thompson, however, has obtained perfect results from the solution No. 5, made by mixing 20 volumes of a nearly saturated solution of magnesium sulphate with 1 volume of a similar solution of cobalt sulphate or chloride. This bath should be used hot, and will then yield a film which, if deposited with all due precautions on iron, brass, or German silver (*inter alia*), is extremely adherent and good. Instead of making up this solution in the manner described, it may be prepared by electrolysis, if the current be passed from a large cobalt anode into a magnesium sulphate solution with a temporary cobalt-cathode, which should be removed as soon as sufficient metal has dissolved in the electrolyte to yield a good deposit.

Anodes of pure rolled cobalt are now obtainable, and should be used like those of nickel, of large superficial area as compared with that of the cathodes; they must not be supported by copper wire. The vats and the general treatment, as well as the character of metal deposited, are, indeed, regulated exactly as in the case of nickel. And because the film is harder even than that of the sister metal, it follows that the same precautions must be taken in regard to previous polishing as well as cleansing. It is even more difficult to impart a smooth polish to a rough cobalt surface than it is to one of nickel. So also the same care is necessary in stripping old deposits previous to plating; corresponding methods may be employed.

CHAPTER XII.

THE ELECTRO-DEPOSITION OF IRON.

THE electro-deposition of iron (or of steel, as it is sometimes wrongly termed) upon the surface of engraved copper plates has long been practised, in order that its superior hardness may enable the printer to obtain a greater number of sharp impressions from the same plate; and the extreme ease with which the worn film may be removed from the copper renders it possible to renew the protective coating again and again. All that is necessary is to suspend the printing operations as soon as the first tinge of the red foundation metal appears through any portion of the iron cover, and then removing the iron by means of dilute acid, to re-immerse it in the iron-plating vat. In this manner the copper need scarcely be appreciably worn, and may be made to yield many thousand impressions without losing any of the sharpness or accuracy in definition of the lines. This, however, is practically the only use for electro-deposited iron; as a metal it is too readily oxidisable to render its use as an external protective coating serviceable in any other way. Its hardness qualifies it for printers' work, and enables copper to compete with steel as a material for steel plate engraving; but even in this field it is outrun by nickel, by which it will probably be superseded in course of time. Nevertheless, for plates which may require alteration from time to time (maps, for example) the iron is preferable, on account of the greater difficulty experienced in removing the nickel coat; yet, if the alterations are likely to occur frequently, it may be better not even to coat the plate with iron, although the stripping is in this case so comparatively simple.

Iron is never deposited by immersion, but always by the separate-current process, for which purpose the Bunsen-cell is best adapted.

The Solution.—Two kinds of iron salts are commonly known—the *ferric* or *per*-salts, and the *ferrous* or *proto*-salts, which are combinations of the metal (Fe) with a greater (Fe_2O_3) or smaller

(Fe O) proportion of oxygen respectively; and, as the heat of formation of any ferric compound is considerably higher than that of the corresponding ferrous body, the proto-salts exhibit a correspondingly greater tendency to absorb oxygen from the air, or from any substance in which it is loosely combined, and thus to become converted into the per-salt of the same class. But the ferrous salts are alone suited for electro-depositing the metal, hence the iron solutions employed must be carefully protected from the action of the air by retaining them in closed vessels when not in actual use. While the current is flowing, it tends to correct any peroxidising action; because the newly-deposited iron upon the cathode is attacked by ferric salts in the solution in its immediate vicinity, and is re-dissolved, while the ferric salt is simultaneously reduced to the ferrous condition, thus—

$$\underset{\text{\textit{Ferric chloride}}}{Fe_2 Cl_6} \quad \underset{\text{with}}{+} \quad \underset{\text{\textit{iron}}}{Fe} \quad \underset{\text{give}}{=} \quad \underset{\text{\textit{ferrous chloride.}}}{3\,Fe\,Cl_2}$$

And newly-precipitated iron or, if we may use the expression, *nascent* iron, exhibits a much greater activity in this respect than metal which has been cast or rolled, and which is usually in a more dense, as well as a more stable, condition. The oxygen which, meanwhile, is being deposited upon the other pole by a moderate current, such as is required for iron coating, acting upon a sufficiently large anode and, therefore, upon an excess of metal, forms only the protoxide. Thus, in course of time, the peroxide originally present will become reduced, and the baths will be brought into the best condition for yielding a good deposit; but in effecting this, a considerable amount of time and of battery-power may have been wasted.

Another effect of oxidation of ferrous salts in neutral solutions is the formation of basic salts (containing an excess of iron), which being insoluble in the liquid give rise to a rust-coloured precipitate or turbidity. This happens because the iron in the ferric condition requires not only more oxygen, but more acid to form salts, than it does in the ferrous state; for example, 2 atoms of iron in the state of ferric oxide requires 3 molecules of sulphuric acid to dissolve them and form the sulphate ($Fe_2 O_3 + 3\,H_2 S O_4 = Fe_2 (S O_4)_3 + 3\,H_2 O$), while in the form of ferrous oxide only 2 molecules of the acid are needed ($Fe_2 O_2 + 2\,H_2 S O_4 = 2\,Fe\,S O_4 + 2\,H_2 O$). The addition of oxygen from the air may thus suffice to produce peroxide, but there may not then be sufficient acid in the bath to combine with it, and it thus appears as a solid precipitate. The following equation sums up the reaction in a general way, and shows how peroxide is formed along with the persulphate, by the action of oxygen upon the protosulphates :—

$$6\,FeSO_4 \;+\; 3\,O \;=\; 2\,Fe_2(SO_4)_3 \;+\; Fe_2O_3.$$

The remedy for this precipitate or cloudiness is obviously the addition of a sufficient quantity of free acid to combine with the peroxide or basic precipitate, and form the soluble persulphate. When much of the ferric compound has formed, just sufficient acid should be added to dissolve the precipitate, and a current of electricity should be passed through it between iron electrodes, until the pure pale green colour of the ferrous liquid has been restored and the metal is depositing well upon the cathode. But it must be remembered that all acid which is added to dissolve the rust-coloured precipitate becomes free again in the solution as soon as the bath is reduced to the ferrous condition, which requires less acid to form its compounds. Prevention in this case is, therefore, better than cure, and the bath should be kept as far as possible out of contact with air; Klein adds a small proportion of glycerin to the liquid to hinder the formation of ferric compounds.

The composition of the principal baths used in depositing is quoted in Table XIX.

One of the best of these solutions is that of Klein (No. 3), which is especially suitable to the production of thick deposits. A solution of ferrous sulphate in water is first prepared in a sufficiently large jar; a solution of ammonium carbonate is now gradually added to it until no further quantity of the precipitate which forms at first is produced; the precipitate is now allowed to subside, the liquid is poured off, fresh water is added, from which the ferrous carbonate is again allowed to separate by subsidence. The water is once more poured away, and sulphuric acid, diluted with twice its volume of water, is added little by little, with constant stirring, until the precipitate is exactly re-dissolved and yet no excess of free acid is present. Towards the end, the acid must be added by a few drops only at a time, after which a few seconds' pause must be made to give opportunity for it to attack the precipitate before introducing a further quantity. A blue litmus-paper suspended in the solution will indicate the condition of the liquid at any time; it should become purple, but never red. Only recently boiled water should be employed to dissolve and wash the ferrous sulphate and precipitate, for natural water contains a quantity of dissolved oxygen, which would peroxidise the iron, but which is expelled by boiling. This solution should be used as concentrated as

TABLE XIX.—SHOWING THE COMPOSITION OF BATHS FOR THE ELECTRO-DEPOSITION OF IRON, RECOMMENDED BY VARIOUS AUTHORITIES.

No.	Authority.	1. Ferrous Chloride.	2. Ferrous Sulphate.	3. Ferric Sulphate.	4. Ammonio-Ferrous Sulphate.	5. Iron Alum.	6. Potassium Ferrocyanide.	7. Rochelle Salt.	8. Caustic Soda.	9. Ammonium Carbonate.	10. Ammonium Chloride.	11. Sulphuric Acid.	12. Water.	Special Method of Preparation.
1	Böttger	…	…	…	150	…	…	…	…	…	…	…	1000	
2	„	…	…	12	…	…	40	80	q. s.	…	…	…	1000	Dissolve **2** in **12**; just precipitate with **9**, and just re-dissolve with **11**; use as concentrated as possible.
3	Klein	…	x	…	…	…	…	…	…	q. s.	…	q. s.	1000	
4	Obernetter	…	30	…	…	30	…	…	…	…	60	…	1000	Stand 2 days; always filter before use.
5	Ryhiner	x	…	…	…	…	…	…	…	…	200	…	1000	Form electrolytically.
6	Varrentrapp	…	10	…	…	…	…	50	…	…	100	…	1000	Dissolve **2** in 250 of **12**, and add to **7** in 750 of **12**; then add **10**.
7	Volkmer	x	…	…	…	…	…	…	…	…	100	…	1000	Form electrolytically.
8	Walenn	…	200	…	…	…	…	…	…	…	…	little	1000	
9	Weiss	…	132	…	…	…	…	…	…	…	100	…	1000	

PARTS BY WEIGHT OF INGREDIENTS.

NOTE TO TABLE.—The black figures in the last column refer to the numbers of the vertical columns indicating the various reagents.

possible and preferably warm; it must never be allowed to become acid. In order to guard against this latter evil, Klein recommended the use of anodes presenting an aggregate area equal to eight times that of the cathode surface, so that any free acid should have every facility for becoming saturated with iron; and he even attached slips of a more electro-negative element, such as platinum or copper to the anode plates, so that a slight local current might be set up, which would assist in the solution of the iron, without affecting the current passing through the bath. The deposit from Klein's solution should not show any tendency to crack upon the surface and peel off in the form of spangles as its inventor observed to happen frequently when other baths were used for producing thick coatings, especially the double chloride of iron and ammonium.

The double sulphate of iron and ammonia, which is obtainable in the market in a very pure condition, is capable of yielding extremely good films upon engraved copper plates. If it be at all acid, a little chalk should be added, or better still, a little washed ferrous carbonate prepared as above described, until no further quantity is dissolved by the liquid.

An excellent solution for the same purpose may be made electrolytically by passing a current between a large iron anode and a copper cathode in a 10 per cent solution of ammonium chloride until the liquid becomes green in colour, exhibits a tendency to form a scum of the red peroxide on the surface, and deposits a clean metallic coating upon the copper plate.

In using any of these iron-baths, the separation of hydrogen with the iron upon the object is to be strenuously avoided, because the gas bells clinging to the surface form pin-holes which are fatal to the impression taken in the press from a plate so affected; and further than this the character of the iron is prejudicially influenced by the absorbed hydrogen. The conditions, therefore, which are least favourable to hydrogen production must be fulfilled; these are mainly—the use of a concentrated solution, the absence of free acid, and the application of a sufficiently weak current.

The Anodes.—The anodes should be made of the purest iron available. Electro-deposited iron would theoretically be most suitable; next to this the softest wrought-iron or so-called mild steel sheet are preferable. Hard steel, and above all cast-iron plates are to be avoided, because they contain comparatively large percentages of carbon and other impurities which, being insoluble in the liquid, remain for a time suspended in the bath and are apt to attach themselves to the objects under treatment.

Cast-iron may contain as little as 93 per cent. of iron (or sometimes even less). The anode surface should be much larger than that of the cathodes (eight times as large) for reasons already given. The anodes should be removed from the vat now and again, brushed to detach the insoluble matter which is left upon the surface in a spongy or pulverulent condition, and returned to their position.

The Vat.—The vat is best constructed of iron which may be heated when the solution is to be used warm; it must be kept thoroughly well cleaned, and for this cause, enamelled iron is to be preferred. It must of course be larger than the work to be plated on account of the increased surface which is given to the anodes.

The Character of the Deposited Metal.—The iron deposited in thick coatings is of a bright grey-white colour, is extremely hard and brittle, and demands the most careful attention if it is to be removed from the matrix. After annealing at a low red heat it becomes softer, and after a fair cherry-red it is as soft as steel which has been similarly treated. The fracture of unannealed deposited iron resembles that of cast-iron; it, of course, contains no carbon but is usually highly charged with hydrogen which it occludes during the process of formation. Cailletet has found as much as 240 volumes of this gas in 1 volume of a sample of iron, which was sufficiently hard to scratch glass, and was extremely brittle. The annealing has the double effect of softening the deposited iron and of removing the hydrogen which it had previously contained; but the inference is not safe that the hydrogen is the sole cause of the brittleness, which is more probably mainly due to the particular arrangement of the molecules of the metal. Its extreme hardness has procured for the application of deposited iron to engraved copper plates the misleading title of *steel-facing*, a designation which implies the presence of combined carbon within it, whereas excepting the hydrogen contained in it (which may be removed by heating), it is the purest form of iron obtainable.

In its relation to magnetism, deposited iron is comparable with mild steel; but Beetz has shown that when deposited under powerful magnetic influence, as between the poles of a strong electro-magnet, and from solutions containing ammonium chloride, it will itself act as a powerful magnet, retaining its magnetism for a considerable period of time.

Iron is so electro-positive a metal that it has a great tendency to combine with oxygen, that is, to rust; the electro-deposited film must, therefore, be dried very thoroughly as soon as possible.

It, however, generally contains within its pores a distinct quantity of the solution from which it was precipitated, and this must be perfectly removed by washing two or three times in boiling water, or the liability to rust will be greatly increased.

The Process of Electro-deposition.—In coating copper plates with iron (which is the chief application of the process), the copper must first be cleaned carefully, so that the sharpness of the lines may not be diminished. Klein dips the plate first into benzene and then into potash to remove grease, but it is usually sufficient first to rinse and then to boil the plate in a solution of caustic potash, then after washing twice or thrice in clean water, to pass it through a bath of dilute sulphuric acid (containing from 2 to 5 per cent. of the acid), and after a second thorough wash to transfer it at once to the iron-vat, without touching the surface at any point. In the bath it is suspended by suitable hooks or by holders such as those mentioned on p. 108 ; two plates may be used with one anode by placing the engraved faces of the copper fronting the latter.

The current from the Bunsen's cell (or cells, arranged in parallel arc, if much work is in hand) is then passed through the solution ; an ammeter and set of resistance-coils should be placed in position. Usually from five to six minutes suffices for the process of deposition ; but if a thicker coating be required, remembering that the last portion deposited forms the printing surface, and hence must be perfect in character, it is advisable to remove the plate, rinse, rapidly examine, and brush it well with a hard brush under water, so that no extraneous matter may cling to the surface, then replacing it for five or six minutes in the iron-bath, the alternation is again and again repeated until the desired thickness is obtained. After the final removal from the bath, the plate is dipped into a large volume of cold water, is then immersed in boiling water for the space of half-a-minute, and is again rinsed in cold water. It may then be lightly rubbed with dilute potash or soda solution, sponged dry, and rubbed with oil, the excess of which is subsequently removed by means of benzene. Thus treated, the plate will not be greatly liable to rust ; but if it is to be stored for any length of time, it must be treated like an ordinary engraved steel plate and covered with a protective film of wax.

Stripping.—When an old iron-coated plate is to be re-plated, the residue of the first coat must be stripped, after removing all grease in a potash-bath, by immersing it in dilute sulphuric acid (5 to 10 per cent.) until the copper surface is left completely bare; the plate, after washing, is then ready for the iron-bath as usual.

Electrotyping **with Iron.**—Thick deposits also may be made, and may be obtained direct from the mould, but in this case the matrix should be first covered with a thin sheath of copper, upon which the iron is precipitated, because iron refuses to deposit well upon the black-leaded surface of the gutta-percha or other non-conducting mould. The copper may be afterwards dissolved away by making the iron sheet the anode in a copper cyanide bath, or (but with greater risk) by simple treatment with the strongest nitric acid, the excess of which must be quite washed away, immediately the copper is removed.

CHAPTER XIII.

THE ELECTRO-DEPOSITION OF PLATINUM, ZINC, CADMIUM, TIN, LEAD, ANTIMONY, AND BISMUTH; ELECTRO-CHROMY.

ELECTRO-DEPOSITION OF PLATINUM.

PLATINUM, one of the most insoluble and acid-resisting metals known, would form an excellent protective coating to metals could it be readily applied; Roseleur, indeed, has stated that he had twenty times evaporated nitric and sulphuric acids alternately in a platinum-plated copper crucible without finding the basis metal to be sensibly attacked until the last operation. But the very insolubility of the metal constitutes one of the difficulties in electro-plating with it; for the anodes resist the solvent action of any solution which may be safely used as an electrolyte, without injuring the objects suspended as the cathode. It is very rarely used, however, as a covering metal.

Platinising.—It is so electro-negative an element that nearly all the other metals are able to decompose its solution, and thus deposit the platinum by simple immersion; but the coating so formed is usually black, granular, and non-adherent. Silver, copper, and brass are the most readily treated, while lead, tin, zinc, iron, Britannia metal, and the like, present great difficulty unless previously protected by a substantial film of copper. For platinising copper by simple immersion, Roseleur recommends the use of a boiling solution containing 10 parts of platinum converted into the neutral chloride, and 120 parts of caustic soda in 1,000 of pure (distilled) water. Another solution may be made by dissolving 25 parts of the double chloride of platinum and ammonium, and 250 of ammonium chloride in 1,000 parts of water; this also is used at the boiling temperature. When silver is platinised, and a simple solution (not too strong) of platinum tetrachloride in water will suffice to effect this, the object should be afterwards rinsed, first in dilute ammonia, and then in water, because silver chloride is formed by the exchange of metals with the platinum chloride; and this silver compound, being insoluble in water, requires the treatment with ammonia, in which it readily dissolves, to effect its complete removal from the plated goods.

Tin, brass, bronze, copper, and tin plate have also been platinised by rubbing upon their surface, with a woollen or linen rag, a solution of 1 part of platinum chloride in 15 parts of spirits of wine and 50 of ether, and then washing well with water, after allowing the ether to evaporate.

Platinum may also be deposited by the single-cell process. Lesmondes' method consisted in placing the articles in a perforated zinc tray and immersing the whole in a solution made by adding sodium carbonate, in the first place, to a strong solution of platinic chloride until effervescence ceases, then a little glucose, and afterwards sufficient sodium chloride to yield a white precipitate. This bath is to be used at a temperature of 140° F., and is most suitable for treating copper and brass, the deposition being mainly due to the current set up by the solution of the zinc of which the tray is composed.

The method adopted by Smee for coating the silver plates required for use in his battery is also a single-cell process. The plate, slightly roughened by mechanical means or by a momentary immersion in nitric acid, is placed in a solution containing dilute sulphuric acid with a few drops of platinum chloride solution added to it. A porous cell containing a rod of zinc standing in dilute sulphuric acid is then introduced into the bath; on making metallic connection between the silver and the zinc, a current is set up, the zinc dissolves, and a proportionate amount of platinum is deposited in the form of a dark-grey or black powder, which, nevertheless, holds fairly tenaciously to the roughened silver surface.

Platinating.—But for general purposes the separate-current process should be employed. Of the various solutions recorded in the appended table, that of Roseleur and Lanaux (No. 4) is the most reliable. To prepare a gallon of the liquid, three quarters of an ounce of platinum is dissolved in aqua regia and converted into chloride, which is then dissolved in a quart of distilled water; in the meantime half a pound of ammonium phosphate should have been dissolved in a quart of pure water in one vessel, and two and a half pounds of sodium phosphate in the remaining two quarts contained in a second vessel. The solution of the ammonium salt is now to be added to the platinum liquid, with which it produces a dense precipitate; disregarding this, the sodium phosphate solution is next added with constant stirring, and the whole bath is boiled until no more smell of ammonia is observed, but on the contrary it is shown to be faintly acid by blue litmus-paper. During the period of boiling, water will have been evaporated, which must be restored before using the liquid.

TABLE XX.—SHOWING THE COMPOSITION OF SOLUTIONS FOR ELECTRO-DEPOSITING PLATINUM, RECOMMENDED BY VARIOUS AUTHORITIES.

PROPORTION OF INGREDIENTS BY WEIGHT.

No.	Authority	1 Platinum as Platinic Chloride.	2 Ammonium Platinic Chloride.	3 Potassium Cyanide.	4 Caustic Soda.	5 Sodium Carbonate.	6 Sodium Chloride.	7 Di-sodium Orthophosphate.	8 Sodium Pyrophosphate.	9 Sodium Citrate.	10 Ammonium Chloride.	11 Ammonium Phosphate.	12 Sulphuric Acid.	13 Acetic Acid.	14 Citric Acid.	15 Water.	Special Method of Preparation.
1	Böttger	...	x	...	...	...	...	...	...	x	...	...	...	...	...	1000	
2	Jewreinoff	10	...	...	...	...	...	...	...	...	...	...	70–80	...	...	1000	Dissolve 1 as platinic chloride in 15; add 12. (For silver.)
3	Langbein	...	15	...	q.s.	...	...	...	...	...	4–5	...	...	...	100	1000	Dissolve 14 and 10 in 400 of 15; neutralise with 4; add 2, and stir; then 600 of 15.
4	Roseleur & Lanaux	5	...	...	...	...	...	250	...	...	...	50	...	...	...	1000	Vide text.
5	Sprague	30	...	...	q.s.	...	18.7	...	...	...	...	...	...	75	...	1000	Dissolve 1 in 15; add first 6, then 13; render alkaline with 4; warm and filter.
6	Wahl	10	...	...	...	400	...	...	...	...	...	...	...	...	...	1000	⎫
7	„	10	...	...	...	...	...	600	...	...	...	...	...	...	...	1000	⎬ Good for thin deposits.
8	„	10	...	...	...	...	...	...	300	...	...	...	...	...	...	1000	⎭
9	Watt	7	...	{slight zz.}	...	...	...	...	...	...	...	...	...	...	...	1000	

NOTE TO TABLE.—The black figures in the last column refer to the numbers of the columns denoting the various reagents.

When in active use, a little of the fresh solution must be added at intervals to supply the place of the platinum which has been lost by deposition upon the cathode; because, as already explained, the anodes are not attacked, and cannot, therefore, replenish the exhausted solution.

A process similar to this, but with the addition of a small proportion of common salt, has been recently patented by Thoms.

The objects to be plated should be well polished before depositing because electrolytic platinum is hard, so that greater difficulty is involved in polishing after deposition than before. The platinum-coated surface may be left dead, or it may be brightened by means of iron-wire scratch-brushes (brass is too soft and, itself becoming rubbed, leaves a yellow stain upon the goods), or by careful rubbing with very finely-powdered pumice.

When old platinum-covered goods are to be re-plated, the stripping of the previous coat presents a difficult problem; it cannot well be removed electrolytically because the baths do not attack the metal, although a long exposure in a gold-stripping bath may sometimes effect the desired object, especially if the platinum be not in its densest and hardest condition; but at best it is a very tedious operation. The chief solvent for platinum is aqua regia, but this cannot be applied because it would vigorously attack the basis metal beneath; and, in fact, as soon as any of the latter metal became uncovered, it would dissolve all the more rapidly on account of its contact with the remainder of the platinum coat, and, being more electro-positive, would even serve to protect the latter from further action. The surest and most rapid method, whenever practicable, is to apply mechanical means and simply rub off the platinum by means of emery-cloth, and then re-polish the metal beneath. This system cannot, of course, be applied when any delicate pattern or design is traced upon the object, in which case the chemical methods must be tried. A greater loss is caused by the use of emery, but this may be minimised by saving the dust produced, and subsequently working it up to recover the platinum.

The name usually applied to platinum-coating by simple immersion is *platinising*, as in the case of the Smee-battery silver plate, while the electrolytically-covered object is said to be *platinated*. Watt, however, raises an objection to this latter term, and suggests the use of the expression "platined" to denominate this class of work. His objections to the other term are doubtless worthy of consideration, but the older word is perhaps more euphonious, and will probably hold its ground; while if it be regarded as a contraction of the compound word "platinum-plating," it is not after all unscientific.

ELECTRO-DEPOSITION OF ZINC.

It is rarely, indeed, that electrolytic zinc deposition is resorted to ; the metal has not a fine colour or lustre, and is readily dulled with the thin film of tarnish, which forms very soon upon exposure to the atmosphere. Its highly electro-positive character certainly renders it suitable to the protection of iron surfaces from destruction by rust ; because when submitted together (in metallic contact) to the same corrosive influence, the zinc is the first of the two metals to become attacked. But, like tin, zinc is more satisfactorily deposited upon iron by dipping the latter into a bath of the molten metal. Such a process yields a perfectly homogeneous and continuous coat, which is applied at such a temperature that it is impossible for water to exist between the two surfaces, or in cavities in which it might previously have been present ; while the electrolytic method deposits a crystalline, and, therefore, to some (however slight) extent, porous cover, in which small quantities of the solution from which it has been deposited may be locked up, and which will tend to facilitate the oxidising action. In every way, then, the dry or fusion method of coating the iron is preferable whenever its application is possible ; the meaningless title *galvanised iron* given to this product is manifestly a misnomer, and is distinctly misleading.

On account of its very electro-positive nature, zinc cannot be precipitated upon ordinary metals by simple immersion, nor is it practically deposited by single-cell methods.

Battery Process.—By separate current it may be obtained from the neutral sulphate, chloride, acetate, or other soluble salt of zinc (except the nitrate), or from the corresponding double salt of zinc and ammonia. The current-strength required for most zinc solutions is considerable, and demands the use of two or three Bunsen-cells.

The solutions are, as usual, various, but good results may be obtained from a 10 per cent. solution of zinc sulphate with a current of from 0·06 to 0·13 ampère per square inch, or 1 to 2 ampères per square decimetre ; other formulæ are given in Table XXI.

Perhaps the best solution is that patented by Watt in 1885 (No. 5) ; 200 ounces of potassium cyanide are to be dissolved in 20 gallons of water ; 80 fluid ounces of the strongest *liquor*

TABLE XXI.—Showing the Composition of Solutions for Electro-Depositing Zinc, recommended by various Authorities.

PARTS BY WEIGHT OF INGREDIENTS.

No.	Authority.	1 Zinc.	2 Zinc Hydroxide.	3 Zinc Chloride.	4 Zinc Sulphate.	5 Potash.	6 Potassium Cyanide.	7 Potassium Carbonate.	8 Alum.	9 Ammonia.	10 Ammonium Chloride.	11 Ammonium Sulphate.	12 Sulphuric Acid.	13 Water.	Special Method of Preparation.
1	Elsner	...	...	5 x	...	...	...	...	...	...	4 x	...	...	1000	(For cast-iron.)
2	Japing	...	...	...	x	excess	...	...	...	...	...	...	...	1000	Dissolve 4 in 13; add 5 until precipitate at first formed is re-dissolved.
3	Lonyet	...	10	...	250	...	...	...	100	...	...	...	little	1000	Precipitate the zinc hydroxide from sulphate by potash.
4	Person & Sire	...	...	...	...	...	...	...	...	...	...	...	...	1000	
5	Watt	18·75	...	...	...	...	62	25	...	25	...	...	...	1000	Dissolve 6 in 13, add 9; mix; add 1 electrolytically; and finally add 7.
6	Weiss	...	...	10	80	...	...	...	...	...	20	...	...	1000	To 5 in 500 of 13, add 3 and 10 in rest of 13; mix well.
7	,,	...	...	...	...	60	...	...	...	...	20	50	...	1000	

Note to Table.—The black figures in the last column refer to the numbers of the vertical columns denoting the various ingredients.

ammoniæ are then stirred into this liquid, and the mixture is transferred to a large vessel containing pure rolled zinc anodes; in this vessel are also several large porous battery-cells which must be filled up with the solution to the level of the surrounding liquid, and in which are cathodes of metallic copper. On passing a current from a Bunsen-battery of several cells, the zinc anode gradually dissolves in the solution until, when it has reached the strength of 3 ounces per gallon (*i.e.*, a loss of weight on the part of the anodes of 60 ounces in the aggregate), the battery is disconnected, and 80 ounces of potassium carbonate are finally added to the liquid little by little, by dissolving each additional portion in a fraction of the liquid, and then returning it to the vat and stirring well. After a final rest of twelve hours, for subsidence, the clear liquid is poured off, and the sedimentary matter at the bottom filtered from the contained liquid; the whole is then ready for use with a battery of three Bunsen-cells or, better, with a dynamo.

The anodes for all solutions should be made of rolled zinc only. It is more likely to be pure than the cast commercial zinc or *spelter* of the market, which frequently contains much lead, iron, or arsenic, as well as other impurities.

Character of Deposit.—The metal should be reguline. If produced by a current which causes a simultaneous separation of hydrogen, it is of course spongy; but Kiliani has made the remarkable observation that with a strong solution of zinc (of specific gravity = 1·38) a very weak current causes an evolution of hydrogen which diminishes in extent as the current increases in volume up to a certain point. For example, a current of about 0·07 ampère per square decimetre yielded 1·6 cubic centimetres of hydrogen for every gramme of zinc deposited; but as the current was increased to 0·2, 0·4, 1·6, and 3·2 ampères per square decimetre, so the hydrogen evolution per gramme of deposited zinc was reduced to 1·5, 0·37, 0·29, and 0·22 cubic centimetre respectively. In all these cases the zinc was in a spongy condition, but less markedly so in the last two instances. When the current-density was increased to 18·5 ampères per square decimetre, the gas separation ceased, and the metal appeared lustrous and adherent. This anomalous action may perhaps be ascribed to the ready oxidisability of the zinc, which deposited by a weak current is attacked by the solution in its freshly precipitated state, with the formation of zinc oxide and evolution of hydrogen ($Zn + H_2O = ZnO + H_2$); but as the current-strength increases so also the quantity of zinc thrown down is increased, until at length the action may

17

perhaps be so hurried that the liquid has no chance of attacking the metal at the moment of deposition.   The current-strength could of course be greatly increased in so strong a solution without causing hydrogen to be evolved to any extent by the simultaneous electrolysis of the metallic salt in contact with the cathode, and a certain proportion of the acid or water.   The same observer also noted that a current of 0·4 ampère and 17 volts, in passing through a one per cent. solution, threw down zinc oxide with the metal upon the cathode.

It is clear, then, that the current-strength must not be too low in dealing with zinc solutions.   The final washing of electro-zinced goods should be in hot water, and the drying, if necessary, effected in a stove, in order that as little time as possible may be given for oxidation of the deposit.

The method of operating in depositing zinc should require no further detailed explanation for those who are acquainted with the matter in the previous chapters of this work.   The same care must be expended upon cleansing and deoxidising at first, and upon washing and finishing subsequently, as has been insisted upon throughout.   The vats, suspension arrangements, and the like call for no special comment.

Electro-Deposition of Cadmium.

Cadmium is a metal nearly allied to zinc but less commonly met with and more costly; its electro-deposition is rarely effected.

Smee found that a liquid made by adding a solution of ammonia to one of cadmium sulphate, until the precipitate at first formed is just re-dissolved, readily yields a good deposit, while the simple solutions of the cadmium sulphate or chloride are difficult to work; Bertrand, however, appears to have been more successful with the sulphate.   This experimenter has also used a bath of cadmium bromide slightly acidified with sulphuric acid.

The only solution remaining to be noted is that of Russell and Woolrich's made by dissolving 40 parts of the metal in dilute nitric acid, adding a ten per cent. solution of sodium carbonate until no further precipitate is produced, allowing the precipitate to subside, pouring fresh tepid water upon it, settling it again and repeating the washing process four or five times, then just dissolving it in sufficient strong potassium cyanide solution, adding ten per cent. excess of the latter, and sufficient additional water to render the total weight of water in the

solution equal to 1000 parts. The bath is used at a temperature
of 100° F., and, with a current of 3 or 4 volts produced by a
like number of Daniell-cells placed in series, may be made to
give a white reguline metal.

ELECTRO-DEPOSITION OF TIN.

The process of coating articles of copper or wrought-iron by
merely bringing their cleaned surfaces into contact with melted
tin is so simple, and, moreover, gives such a sound and perfect
covering, that there is but little room for an electro-tinning
method which must usually entail greater trouble and expense.
Notwithstanding this, the wet deposition is to some extent
practised, largely indeed for whitening small objects of brass
wire, such as pins, hooks and eyes, and the like, for which it is
to be preferred ; and for giving a preliminary coating to certain
metals, like cast-iron, which are subsequently to receive an
electro-deposit of any more electro-negative metal that may not
be deposited directly upon them. Like many other metals, tin
may be deposited either by simple immersion, by single-cell, or
by separate-current methods. This is of course a consequence of
its relatively low position in the electro-chemical series of
metals.

Tinning by Simple Immersion.—This is the commonest method
of dealing with small copper or brass objects. Of the various
processes, the oldest and simplest is to place them in a strong
solution of potassium bitartrate (cream of tartar), with which
may be mixed a small proportion of stannous chloride (tin proto-
chloride) to quicken the action. They are then covered with a
layer of pure tin, either in small pieces, in the form of foil, or as
cast plate; a second layer of brass pieces is now placed above
this, then a fresh stratum of tin, and so on. On boiling for a
short time, it will be found that an exchange has taken place
superficially upon the articles, so that they have become covered
with a white film of tin. Iron and steel articles to be treated
in this way must first receive a coating of copper, which will
enable them to take the tin satisfactorily. This coating of tin is
very thin—a mere wash—and will not withstand much wear, but
suffices for the purposes for which it is required. On removal
from the bath, the articles are thoroughly washed, and are
polished by shaking with bran in a revolving drum (*e.g.*, fig. 81),
or by any other mechanical expedient. Large objects may be
polished by scratch-brushing. An alternative method is to
dissolve stannic oxide (putty powder) in caustic potash solution,

TABLE XXII.—SHOWING THE COMPOSITION OF SOLUTIONS FOR TINNING BY SIMPLE IMMERSION, RECOMMENDED BY VARIOUS AUTHORITIES.

No.	Authority.	Special Application of Bath.	PARTS BY WEIGHT OF INGREDIENTS.								Special Method of Preparation.
			1	2	3	4	5	6	7	8	
			Tin Binoxide.	Tin Bichloride.	Tin Tetrachloride.	Caustic Potash.	Potassium Bitartrate.	Ammonia-Alum.	Sodium Pyrophosphate.	Water.	
1	Gore, . . .	Copper, Brass, or Bronze	x	...	...	y	...	...	...	1000	
2	,, . . .	Iron or Zinc	...	0·5	...	...	...	15*	...	1000	
3	Lüdersdorf, .	Zinc	...	...	400	...	400	...	...	1000	Dissolve 2 in 330 of 8, cold; and 5 in 670 of 8, warm; mix. (*Use with zinc contact*).
4	Roseleur, .	...	...	1	...	...	10	...	...	1000	
5	,, .	...	...	1·3	...	...	...	...	20	1000	To be used with zinc contact.
6	,, .	Zinc	...	3·3	...	...	...	...	16·6	1000	
7	Wahl, . .	...	...	3	...	...	...	50	...	1000	

NOTES TO TABLE.—The black figures in the last column refer to the numbers of the vertical columns denoting the various reagents.

* For zinc, potash- or soda-alum is recommended in lieu of ammonia-alum.

and immerse the articles in it, in contact with fragments of tin, the subsequent processes being identical with those practised in the first process.

Zinc, placed in a solution of stannous chloride, precipitates the tin in a spongy and useless condition (analogous to that of the "lead tree"); but by using a very dilute solution, to which a certain proportion of alum (potash- or ammonia-alum) is added, a good adhesive deposit may be obtained. A similar solution of stannous chloride and ammonia-alum may be applied to the tinning of iron by simple immersion. The whole process is very simple; and the principal solutions are enumerated in the preceding table.

Deposition by Single-Cell Process.—Solutions recommended by Roseleur are given in the above table (Nos. 4 and 5). In use, small articles are most conveniently placed upon a tray of perforated zinc and lowered into the liquid, which should be kept warm. The tray should be lightly shaken from time to time in order to bring different points into contact with the zinc, and the surface of the latter must be scraped at intervals to remove the white incrustation which forms upon it and destroys metallic contact. Large objects are immersed in the liquid in contact with fragments of zinc, which should have a surface equal in the aggregate to about the one-thirtieth part of that which is to be coated with tin. At the end of an immersion lasting from one to three hours, Roseleur recommends that the pieces should be removed and scratch-brushed, while at the same time the bath, which has been impoverished by the deposition of tin, is regenerated by the addition of fresh tin and alkaline bitartrate or pyrophosphate. After a further immersion of about equal duration, the goods are well washed, scratch-brushed, and dried.

Deposition by Separate Current.—For this purpose the current should have a fairly high electro-motive force, such as would be yielded by two Bunsen-cells in series. Of all the baths quoted in the following table, that of Roseleur will probably be found to give the most universally good results.

This bath of Roseleur's (No. 8) is made by placing the pyrophosphate and water in a tin-lined wooden tank, the lining serving also as anode; afterwards the stannous chloride is suspended in the liquid in a copper sieve. The first result is to produce a cloudiness in the liquid, which, however, subsequently becomes clear; and after complete solution of the tin salt is ready for use. The liquid may have a yellowish colour, but should be perfectly transparent and clear. This bath requires an addition of tin from time to time, because the metal is de-

TABLE XXIII.—Showing the Composition of Electro-Tinning Baths, recommended by various Authorities.

No.	Authority.	1 Tin Binoxide.	2 Fused Tin Bichloride.	3 Tin Tetrachloride.	4 Caustic Potash.	5 Potassium Carbonate.	6 Potassium Cyanide.	7 Potassium Bitartrate.	8 Caustic Soda.	9 Sodium Carbonate.	10 Sodium Pyrophosphate.	11 Zinc Acetate.	12 Sulphuric Acid.	13 Tartaric Acid.	14 Water.	Special Method of Preparation.
																PARTS BY WEIGHT OF INGREDIENTS.
1	Elsner	…	…	24	q.s.	…	…	…	…	…	…	…	…	…	1000	Dissolve 3 in 14 and add 4 till precipitate at first formed is re-dissolved.
2	Fearn	…	1·2	…	111	…	111	…	…	…	111	…	…	…	1000	Add 4 in 740 of 14 to 2 in 37 of 14; add 6 in pieces; and finally 10 in residue of 14.
3	,,	…	1·3	…	71	…	…	…	…	…	…	…	…	25	1000	Dissolve 13 in 480 of 14; and 4 in 480 of 14; mix with 2 in remainder of 14.
4	Hern	…	30	…	…	…	…	…	30	…	…	…	…	…	1000	Dissolve 2 and 6 in caustic soda lye of 3° Baumé.
5	Lobstein	…	0·3	…	…	…	0·4	…	18	…	…	…	…	21	1000	
6	Maistrasse	…	0·1	…	…	…	0·3	…	q.s.	…	…	…	…	…	1000	
7	Munro	…	x	…	…	…	…	…	…	q.s.	…	…	q.s.	…	1000	Dissolve 2 in 12; add sufficient 9 to precipitate it; and just re-dissolve in 12.
8	Roseleur	…	1	…	…	…	…	…	…	…	10	…	…	…	1000	Form electrolytically.
9	Sartoria	…	x	…	…	…	…	125	…	…	…	…	…	…	1000	
10	Steele	17·6	…	…	5·5	16·5	0·9	…	…	66	…	0·9	…	…	1000	Dissolve 4, 5, and 9 in water; filter; add 1 and 11; stir till dissolved.
11	Weiss	…	1	…	…	…	…	…	…	…	5	…	…	…	1000	

NOTES TO TABLE.—The black figures in the last column refer to the numbers of the vertical columns denoting the various reagents. No. 2 is intended to give thick deposits; No. 7 especially to coat lead; and No. 11 to tin lead surfaces.

posited from it at a greater rate than that at which the solution of the anodes replenishes it, in spite of the relatively large surface of the latter exposed. The solution already mentioned as being made by dissolving the binoxide of tin in caustic potash may be used as an electrolytic bath for coating iron. Maistrasse ensures the complete continuity of the tin-covering given by his bath (No. 6) by heating the coated object to the melting point of the tin, thus causing the latter to fuse over the surface and alloy with the basis metal; this solution is said to be especially applicable to the tinning of cast-iron.

No special remarks are called for upon the practical electro-deposition of tin. A current of moderate volume at from 3 to 5 volts pressure is required. The time to be expended varies with the process and with the thickness of metal to be deposited, from one or two up to twenty-four hours, which latter period is recommended for Maistrasse's solution. The objects, as usual, are carefully cleaned before immersion, and are subsequently most thoroughly washed. They may be left in the "dead" condition, if preferred, but are more generally polished, either by scratch-brushing or by friction with bran.

The anodes should be of pure metal. So-called *tin-plate*, which is only sheet-iron covered (in the dry way) with the thinnest possible coating of metallic tin, is, of course, useless; and tin-foil must be used with caution, for many samples of the foil are made from alloys of lead and tin, but these are generally duller in their outer aspect; while others may consist of the thinnest lead-foil, covered on one or both sides with tin, the two surfaces being united together by rolling; and such samples in external appearance have all the characteristic appearance and brightness of the pure metal. Only pure tin-foil, or plates cast from the best *grain tin*, should be employed.

Electro-Deposition of Lead.

With lead, as with tin, the low fusing-point renders the coating of an object more simply effected by immersion in the melted metal than by electro-deposition. The old experiment of growing a "lead tree" by suspending a fragment of metallic zinc in a dilute solution of lead acetate (sugar of lead) is simply a case of deposition by "simple immersion;" the peculiar, largely-crystalline, spongy formation of the resulting lead illustrates very well the difficulty of getting a good solid adherent metal by simple exchange with a more electro-positive element. When,

however, it is necessary to deposit lead in the wet way, a simple
dilute solution of the acetate may be electrolysed by separate
current; but the alkaline bath, prepared by boiling 5 parts of
lead oxide (litharge) in a solution of 50 parts of caustic potash
in 1,000 of water until it is completely dissolved, is preferable.

With either liquid, lead anodes are used, and the objects are
carefully prepared for the bath, and polished afterwards as usual.
The methods cannot, however, be relied upon to give a thick
deposit, nor are they largely used in practice.

Many lead solutions tend to form an insoluble higher oxide
(a peroxide $= Pb O_2$) at the anode, which thus receives a coating
as well as the cathode, but of a different kind; and the principal
interest attaching to the process of lead electrolysis centres in
the possibility of producing films of oxide which present different
colours to the eye by reason of their extreme tenuity.

ELECTRO-DEPOSITION OF ANTIMONY.

The deposition of antimony, again, is a process of no commercial
importance, although the metal, which has a fairly bright lustre
when polished, but is rather grey in colour, resists well the
tarnishing action of the atmosphere. It is a very brittle metal,
and would be useless as a coating upon any thin article, or
upon one which is liable to be bent in any degree when in
use.

Immersion Process.—It is fairly electro-negative, and will,
therefore, give a deposit upon many metals by simple immersion.
For example, brass will receive a lilac-coloured surface tint,
varying in depth of shade according to the time of immersion,
by dipping it in a boiling dilute solution of antimony terchloride
(butter of antimony), made by adding much water to a little
of the antimony compound, and boiling until the dense white
precipitate formed on mixture has re-dissolved; and then, after
a further addition of water and a second boiling, filtering and
heating for use. The coated pieces must be well dried in hot
sawdust or in a stove, and must be protected by a varnish of
lacquer, if the lilac colour is to be preserved.

Battery Process.—But, as usual, the separate-current process
s to be preferred, for which the following solutions, *inter alia*,
have been recommended :—

TABLE XXIV.—Showing the Composition of Solutions for the Electro-Deposition of Antimony, recommended by various Authorities.

No.	Authority.	Antimony.	Antimony Terchloride.	Antimony-Potassium Tartrate.	Antimony Tersulphide.	Sodium Carbonate.	Ammonium Chloride.	Hydrochloric Acid.	Tartaric Acid.	Water.	Special Method of Preparation.
										1 2 3 4 5 6 7 8 9	
					PARTS BY WEIGHT OF INGREDIENTS.						
1	Gore . .	...	x	...	...	...	y	...	...	1000	
2	„ . .	q.s.	...	...	...	...	...	q.s.	...	1000	Electrolyse pure strong hydrochloric acid with antimony anode, until solution is charged.
3	„ . .	...	...	4000	...	...	...	2000	...	1000	
4	„ . .	...	...	83	...	...	...	124	83	1000	
5	Roseleur	...	...	...	50	100	...	...	...	1000	(Use boiling; it deposits kermes mineral on cooling.)

Of these solutions No. 3 will probably give the best results in workshop-practice. It is made by dissolving four pounds of the double potassium-antimony tartrate (tartar emetic) in a mixture of two pounds of strong hydrochloric acid with one of water. This solution is particularly useful for producing thick deposits, as considerable latitude in current-strength is permissible. A current of 0·06 to 0·1 ampère per square inch, or 1 to 1½ ampères per square decimetre, will be found most suitable; but the volume may be greatly increased, and the rate of deposition correspondingly hurried, without danger. Tartar emetic itself is a feeble conductor, and cannot alone be made to give good deposits, for, as Gore has shown, even a very weak current brings down the metal in a pulverulent form; hence the addition of hydrochloric acid, which sufficiently increases the conductivity. The other solution containing tartar emetic has less hydrochloric acid, is a poorer conductor, and must be used with a weaker current, not exceeding 0·013 ampère per square inch, or 0·2 ampère per square decimetre. The bath prepared by Roseleur by boiling together for the space of one hour, and subsequently filtering the solution, 1 ounce of antimony tersulphide, 2 of sodium carbonate, and 1 pint of water, tends to deposit antimony oxysulphide (kermes mineral) on cooling, as above noted; and must, therefore, be always used hot. The pale orange precipitate of oxysulphide is soluble in the mother-liquor as soon as the boiling-point is reached.

The anodes may be made of platinum, but preferably of antimony, which must be cast into slabs of the required shape and size, as it is far too brittle to allow of mechanical work, such as rolling or hammering.

No special treatment is required in depositing antimony; the pieces must be cleaned thoroughly, and the strength of the bath must be maintained by adding a further quantity of the solution from time to time. After coating, the pieces are rinsed, dried in a stove (or in hot sawdust), and brightened in the usual way. But if the chloride solution has been employed, the object must be rinsed once or twice in hydrochloric acid immediately it is removed from the bath, and then in water, because water added to the original solution produces a dense curdy-white precipitate of antimony oxychloride. So that, if the pieces, with a portion of the bath liquor clinging to them, were dipped into water at the outset, they would be covered with this white deposit; but if first washed in a menstruum with which the solution mixes without decomposition, such as hydrochloric acid, the original liquid is safely removed, and the final cleansing may be effected in water without risk.

Character of Deposit.—The metal deposited by too strong a current is, as usual, black, powdery, and non-adherent; and that yielded by some solutions may be so, even when a weak current is employed. But the metal is capable of being thrown down in two different modifications of the reguline or solid form—one of a grey-slate colour in the dull condition, but taking a good polish and resembling cast-iron when scratch-brushed, having a crystalline fracture, and being hard and very brittle; while the other is darker and more steely, but somewhat softer, non-crystalline, or amorphous, and with a lustre which resists atmospheric influences for quite a lengthened period.

Explosive Antimony.—The most curious and interesting phenomenon in connection with antimony deposition is the production of an *explosive* variety, which has been fully studied and described by Gore. He found that the amorphous antimony deposited from a solution of 1 part of antimony terchloride in 5 or 6 parts of hydrochloric acid (of specific gravity 1·12), or in 10 of hydrobromic acid (specific gravity 1·3), or in 15 parts of hydriodic acid (specific gravity 1·25), would, under certain conditions, undergo a physical change and become crystalline; and that this change was attended by an increase of density, and with an evolution of heat so considerable that, if evolved instantaneously by a considerable mass, it may develop almost explosive violence. The heat is so great that, if a sufficient body of metal undergo the change, paper in contact with it is burned, and wood is scorched brown; the "explosion" is often accompanied by a flash of light, but always by a slight cloud of vapour expelled from the interior.

The three varieties (from the chloride, bromide, and iodide) differ in their sensitiveness, as well as in other particulars. None of them are pure, but retain respectively within their pores about 6, 20, and 22 per cent. of the depositing liquor, which may be expelled by heating, as, for example, at the moment of explosion; the cloud of vapour observed is thus accounted for. The presence of this liquor in the metal gives rise to an apparently abnormal excess of deposited metal over that which should be yielded according to the electro-chemical equivalent. The alteration of condition proceeds gradually on keeping, but more quickly in the case of powder or of thin pieces, than with larger masses of metal; and the heat is then evolved almost imperceptibly. But freshly-deposited material may be caused to undergo the change, in a rapid or explosive manner, by any physical means, which is capable of sufficiently affecting the molecular arrangement of the body. With the chloride variety the action begins when it is heated to 170° F., becoming sudden and complete at about 205° F.; with the bromide deposit the explosion occurs at 320° F.; with that from the iodide at a still higher temperature. A similar descending order of sensitiveness was observed when other means were employed to initiate the action; a touch with a red-hot wire caused immediate conversion of the chloride variety, while the bromide metal was merely locally affected by contact with the hot wire, the action only spreading through the whole when it was raised to 250° F. throughout, and through the iodide specimen when it was heated to 338° F. The heat developed by the alteration in the first-named case was so great that a tin rod $\frac{1}{8}$ of an inch in diameter, upon which the amorphous antimony was electrolytically built up to a total diameter of $\frac{1}{2}$ an inch, melted, flowed away from the antimony, and remained fluid for some time.

A sudden blow, or even rubbing with glass or metal, is liable to convert the amorphous into the crystalline variety, so that if it is required to break up the unexploded metal into smaller pieces, it should be fractured under cold water by a comparatively soft material such as wood. Provided that they are kept under iced-water meanwhile, very thin pieces may even be crushed to a fine powder in a mortar; and this powder may be dried in the cold over sulphuric acid; and, remaining in the original condition, will evolve subsequently the same proportion of heat as the thicker untouched deposits.

The Electro-Deposition of Bismuth.

The deposition of this metal possesses at present little beside a scientific or theoretical interest.

It may be thrown down from a weak and very slightly acidified solution of the nitrate, either by simple immersion upon certain more electro-positive metals, such as tin, or by the separate-current process. Bertrand uses for the latter method a solution of 30 parts of the double chloride of bismuth and ammonium in 1000 of water, containing a small proportion of hydrochloric acid. With one Bunsen-cell he succeeded in obtaining a coat which, although black exteriorly, exhibited the well-known slightly-pink shade of the metal, and was susceptible of a very high polish. Like antimony, the brittle nature of the metal renders it unfit for coating objects which are at all strained or altered in shape subsequently.

Of the remaining metals there are none which render necessary in this work a description of the means by which they may be electrolysed. With regard to some of them, indeed, many published processes would appear on thermo-chemical grounds to be visionary.

In regard to **aluminium** especially; the extreme popularity of this metal combined with a great want of knowledge, on the part of the public, as to its properties, have led to a demand for its electro-deposition. Many solutions have been proposed which it was claimed should give good deposits of the metal, but have been found by various experimenters to be worthless. In our own experience, the brilliant grey deposit, which has been afforded by some of these methods, but which has never exceeded in thickness that of a mere film, has consisted principally of iron, a metal which is almost universally present in commercial aluminium compounds. The deposit has been often found, on testing, to contain aluminium; this may have been due to traces of the solution remaining in the pores of the coat, or it may have resulted from aluminium which had actually been deposited with the iron as an alloy; but in all cases the iron was found to be vastly preponderating. That it is possible to deposit aluminium by the electrolysis of fused compounds is no doubt true, but further analytical evidence is necessary to prove the satisfactory deposition of the pure metal from aqueous solution.

COLOURING OF METALLIC SURFACES.

Advantage has been taken of the fact that lead and certain other metals tend to deposit as peroxide upon the anode, instead of, or sometimes in addition to, precipitating as metal upon the cathode, to obtain certain colours upon metal surfaces. The most interesting application of this is to be seen in the formation of metallo-chromes by the deposition of an infinitesimal film of lead peroxide upon a polished steel surface. It has long been known that colourless transparent substances, if sufficiently thin, are capable of displaying a series of colours by reflected light by the optical phenomenon known as the interference of luminous waves (where the wave of light reflected from one side of the film passes so near to that reflected after refraction from the other, that their respective vibratory influences interfere with one another). The play of colours upon the soap-bubble or upon oil floating on water are instances of this phenomenon, which was first studied by Newton. A momentary immersion of a bright steel or platinum plate as anode in a lead solution suffices to deposit a film of peroxide, which answers the requirements for the production of these iridescent colours. Nobili was the first to observe this action with acetate of lead; Becquerel's solution is now used for this purpose; it is made by dissolving 14 ounces of caustic potash in half a gallon of water, adding to this $10\frac{1}{2}$ ounces of lead oxide (litharge) and boiling for from half-an-hour to an hour, allowing it to stand for some time, then decanting the clear liquid from the subsided precipitate, and making up the whole to a gallon in volume. The electrolytic action must be continued for exactly the right period of time, an insufficient exposure does not give time for the development of sufficient thickness to allow of interference, while an excessive action causes an opaque dirty brown deposit; intermediately between the two, a very beautiful play of colours may be secured. The cathode may be of copper-sheet. Gassiot produced patterns upon the anode by interposing a cardboard disc, with a perforated design upon it, between the electrodes, so that the deposit chiefly occurred on the portions unshielded by the solid portions of the card. Watt uses copper wire bent into various shapes; this system has the advantage that there are varying distances between the different portions of the anode and the cathode, and, therefore, a varying thickness of film is produced, which adds to the beauty of the iridescence. The film is fairly adhesive, but should not be handled more than is necessary.

CHAPTER XIV.

THE ELECTRO-DEPOSITION OF ALLOYS.

THE principles upon which the possibility of electro-depositing alloys may be said to depend has already been explained in Chapter II. (p. 34). Brass, bronze, and German silver are practically the only alloys deposited, if we except the mixtures used in producing coloured gold coatings, and of these the first-named alone has any widespread use.

THE ELECTRO-DEPOSITION OF BRASS (COPPER AND ZINC).

A brass coating may be given to a copper article by covering it electrolytically with a thin film of zinc, and then, after washing and drying, applying to it a heat just sufficient to cause the two metals to form an alloy superficially; and similarly an object made of any other material, which will withstand the necessary heating, may be brass-surfaced by depositing alternate layers of copper and zinc, and alloying them *in situ* as before. But in practice this could not well be done. Nor is brassing usually effected by simple immersion, although Watt has shown that a zinc rod, dipped into a mixed solution of copper and zinc acetate, becomes covered with a yellow deposit of the alloy. But for practical purposes the production by the separate-battery process is alone adopted. The Bunsen form of battery is the best, and should generally be arranged with two cells in series.

The Solution.—The solution may be greatly varied, and, indeed, no absolute and unalterable rules can be laid down as to its constitution. The basis of most of the liquids is the mixed cyanides of copper and zinc, as in this combination zinc does not readily displace copper from its solution, and there is in consequence a better chance of obtaining a simultaneous coating of the two bodies; but the relative proportions of these may require to be varied in working, according to the behaviour of the solution, which depends upon several inconstant quantities—strength of current, resistance of solution, and the like. The following table (XXV.) summarises the principal electro-brassing solutions.

One of the best of these solutions is that recommended by Roseleur (No. 8) which is somewhat complex, but will be found to give good results—to prepare one gallon, $2\frac{1}{2}$ ounces of copper sulphate and a like weight of zinc sulphate are dissolved in water; and a solution of $6\frac{1}{2}$ ounces of sodium carbonate in a convenient quantity of water is added to the mixture. A dense precipitate of copper and zinc carbonates is thus formed. It is allowed to settle, water is poured on, and it is thus washed several times by decantation; the clear liquor is finally siphoned away from the precipitate, which is then filtered off from the residual liquid, and mixed with a solution of $3\frac{1}{4}$ ounces of sodium carbonate, and $3\frac{1}{4}$ ounces of sodium bisulphite, in $7\frac{1}{4}$ pints of water. To this is added $3\frac{1}{4}$ ounces of potassium cyanide and 20 grains of arsenious acid (white arsenic) in $\frac{3}{4}$ pint of water. After filtering, the clear liquid contains the copper and zinc, and should be quite decolorised; any blue colour indicates the presence of unaltered copper salt, and calls for the addition of a further quantity of potassium cyanide solution. It may be taken as a general rule that all the cyanide brass-baths should be free from blue colour; and that if they are not so initially, they must be made so by the introduction of sufficient extra potassium cyanide, while if they become blue when in use, the same remedy is to be applied. It is to be recommended that a 10 per cent. solution of the cyanide should be kept for this purpose, so that a little may be added whenever necessary. The bath (Roseleur's) is used cold and with brass anodes, but as the composition is liable to variation by unequal solution of the brass, a further addition of copper or zinc may be required from time to time. This should be effected by adding separate solutions of copper or of zinc cyanides, made by dissolving their carbonates in solutions of potassium cyanide. Sodium or potassium arsenite may be used in place of arsenious acid, but a proportionately larger quantity must of course be employed. A small proportion of arsenic is found to give a brighter deposit than is yielded by the copper-zinc solution alone; a large quantity on the other hand is found to give to the deposit a temporary increase in whiteness, which is objectionable.

Baths may be prepared electrolytically either, as in Gore's solution (No. 3), by passing the current through a suitable solution from a brass anode, so that the constituents of the anode alloy may dissolve simultaneously into the liquid, or, as in Volkmer's bath (No. 14), by passing the current from a copper anode alone at first, until sufficient of that metal has been taken up; and next from a zinc anode alone, until the requisite shade

TABLE XXV.—Showing the Composition of Electro-

No.	Authority.	Brass.	Copper Acetate.	Copper Carbonate.	Copper Chloride.	Copper Cyanide.	Copper Sulphate.	Zinc Acetate.	Zinc Carbonate.	Zinc Chloride.	Zinc Cyanide.	Zinc Sulphate.	Caustic Potash.	Potassium Carbonate.
		1	2	3	4	5	6	7	8	9	10	11	12	13
							PARTS BY WEIGHT							
1	Brunel	...	...	...	10	...	...	...	...	...	...	20	...	250
2	,, ...	...	...	...	5	...	...	...	...	...	...	10	...	80
3	Gore	x	...	...	...	...	...	...	..	...	...	...	...	...
4	Heeren . . .	...	...	...	...	...	3	...	...	...	...	26	...	...
5	Hess	x	...	...	...	...	...	...	...	...	...	...	...	...
6	Japing	...	...	...	...	...	...	...	...	...	...	...	...	...
7	Morris & Johnson .	...	...	...	...	12·5	...	...	...	...	6·2	...	...	...
8	Roseleur . . .	...	...	10	...	...	...	...	10	...	...	...	...	...
9	,, . . .	...	12·5	...	...	...	...	...	...	10	...	...	...	...
10	,, . . .	...	14	...	...	...	...	...	...	14	...	...	...	...
11	Russell & Woolrich	...	150	...	...	...	...	15	...	...	...	...	...	...
12	De la Salzède . .	...	...	...	5	...	...	...	...	...	...	9·5	...	120
13	,, . .	...	...	...	3	...	...	...	...	...	...	7	...	100
14	Volkmer . . .	...	...	...	...	...	...	...	...	...	...	...	...	...
15	Watt	...	4	...	...	...	...	...	...	...	...	8	56	...
16	Weiss	...	4	...	...	...	...	...	...	...	...	6·8	40	...
17	Wood	...	...	...	...	14	...	...	...	...	7	...	...	...

NOTES TO TABLE.—The black figures in the last column refer to the numbers of the vertical
Nos. 9 and 17 to iron and steel. No. 8 is recommended to be used cold No 14 at 86° to 100°;

BRASSING SOLUTIONS, AS RECOMMENDED BY VARIOUS AUTHORITIES.

14 15 16 17 18 19 20 21 22 23 24

OF INGREDIENTS.

Potassium Cyanide.	Potassium Acetate.	Sodium Carbonate.	Sodium Bicarbonate.	Sodium Bisulphite.	Ammonia.	Ammonium Carbonate.	Ammonium Chloride.	Ammonium Nitrate.	Arsenious Acid.	Water.	Special Method of Preparation.
...	...	...	...	...	...	...	...	125	...	1000	
12	...	...	...	...	...	...	...	...	...	1000	Dissolve together in 24; add 14 last.
125	...	...	...	...	62·5	...	...	...	...	1000	Form electrolytically.
60	...	...	...	...	...	...	...	...	...	1000	Dissolve 6 in 12 of 24; 11 in 52 of 24; 14 in 120 of 24; mix; add rest of 24.
6·5	...	...	42	...	...	...	27	...	...	1000	Form electrolytically with brass anode.
q.s.	...	...	...	...	...	...	...	...	...	1000	Form electrolytically, with brass anode supplemented by copper or zinc anodes, if necessary, to improve colour.
100	...	...	...	...	...	100	...	...	...	1000	
20	...	20	...	20	...	...	...	...	0·2	1000	Precipitate 3 and 8 from $CuSO_4$ and $ZnSO_4$ with Na_2CO_3, and wash; add 16 and 18 in 900 of 24; add 14 and 23 in 100 of 24; filter. Add more 14 if solution remain blue.
40	...	100	...	20	...	...	...	...	...	1900	Dissolve 14, 16, and 18 in 800 of 24; add 2 and 9 in 200 of 24.
30	...	...	28	16	...	...	...	...	...	1000	Dissolve 14 and 18 in 800 of 24; add to solution of 2, 9, and 19 in 200 of 24.
excess	150	...	...	...	...	...	...	...	...	1000	Add 14 last, until precipitate re-dissolves.
2·5	...	...	...	...	...	...	...	61	...	1000	Dissolve 14 in 24 of 24; dissolve 4, 11, 13 in rest of 24; add 22; stand for some days and decant.
10	...	...	...	...	...	...	...	...	...	1000	
125	...	...	...	...	...	125	...	...	...	1000	Form electrolytically; with copper anode till saturated; then with zinc anode till deposit is brass colour.
6·5	...	...	...	...	31	...	...	...	...	1000	Dissolve 2 in 65 of 24 and add 15 of 19; dissolve 11 in 125 of 24 and add 16 of 19; mix; add 12 in 125 of 24, then 14 in 125 of 24. Stir, add rest of 24; stand and decant.
3-4	...	...	...	...	24	...	...	...	...	1000	Dissolve 2 in 19; add 12; then add 11, 14 and 24.
82	...	...	...	...	...	...	14	...	...	1000	

columns indicating the various reagents. Solutions Nos. 6 and 10 are especially applicable to zinc; No. 7 at 150°; No. 17 at 160° F.; and Nos. 3 and 4 boiling.

of brass-colour appears on the sheet metal cathode; a brass anode is then substituted for that of zinc, and the plate at the cathode pole is replaced by the object to be coated.

Anodes.—The anode used for the various solutions above given may be made of brass, and this should have approximately the same composition as the metal which it is proposed to deposit, and should be prepared from the pure virgin metals; but instead of using the copper and the zinc combined together in the form of an alloy, they may be used separately by suspending alternate strips of the two metals from the anode rod; and this method, although less usually adopted, presents the advantage that the composition of the bath may be controlled by altering the relative numbers of the two kinds of strips, so that a greater area of copper anode surface may be presented to the liquid when the deposit is becoming too pale in colour, or of zinc when it grows too red or yellow. The anodes are supported in the usual way, either completely immersed and supported by stout brass hooks, or partially immersed only. The vats for cold and for hot solutions are similar to those used for the copper (cyanide) depositing process.

Nature of Deposit.—The character of the metal deposited is entirely dependent upon the conditions of current and of bath, but chiefly upon the composition of the solution. If the liquid contain an excess of either constituent beyond the normal, the deposited metal will also contain an excessive percentage of that metal, and its properties and colour will be influenced accordingly. The composition of the anodes vastly influences the deposit yielded by a solution, inasmuch as it modifies the constitution of the bath itself.

A weak current, or the imparting of motion to the suspended objects (which is in some respects equivalent to a reduction in current-volume), tends to produce a deposit containing a greater proportion of the more electro-negative element copper, while a stronger current yields a larger percentage of zinc. Then, again—given the same battery, solution, and anodes—heating the bath, by increasing its conductivity, raises the current-strength, and thus also tends to increase the percentage of zinc. The deposition of hydrogen must as usual be avoided, in order to obtain a good adherent deposit. But, to sum up the preceding observations, within the limits of current-volume that permit the production of a good coat, the conditions which favour the precipitation of an alloy rich in copper, and, therefore, a red or yellow colour, are—a solution and anode containing a high percentage of copper, a weak current, a cold bath, and the movement of the articles under treatment; while the opposite effects of a greater proportion of zinc with a corresponding whitening of the deposit, are yielded by a strong current, by anodes and solution richer in zinc than in copper, and by maintaining the objects motionless in a hot liquid.

It will now be understood that the various relations of current-strength and other conditions of work are mutually so inter-dependent that it is impossible to lay down inviolable rules for working brass solutions; but with the requisite constant obser-vation and care, there is no difficulty in obtaining any desired nature of deposit. If the colour of the metal is too red, an excess of copper is indicated, which may be rectified by increasing the current-volume or adding more zinc to the liquid; if it is too white, showing an excess of zinc, the current is reduced or copper is added to the bath. The alteration of current may often be conveniently effected, as mentioned in another chapter, by increasing or decreasing the surface of the anodes according as its volume is required to be greater or less. It is evident, therefore, that baths of even comparatively widely-differing compositions may be caused to yield deposits of the same alloy by adjusting the various conditions of work. But since the proportion of the constituents in the deposited metal, and hence its colour, are so susceptible of alteration by a change in any of the conditions, it becomes necessary to watch the progress of the electrolysis most carefully, not only to prevent the variation of the alloy, but to prevent local alterations, which may give rise to spotted or unevenly coloured deposits, such as would be produced if any portion of the cathode object were receiving more or less current than the remainder, or if the solution were not properly mixed. It is, therefore, advisable to stir the solu-tion very thoroughly before commencing work, and at intervals while the deposition is in progress; and again to observe that the pieces forming the opposite electrodes are as nearly as possible equidistant from one another. The observance of this latter precaution, which is necessary enough in depositing single metals, where the main question is one of thickness of deposit, becomes greatly exalted in importance when it is seen that not only the thickness but also the colour of the coating is influenced, the portions more remote from the anodes receiving less cur-rent, and, therefore, having a redder shade than those in closer proximity. Similarly, the difference in specific gravity of the various strata in unmixed liquids produces a variation in colour between the deposit at the top and that at the bottom of an article.

The Process.—In the practical application of the process, the objects to be brassed must be first carefully cleansed and polished after the orthodox fashion; they are then immersed in the electrolytic bath until the required thickness of metal has been attained, and, provided that it be of the right colour, it is then rinsed, scratch-brushed, well washed in hot water, and dried in hot sawdust or in a stove. In some of the liquids used, the cyanide solution does not exert sufficient solvent action upon zinc oxide, which thus becomes separated in the solid form upon the surface of the anode, finally crumbling away and collecting on the bottom of the bath. Such a formation is objectionable, firstly, because it impedes the action by yielding a film upon the electrode surface; then, because it becomes detached, and so introduces into the bath solid matter that may be held for a time in suspension in the liquid, which is always dangerous, because fragments are liable to become attached to the surface of the object being plated; and, thirdly, because a deficiency of zinc passing into solution from the anode is equivalent to a relative increase in the amount of copper in the bath. In cases where such a precipitate is observed upon the anode (which should, therefore, be inspected from time to time), a small quantity of liquor ammoniæ mixed with a slight excess of potassium cyanide should be added to the liquid; any oxide already formed will thus be dissolved, and the anode surface will be kept clean, owing to the non-formation of fresh quantities of the substance. When ammonia is one of the constituents of the bath, a further quantity must be added at intervals, because it is constantly evaporating by exposure to the air.

When an object is found to be taking a bad deposit at first, it should be removed from the vat, scratch-brushed and returned, the defect in the electrolytic arrangements having been made good in the meantime. A dirty yellowish or earthy-looking deposit is often the result of insufficiency of cyanide, and may be corrected accordingly.

The chief use of brass composition is to coat zinc or iron surfaces with a rich coloured material that may be subsequently bronzed or otherwise ornamented, or which may be simply lacquered and left with the original brass colour. It is sometimes employed for facing typographic matter, but presents no advantage over nickel for this purpose; it is especially applicable to coating bookbinders' type, which should have a hard face, and which must bear heating, as these tools are frequently used hot.

TABLE XXVI.—Showing the Composition of certain Electro-Bronzing Solutions, recommended by various Authorities.

No.	Authority	PARTS BY WEIGHT OF INGREDIENTS.														Special Method of Preparation.
		1	2	3	4	5	6	7	8	9	10	11	12	13	14	
		Copper Sulphate.	Cupric Chloride.	Cuprous Chloride.	Cuprous Cyanide.	Tin Tetrachloride.	Tin Bichloride.	Tin Binoxide.	Sodium Stannate.	Caustic Potash.	Potassium Carbonate.	Potassium Cyanide.	Soda-Lime.	Rochelle Salt.	Water.	
1	Elsner . .	70	...	...	...	8	...	...	...	q.s.	...	...	...	...	1000	Dissolve 1 in 14; add 5 in a little strong solution of 9.
2	Ruolz . .	...	...	...	5	...	...	2	...	...	...	q.s.	...	...	1000	Dissolve 11 in 14 until sp. gr. = 1·03; dissolve in this, 4 and 7 at 125°-150° F.
3	De la Salzède	...	...	1·5	...	...	1·2	...	...	...	100	10	...	...	1000	
4	Weiss . .	...	20	...	...	...	10	...	...	...	100	10	...	...	1000	
5	,, . .	64	...	...	...	8-10	...	...	...	q.s.	...	128	...	...	1000	Dissolve 1 in 14; add 11; when dissolved, add 5 in little solution of 9.
6	Weil . .	35	...	...	...	...	...	...	q.s.	...	...	...	80	150	1000	Add sufficient 8 to rest of solution, to give required deposit.

NOTE TO TABLE.—The black figures in the last column refer to the numbers of the vertical columns indicating the various ingredients.

THE ELECTRO-DEPOSITION OF BRONZE (COPPER AND TIN).

Maistrasse has obtained a superficial bronze coating upon copper objects by first electro-tinning them, and then heating them for some time above the melting-point of tin, so that it may fuse and alloy with the base metal upon the surface. But, in regard to the more purely electrolytic methods, the general remarks which were made in reference to brass apply with equal force to the less commonly employed deposition of bronze.   The principles underlying the two processes are identical, and the methods similar The preceding short table of solutions will show that there is practically only a substitution of tin for zinc as compared with brassing liquids.

Bronze anodes are used, varying in composition with the character of the metal to be deposited.   A ternary alloy of copper, tin, and zinc may be obtained by suitably combining the brassing and bronzing solutions ; but, like bronzing itself, this process has but little practical utility.

THE ELECTRO-DEPOSITION OF GERMAN SILVER (COPPER, NICKEL, AND ZINC).

This alloy is recommended by Watt as a substitute for nickel in coating certain small articles such as revolvers, on account of the colour, which has a somewhat redder shade than the pure white of nickel, and is sometimes considered to be more pleasing. The bath which he recommends is made by dissolving 1 ounce of German silver, of the required composition, in nitric acid, diluted with an equal volume of water, taking care that no excess of the acid is present by adding it very gradually, and so that a small portion of the metal remains undissolved in the liquid when all chemical action has ceased (this is indicated by the cessation of gas evolution) ; about 4 ounces of potassium carbonate are dissolved in a fairly large quantity of water, and this solution is added to that of the alloy, until no further precipitate is produced by the addition of another drop of the carbonate.   The precipitate is then allowed to subside, the liquid is poured away from it, and fresh water is added several times successively.   A strong solution of potassium cyanide with 1 ounce of the strongest ammonia solution (*liquor ammoniæ fortiss.*), filtered clear, if necessary, is now added with constant stirring, until the precipitated carbonates are just dissolved ; a slight excess of potassium cyanide is now introduced, with sufficient water to make up the bulk of the liquid to 1 gallon. After filtering, the liquid is allowed to stand for twelve hours, the clear liquid is decanted off, and is used as an electrolyte with a German-silver anode, similar in composition to that from which the bath is prepared, the current being produced by a Bunsen-battery.

Morris and Johnson, whose electro-brassing bath has already been given, protected, by their patent of 1852, the deposition of German silver from a solution of the cyanides of copper, zinc, and nickel (in the correct proportions of the alloy) in one of ammonium carbonate and potassium cyanide dissolved in 10 parts of water. An excess of potassium cyanide produces an alloy containing a larger percentage of copper, and imparts to it a reddish shade in consequence ; but this may be rectified by the addition of ammonium carbonate, while on the other hand, if the deposit be too white, fresh cyanide is added by degrees until the desired shade is produced. This bath is to be used with a powerful current at a temperature of 160° F.

Other alloys may also be deposited by adopting suitable solutions prepared to satisfy the requirements set forth in an earlier chapter. At present there is no demand for any of these, and they have generally been designed to imitate the qualities of a more costly unalloyed metal. For example, Round proposed to precipitate an alloy of tin and silver to take the place of the latter metal, by electrolysing a clear solution of 4 ounces of potassium cyanide, 5 of the strongest *liquor ammoniæ*, and ½ ounce of silver, to which a suitable proportion of any soluble tin compound, and subsequently about 2½ ounces of potassium carbonate, had been added, and from which sediment had been separated by decantation. He used a dual anode, consisting of a large sheet of tin with a small plate of silver in juxtaposition.

Alloys have also been prepared to resist the action of acids or other corrosive fluids, such as that of platinum and silver, which Campbell has patented, and made by depositing from a solution containing silver and platinum in the proportion of 70 : 30, with an anode made from an alloy of similar percentage composition. The solution was prepared by dissolving the mixed chlorides, obtained by the addition of ammonium chloride to the conjoint solution, in potassium cyanide solution.   An alloy of aluminium and nickel is said to have been deposited in Philadelphia and denominated *alu-ni*, to take the place of the nickel, but it is more costly while presenting no commensurate advantages.

Another group of alloys sometimes prepared includes those of magnesium and aluminium, which it is difficult, if not impossible, to deposit alone.

CHAPTER XV.

ELECTRO-METALLURGICAL EXTRACTION AND REFINING PROCESSES.

Preliminary.—As we have already explained in the first chapter, the term electro-metallurgy is commonly applied to all classes of work which necessitate the deposition of metals by means of the electric current. Some authorities are disposed to reserve it for those processes by which metals are extracted from the ore with the aid of electricity. It would seem that technically the first and wider definition is admissible, and it is certainly more convenient to group together under one name all the kindred processes of electro-extraction, refining and plating. The term electro-metallurgy may, then, embrace all these, and the different subsections may receive special designations such as electrotyping, electro-plating, and the like.

In the more restricted sense of the term there are two principal divisions—the extraction of metal from the ore, and the refining of metals already produced, either in this way or by other methods. Only comparatively few ores and metals are thus dealt with at present—of these the principal are copper, zinc, and lead, although incidentally gold and silver are produced from other metals, in which they had previously existed in minute quantities ; and, indeed, one of the advantages of electrolytic refining is that usually the precious metals are completely and readily recovered by its application.

The laws which govern the electro-deposition of metals in plating or typing are equally applicable in this case, and well repay the closest study ; but there are financial questions entering into the problem as here presented which are not so clearly placed before the electro-plater. This is due to the fact that the principal object in plating or typing is to produce a good and reliable deposit, and failing this, all attempts at economy result in extravagance; whereas a slight additional expenditure may repay itself again and again by the enhanced excellence of the work ; thus, although there may be great competition, and, therefore, necessity to minimise expenditure, yet the competition is in the same field, and all rivals are working under like conditions. On the other hand, the electro-reduction of ores or refining of metals has to be measured against other purely metallurgical or chemical processes, both as to purity of product and cost of production. And although there may occasionally be a demand for a pure article at any price for certain specific purposes, and although the electrolytic refining is perhaps best fitted to meet this, yet the demand is comparatively small. If, therefore, the field is to be widened so that the electrical system enters into competition with metallurgical methods upon common ground, the greatest attention must be paid to every detail of the work, and every effort must be made to ensure the highest efficiency at the least cost. But as the main object in electro-refining is the production of a metal having the maximum purity, it may often be necessary to sacrifice the false economy of saving in the conduct of the process, for the true economy of obtaining a pure metal, although at a slightly increased expenditure.

THE ELECTRO-REFINING OF COPPER.

The refining of crude metallic copper produced by other processes is one of the chief applications of this branch of the electro-metallurgist's work. The impure metal may contain, in addition to the copper which is generally present to the extent of 95 per cent. and upwards, such substances as gold, silver, lead, bismuth, tin, zinc, iron, manganese, nickel, antimony, arsenic, sulphur, and siliceous matter or slag in varying proportions, and the problem is, so to remove these that perfectly pure copper is obtained. Many experiments have been tried and recorded, both on a small and on a large scale, by such authorities as Sprague, Gramme, Becquerel, Kiliani, Keith, and others, whose observations have borne results of extreme value, in fixing the most suitable conditions under which the copper-baths should be worked. They have been freely quoted in this chapter, more especially those of Kiliani.

Principles.—The universal principle underlying the various processes in use for this purpose may be said to consist in passing a current of electricity of suitable strength, from an anode of the metal to be refined, through a solution of copper sulphate, acidulated with sulphuric acid, and collecting the pure metal upon a convenient cathode; at the same time allowing for the removal of such insoluble impurities as gradually form a slimy sediment upon the bottom of the vat, and guarding against the excessive accumulation of soluble impurities in the liquid itself. It is clear that the precise manner of applying this principle is capable of almost infinite variation; and so it is scarcely a matter for surprise that almost every electro-refinery has its own special method (concerning which in many cases secrecy is most jealously preserved), differing from others either in the current-strength or in the disposition of the baths, or in some other matter at first sight trivial, perhaps, but in reality vastly influencing the economical use of the system for good or for evil.

In order to obtain a complete understanding of the process itself, it is desirable that the behaviour of each of the elements likely to occur in the crude copper should be examined, both at the anode in relation to the action of the solvent employed, and at the cathode in regard to its tendency to deposit from copper solutions of various degrees of dilution, and with different volumes of current.

In accordance with the principle explained in an earlier chapter, given an alloy of all known metals in equal proportions, a suitable electrolyte, and only a weak current to effect the electrolysis, the most electro-positive metal should first dissolve, then the next on the scale, and so on, until at last only the most electro-negative element remains, and this would itself dissolve in turn; or, on the other hand, if all were together in solution, the order would be reversed, and the most electro-negative element would first deposit upon the application of a current. In the copper anode, however, and with the strength of current employed, such a reaction is not likely to occur, because of the comparatively small amount of the total impurities present in it and the still smaller proportion of any one taken singly: there is only the *tendency* to follow in this order of sequence. Actually, there is an attack at first on such molecules of the most electro-positive metals as are upon the surface of the anode, to an extent proportionately greater than on those of copper; but the current-volume is so intense as compared with the amount of these impurities, that the last-named metal also dissolves. In this manner a fresh series of molecules is exposed by the removal of the outermost layer, and these are in turn attacked, the more positive metals in greater proportion than the copper (in relation to their respective percentages) and so forth; thus, there is a tendency to form corrosion-pittings or hollows, due to excessive action locally upon superposed electro-positive molecules; then, as the acid penetrates these, the action presses more and more unevenly into the body of the anode, until, in course of time, the latter becomes deeply honeycombed, or it may be even a mass of sponge, if originally it were very impure. This sponge consists mainly of copper and of metals more electro-negative than itself, together with any oxidised bodies which may be insoluble in the liquid, the greater proportion of the more electro-positive elements having been removed by solution. This sponge being very frangible, is constantly becoming broken up to a slight extent by any motion of the liquid, and the detached portions fall to the bottom in the form of mud. When the percentage of impurities in the copper is not excessive, the anode is less readily penetrated by the solution, the action proceeds more evenly, a greater proportion of copper dissolves superficially, and the spongy deposit on the surface consists almost entirely of the bodies which are not oxidisable in the bath, or whose oxides are insoluble in it; and this being very limited in quantity is not coherent, and is more readily detached, forming a mud comparatively poor in copper.

Dealing first with the solution of the anode, copper is a very electro-negative element (see p. 21), and thus nearly every metal, other than the precious metals, *tends* to become oxidised before it.

Behaviour of Individual Impurities.—Zinc, iron, nickel, cobalt, and manganese dissolve and remain in the solution, and by taking the place of the equivalent of copper, which should have dissolved in their place had the anode been pure, bring about a gradual impoverishment of the liquid in regard to copper; and ultimately, by continued solution of small quantities, they accumulate to so great an extent as to render the bath unworkable. This relates to the *electrolytic* solution of the anode; but it should be remarked that acid baths tend also to act upon these bodies by *simple solution* (or by *local action*—see p. 39—with the more electro-negative copper) when they are present in any quantity, and thus the acidity is reduced, whilst the weight of metallic substances present is increased, and this is objectionable because neutral baths afford less satisfactory results than acid solutions. With a neutral liquid, silver must be added to the list of the substances which dissolve in the bath.

Bismuth and antimony in part dissolve in the solution, in part remain upon the anode, and thence pass into the mud as insoluble basic sulphates (salts containing less than the due proportion of acid), but even the portion which dissolves is more or less completely precipitated on standing. Tin behaves similarly, but a greater proportion remains as an insoluble oxide at the anode when the latter is rich in this metal. Arsenic also dissolves at first, but as soon as the liquid is saturated with it the remaining proportions are found in the slime.

Gold, platinum, lead, and, in acid solutions, silver remain in the anode mud, the two former as undissolved metal, and the latter as lead sulphate, a compound which is quite insoluble in the acid liquid.

Crude copper generally contains a certain percentage of admixed slag, cuprous oxide (*i.e.*, suboxide of copper), and cuprous sulphide ; and these, with the usual intensity of current, being but feeble conductors of electricity, pass unchanged into the slime, from which, however, the oxide may gradually be dissolved out in the free acid of the bath by a purely chemical reaction.

Thus, the composition of the mud varies with the nature and quantity of the impurities in the unrefined copper, and may embrace gold, platinum, silver, lead sulphate, basic sulphates of bismuth, tin, and antimony, arseniates and antimoniates, slag (chiefly silicate of copper and iron), cuprous oxide and sulphide, and metallic copper.

The solution may contain besides copper, iron, zinc, cadmium, nickel, cobalt, manganese and small proportions of arsenic, tin and antimony, together with silver if the solution be neutral. It is evident that, so far as regards the refined copper at the cathode, the bodies which remain undissolved at the anode are *ipso facto* removed, as it were, from the sphere of action, and are perfectly eliminated ; unless, through bad working, minute fragments obtain access and cling to the cathode surface. The only impurities which can possibly be deposited upon the cathode, and thus contaminate the refined metal under normal working, are those which have become dissolved in the solution, and diffused through it, so that they are actually brought into contact with the surface of the cathode plate. Of these, silver is always precipitated with the copper, together with gold and platinum if, from any unusual cause, either of them be present in the liquid. In neutral solutions, and even in acid solutions, if they be weak in copper, antimony and arsenic are deposited, and thus tend to make the refined metal brittle and unserviceable. The other constituents of the bath, being far more electro-positive, do not precipitate when the electrolytic conditions are satisfactory, and require no consideration, except in so far as they take the place of copper in the solution, and thus reduce the strength of the bath in regard to its principal element.

Current Regulation.—The source of the current was originally the voltaic battery, but the high price of the zinc was prohibitive, so that the electro-refining or extraction of metals remained merely an interesting laboratory experiment, the practical application of which was beyond the range of economical solution until the invention and multiplication of dynamo-electric machinery introduced the possibility of unlimited electric power at a moderate cost. Now, however, the readiness with which other forces may be converted into electricity renders the electrolytic process a formidable competitor with existing methods, especially when natural power is available, as, for example, in the neighbourhood of hill torrents or continually-running water of any kind, and the loss of energy inseparable from the use of the steam-engine as a motor is thus excluded. The dynamo used is of low electro-motive force, and similar to those which are employed for electro-plating, though usually, of course, of increased size and output (see Chapter III.).

The current-volume which has usually given the best results averages 0·013 to 0·020 ampère per square inch, or from 0·2 to 0·3 ampère per square decimetre, but in some installations it is, we believe, as high as 0·065 ampère per square inch (1 ampère per square decimetre). The electro-motive force or pressure required depends upon the circumstances of the case—the nature of the copper under treatment, the distances between the electrodes, and the number of baths or pairs of plates in series. Thus, two identical sheets of pure copper, opposed to one another, and joined by metallic connection in a homogeneous depositing solution, show no difference of potential; but any variation between them, even in the degree of compression of the metal, though more markedly, of course, in the chemical composition, gives rise to a certain development of electro-motive force; and thus, by producing a current (however feeble), either aids or retards the electrolysing current, according to the direction in which it is flowing. It is obvious that the crude copper anode and the pure refined metal cathode thus set up an electro-motive force between them, which must to a small extent influence the decision as to the pressure that is required from the electric generator. In the second place, an increase in the distance between the electrodes involves also an increase in the resistance of the bath, and demands a corresponding development of current-pressure in order to overcome it. While, thirdly, the multiplication of the number of baths in series intensifies the interfering action of the unlike electrodes, and of the varying distances between them.

The Electrodes.—The space most usually allowed between anode and cathode varies from 1½ to 2 inches; less than this is not advisable, because of the facility with which detached fragments of copper and mud from the anode, or excrescent growths, which are common enough on the cathode—especially when powerful currents are employed—tend to bridge over the space and short-circuit the current (see p. 42), thus entirely stopping the action, while at the same time wasting energy, and even perhaps damaging the dynamo. With the inter-electrode distance above recommended, an allowance of 0·3 volt E.M.F. for every pair of electrodes placed in series will be found suitable.

The cathode plates, on which the metal is to be deposited, may be of thin sheet copper, but in this case the surface should be slightly greased and well covered with black lead, to prevent the uniting of the surface with that of the precipitated metal; but it is in many respects better thus to form a plate up to a thickness of $\frac{1}{18}$ of an inch, and then stripping it from the original sheet, to use it in turn as cathode for the remainder of the process. The anode consists usually of a cast slab of the metal to be refined, of convenient size for the bath employed; it may range, for example, from 2 to 2½ feet in length, from 6 to 9 inches in width, and from $\frac{3}{8}$ to $\frac{5}{8}$ of an inch in thickness. The distance between the two plates should not be less than 1½ inches, as shown above; any great increase beyond this lessens, it is true, the risk of short-circuiting, but it also increases the resistance of the bath, and, in consequence, not only adds to the time required for depositing a given weight of metal, but also gives rise to an increased loss of electrical energy, by conversion into useless heat, in the bath, so that the efficiency of the installation is diminished. The distance which best strikes the mean between the two opposing sources of danger may be taken to be 2 inches; but with very strong currents, with which there is a greater risk of projections forming upon the cathode, or with very impure copper, which yields unduly large volumes of spongy matter, this distance may with advantage be extended to 2½ inches.

The Solution.—The solution employed as electrolyte should contain from 1½ to 2 pounds crystallised copper sulphate, and from 4 to 10 ounces of sulphuric acid, in each gallon of water. The baths are made after the manner already described, and the fittings of each bath may be conveniently arranged like those in use for electro-plating or electrotyping, all plates in the same bath being placed in parallel arc, not in series, so that all the anode plates are in direct connection with the positive lead from the dynamos, and all the cathodes with the negative lead.

Arrangement of Baths.—The disposition of the baths, however, is a different, and indeed a critical, matter, because upon the method of dealing with this problem may depend the financial success of the installation. In order to make this point quite clear, the theoretical principles already laid down must be thor-

oughly understood, in so far as they refer to the effects of
increasing the number of plates in series or in parallel arc. It
will be remembered that each ampère of current in electrolysing
a simple solution in a single cell, invariably reduces a constant
quantity of the same metal in an equal period of time, without
regard to the size of the electrodes; and that an increase of plate-
area affords no gain so long as the total current-volume remains
unaltered, while an addition to the number of parallel anodes is
only equivalent to an enlarged electrode surface. Thus the
multiplication of parallel pairs of electrodes in the same bath with
the same strength of current, has only the effect of causing the
same weight of deposit to distribute itself over a larger surface,
and thus to produce a thinner film at any given point; the total
current is constant, but the current-intensity per square inch is
diminished, by reason of the enlarged superficial area. But
there is another side to the question—with the increase in the
size of the plates there is an increased area of solution traversed
by the current, and, therefore, in inverse proportion a reduction
in the resistance, so that a greater volume of current is allowed
to pass, current-volume being proportional to the electro-motive
force (which is constant in this case) divided by the resistance
(which is less).

But by increasing the number of pairs in series, the amount
of metal precipitated is increased in direct proportion, if the
current be constant in strength, because it deposits upon each
cathode in series a mass equal to that which would be thrown
down upon one alone. On the other hand, the resistance is
increased in equal proportion, because the current has to
traverse a greater length of solution in passing from plate to
plate, while it has only the same area of solution as before,
because the size of the electrodes is supposed to be unchanged.
This idea may be grasped more readily by comparing the cells
placed in parallel arc to a number of equal lengths of copper
wire placed side by side and bound together into a compound
rod or cable of proportionately greater diameter; and those
placed in series to the same lengths of wire united, end to end,
to form one continuous piece of only one thickness; it is evident
that the former will present far less resistance than the latter to
the flow of a given volume of current.

By a judicious combination of baths in parallel and in series,
the resistance and the amount of the deposit afforded by a given
current may be varied indefinitely. For example, given four
baths under precisely similar conditions; if all are placed
parallel, the total external- or bath-resistance will be one-fourth

of that caused by a single vat; this lower resistance will allow of a larger current, but the amount of metal deposited per unit of current will remain constant. If now they are arranged in series, the bath resistance will be four times as great as that of one cell, and a smaller volume of current will flow through them, but the quantity of metal deposited per ampère will be four times as great as in the first instance. Thus, in the former case, a greater amount of copper will be precipitated than would have been yielded by a single cell, because the current is increased; yet not by four times, because it is only the external resistance which is reduced, the internal resistance of the battery or dynamo remaining unaltered. In the latter case, the amount of copper separated is greater, although the actual current-strength is reduced, because the same current gives the unit weight of metal in each bath.

By arranging two vats parallel, and coupling them in series with the other two cells, which are also placed parallel, the resistance of each parallel pair is equal to half that of one bath, and that of the two groups of cells in series, being twice that of one group, is equal to the resistance of a single vat. Thus the resistance of (and, therefore, the total current flowing through) the four cells so arranged is the same as that of (or through) one only, but the weight of copper deposited is exactly double, because there are two baths in series. Similarly, by placing in series three groups, each of three parallel cells, the mass of copper refined will be three times as great. Thus apparently at first sight, a current of 1 ampère might be economically made to yield any given weight of copper per unit of time, for it might be argued that since 1 ampère of current deposits 18·26 grains of copper per hour, it could be made to precipitate 18·26 × 383 grains or 1 pound per hour, by coupling up a series of 383 groups of cells containing 383 parallel cells in each group. Theoretically this is true, but there are several circumstances which limit this apparently infinite extension of duty. In practice, the amount of profit obtainable would of course be entirely swallowed up in the immensely increased cost of plant and space, and the vast stock of copper anodes required for so gigantic an undertaking (383 groups of 383 vats in each, for example, means the use of an aggregate of 383 × 383 or 146,689 baths!); and in a strictly commercial process, such as that of copper-refining, the arrangements are necessarily primarily limited by financial considerations. It is of no avail to insist that the scientific efficiency is higher, if the cost of obtaining it be in excess of the small margin which is allowable for working

19

charges in these days. Thus many variable items have to be dealt with, which cannot well be discussed in this place; for example, the power available, the cost of labour and of land, the capital at command, the position of the works, and conveniences for effecting repairs as well as the charge for freightage of raw and finished material, and the like. All these must be taken into account, more or less, in determining the exact disposition of plant.

But there are other limiting conditions inherent in the process itself. First, in regard to the number of groups in series. Since each group requires a certain expenditure of electro-motive force, which we have taken at 0·3 volt, it is clear that the maximum number of vats, which it is convenient to place in series will be found by dividing the electro-motive force of the generator, expressed in volts, by 0·3. Thus, if the generator can give only 6 volts, the highest number of baths permissible in series is $\frac{6}{0·3}$ = 20. If a larger number were grouped in series, the voltage per group would be less than that recommended, and although metal might be deposited (though much more slowly) at a certain point the further increase of the number of groups placed in tandem fashion would entirely stop the current from flowing through them at all. By using a dynamo, the brush potential of which is higher than 6 volts, the number of baths in series may be proportionately increased. In the second place, in regard to the number of parallel cells in each group, if these be multiplied, the surface of plate is increased to a corresponding extent, and with only a given current available, the proportion of current-strength to superficial electrode-area will be reduced to a point, at which the character of the deposit would be greatly deteriorated, and at which the thickness deposited in a given time would be so small that the duration of the process would be abnormally extended. These two factors then—the current density per unit of electrode surface and the electro-motive force—form the scientific limits to the extension of plant in the directions shown; and within these limits the refiner must be guided by the considerations of time, space, and capital, under which he is called upon to work.

In illustration of the practical application of these laws, the following table has been compiled from accounts, which have been frequently republished, of the work carried on in certain large electro-refineries; the figures given are transcriptions of those which have been published, or are deduced by calculation from them. This tabular statement is self-explanatory and calls for no further comment.

TABLE XXVII.—SHOWING SYSTEMS OF COPPER-REFINING AT CERTAIN ELECTRO-METALLURGICAL WORKS.

No.	Name of Works	Locality of Works	Number of Dynamos	Type of Dynamo	Total Number of Baths	Number of Baths in Series	Number of Anodes per Bath	Number of Cathodes per Bath	Area of Electrodes per Bath (sq. ft.)	Thickness of Anodes (in.)	Distance between Electrodes (ins.)	Specific Gravity of Bath	Total Ampères used (calculated)	Total Copper Deposited per Hour (lbs.)	Copper Deposited per Bath per Hour (lbs.)	Ampères per Square Foot	Ampère per Square Inch	Thickness of Metal Deposited per Hour, in Inches	Horse-Power Absorbed
1	North-German Refinery	Hamburg	1	Gramme No.1	40	20	...	...	325	...	5	Hydrometer.	1227	64	1·6	1·88	0·0130	0·000105	16
2	" "	"	2	Gramme No.1	240	120	...	...	160	...	...	...	266	83·3	0·3	0·83	0·0057	0·000046	12
3	Œchsger Mes-dach	Biache	1	"	20	20	88	69	540	0·4	2½	10° Bé.	700	36·5	1·8	1·3	0·0090	0·000073	...
4	Hilarion Roux'	Marseilles	1	"	40	40	115	...	215	0·4	2	16°-18° Bé.	2·0	23	0·6	1·0	0·0069	0·000056	5
5	Oker	...	1	{Siemens Halske C1}	12	12	...	...	...	...	...	...	862·5	27	2·2	...	...	...	...
6	Selly Oak	Birmingham	1	Wilde	48	48	16	10	30	0·5	3½	16° Bé.	240	30	0·6	8·0	0·0555	0·000450	...

NOTE TO TABLE.—The total current-pressure in No. 2 is 27 volts, and resistance of each bath = 0·00084 ohm.

Precautions.—In working this process the usual attention must be paid to perfect cleanliness throughout. All connections and all cathode surfaces (unless the deposit is to be detached afterwards) must be clean initially ; the mud should be frequently removed, since it becomes slowly attacked by the acid liquor of the bath and thus gives rise to an alteration in the composition of the latter. But as the slime contains practically the whole of the precious metals originally present in the copper it must be preserved, so that they may be subsequently extracted, for this recovery of gold and silver may form no inconsiderable source of profit. By the gradual accumulation of impurities in the liquid, the percentage of copper is greatly reduced. This should be rectified by partially evaporating the solution, and collecting the crystals of iron sulphate which separate out, and which constitute the chief foreign matter introduced during the process. The residual liquid, still containing iron, however, but in reduced proportion, may be diluted sufficiently, and mixed with a fresh quantity of copper sulphate, and is then ready for further use. The copper may be deposited in thick or in thin sheets, but conveniently about $\frac{1}{16}$ to $\frac{1}{8}$ of an inch in thickness, which may require from two to four weeks to produce ; it should be almost absolutely pure, and may be sent into the market as electro-deposited plate, or it may be remelted and poured into ingots of the ordinary size and shape. It should be finely crystalline, with clear close grain, solid and free from pin-holes or blisters in the sheet form, and as such is the purest copper known.

The Electro-Extraction of Copper from Ores and Products.

Progress.—The first step in the direction of electro-extraction was, no doubt, the discovery alluded to in the first chapter of this treatise, that metallic iron dipped into liquors containing copper became covered with a deposit of the latter metal, a reaction known, but misunderstood, even in the fourth or fifth centuries of the Christian era, and applied in the fifteenth century to the extraction of copper from the cement water at Schmöllnitz in Hungary. This precipitation by a system of "simple immersion" is still very largely used in the treatment of many thousands of tons of copper ore annually, and is especially applicable to the recovery of the small proportion of this metal (2 to 4 per cent.) present in the burnt Spanish pyrites, which has been used and discarded by the vitriol maker.

The next step was that taken in 1846 by Dechaud and Gaultier, who mixed crude copper oxide (either the natural ore, or a roasted sulphide) with sulphate of iron, and strongly heated the two in an air-current until the copper had become converted into sulphate at the expense of the iron salt, which thus became changed to peroxide. After cooling, the burnt charge, on treating with water or *leaching*, gave up its copper sulphate into the solution, which was then ready for precipitation with iron. Instead, however, of merely inserting scrap-iron, which deposited the copper in a spongy condition, and left it commingled with all other metals which iron was capable of precipitating by simple immersion, and with all the carbon and other impurities contained in the iron itself, plates of the last-named metal were placed in the solution, parallel with copper plates, and joined to them externally by copper wires, but separated from them in the liquid by cotton cloth, to prevent the contamination of the copper by impurities from the precipitating element. Thus the iron dissolved, and, producing a current of electricity, caused the deposition of the copper to take place upon the copper plates (or upon the lead plates which were sometimes substituted for the latter). To compensate for the gradual exhaustion of the liquors in regard to copper, a continuous gentle flow of saturated extract from the roasted product was introduced into the bottom of the bath, while simultaneously the decopperised liquid, rich in ferrous sulphate, was removed from above.

So also in 1867, Patera treated the Schmöllnitz cement liquors by immersing in the copper fluid porous clay cells, containing fragments of iron in a strong solution of sodium chloride, and connecting this metal with pieces of coke immersed in the cement water; here, too, a "single-cell" arrangement resulted, and the solution of the iron in the porous vessels caused the deposition of the copper upon the coke fragments without. Later still, the copper liquors obtained from an ore by any treatment such as that above alluded to, or by attacking oxidised copper products with sulphuric acid, have been treated with the aid of dynamo-electric machinery.

Yet another method of procedure has been employed. A regulus of copper (that is, an impure cuprous sulphide) is obtained in the usual way by smelting sulphide ores in the furnace; and this regulus is cast into slabs, which are then used at once as anodes, in a solution of copper sulphate, a powerful dynamo-current being employed to effect the electrolysis. The compound becomes broken up, the copper of the copper sulphide is oxidised and dissolved, and the sulphur, which is separated and remains practically unaltered even in presence of the nascent oxygen at the anode, together with the precious metals, and all the other bodies that were mentioned on p. 285 as being left in the slime during refining, remain undissolved at the anode, which is usually enclosed in a bag of porous material, so that the pulverulent deposit may not become mixed with the solution and the deposited copper.

The Processes Available.—Thus the general methods may be classified as those in which—1. The ore is treated by ordinary furnace processes with the object of rendering certain constituents (chiefly copper) soluble, and these are extracted electrolytically from their solution in water by processes of simple immersion, single cell, or separate current with insoluble anode. 2. The ore is simply melted, or is partly refined, and then run into slabs, which are used as anodes in a bath of a suitable copper salt with a separate current for electrolysis.

By the first method any copper ore may be successfully treated, provided that due attention be paid to metallurgical details in the roasting, into which it would be out of place to enter here. The second is applicable to native copper ores, such as those of the Lake Superior district, which are already in the metallic state, and, therefore, need refining only, and to sulphide ores, or to those in which the copper may be converted metallurgically into regulus or sulphide.

Becquerel in 1836 made a series of experiments on the treatment of complex ores containing copper, lead, silver, and iron, by converting them into chlorides or sulphates, and subsequently electrolysing these solutions; but although they were successful enough in their result, the high cost of producing electrical energy at that time proved an insuperable bar to their commercial adaptation.

In 1871 Elkington dealt with copper mattes and solutions with the aid of dynamo- or magneto-electric machinery, and thus placed the process at once upon a practical footing. Cobley in 1880 electrolysed copper solutions obtained by lixiviating roasted copper ores with water or sulphuric acid, obtaining, if possible, a solution containing 18 per cent. of copper salt and 1 per cent. of free acid. A year later Deligny described experiments in which he had successfully used sulphide ores of copper packed around carbon rods as anodes in a weak solution of copper sulphate, with a copper cathode, and with two Bunsen-cells as the operating force. In carrying out this process he took advantage of the comparatively high electrical conductivity of many natural sulphides, such as those of copper, iron, lead, silver, &c., conductivity being quite essential to the success of the process in order that the current may be transmitted through the anode at all. It may be noted that most oxide ores and oxidised compounds, and the sulphides of zinc and antimony, are bad conductors, and are, therefore, unsuited to such treatment.

In 1882, Blas and Miest elaborated a process intended to deal with many complex ores, but especially with those containing copper. Firstly, in order to ensure better conductivity than may be obtained with a loose powder, the crushed mineral is moulded into cakes at a pressure of about 40 atmospheres (600 pounds per square inch), and is heated to a temperature of from 950° to 1100° F., bad conductors being formed into slabs of greater thickness than those which exhibit a superior conductivity, in order that a greater area may be allowed for the passage of the current. The plates are then used as conductors in a suitable solution—copper sulphate for copper ores, lead nitrate for lead ores, and so forth—when the metals soluble in the electrolyte are gradually removed under the action of the current, and certain of them are deposited upon a sheet metal cathode. The whole of the sulphur remains with the gangue or earthy matter, and other insoluble constituents of the anode; and these, after a time, being weakened by the honeycombing process of solution, tend to fall to the bottom of the bath; it is well, therefore, to remove and re-agglomerate them from time to time before this can take place. The cathode metals are finally electro-refined, while the spent anodes, containing the sulphur in a recoverable form, are treated accordingly. In this process, and in that of Marchese proposed a year later, the number of soluble constituents, and principally the amount of iron, which are constantly passing into the solution, and thus altering its composition, militate greatly against the economical success of the treatment. It is, indeed, very questionable whether the use of raw ores as anodes can ever really compete with a mixed process of preparing a less impure compound—for example, blister copper, by metallurgical, or furnace methods, and then refining this product by electrolytic means. In either case, the behaviour of the various metals which are likely to be present will now be understood, so that an analysis of the ore or of the crude compound to be electrolysed should give a sufficient indication of the probable result of applying this method of treatment. The deposited metal may or may not be sufficiently pure to use as electrolytic copper; if not, it must be refined as previously explained.

THE ELECTRO-REFINING OF LEAD.

Base Bullion.—The principal lead product which offers itself for electrolytic treatment is that known as *base bullion*, in which the lead usually exists to the amount of 90 per cent. or more, and is accompanied by varying proportions of silver and gold with antimony, arsenic, zinc, copper, iron, tin, &c., the main object being the recovery of the precious metals, together with the production of a good soft lead, free from antimony and arsenic. This problem was taken in hand by Keith; difficulties were encountered at the outset in the selection of an electrolyte which, while dissolving the lead, should not at the same time dissolve silver, as lead nitrate tends to do, nor encourage the formation of crystalline growths of spongy character resembling the "lead tree," which may form a metallic bridge between the electrodes, as lead acetate would do; the simple sulphuric acid solution is useless, because lead sulphate is insoluble in it, and there is also a tendency to form lead peroxide upon the anode, which is still less soluble than the sulphate in this menstruum. A solution of lead sulphate in sodium acetate is found, however, to satisfy the requirements of the process.

The bullion is fused and run into thin plates (about $\frac{1}{8}$ of an inch thick), which are enclosed in muslin bags and immersed as anodes in a solution of $1\frac{1}{2}$ pounds of sodium acetate and $2\frac{1}{2}$ to 3 ounces of lead sulphate in every gallon of water. The solution must not be allowed to become alkaline or the anode will be covered with a film of lead peroxide, which will protect it from further action; it should be gently agitated to ensure thorough admixture and guard against polarisation due to unequal distribution of lead, and it may with advantage be heated to about 100° F. with the same object in view.

By the application of 0·016 ampère per square inch (= 0·25 ampère per square decimetre, or 2·3 ampères per square foot), the greater proportion of the lead will be gradually transferred to the cathode plate, where, however, it may be mixed with mere traces of silver, copper, tin, antimony, and zinc, and a somewhat larger proportion of the total bismuth present in the bullion. The percentage of lead, however, in the refined metal should exceed 99·99 per cent., indicating a total impurity of less than 0·01 per cent. The remaining substances are usually found in the anode bag, often in the form of a spongy mass retaining the shape of the original anode. After drying this sponge, it is fused with nitre and poured into a small ingot-mould, by which means the silver and gold are obtained in the form of a button of alloy, the other constituents having passed into the slag, from which they may be more or less readily recovered. The lead frequently crumbles into the solution from the cathode plate, but is readily collected and fused; the use of a containing bag for the anode prevents any of the slime becoming mixed with the detached lead in the bath.

A modification of the process has been suggested by Keith, by which the bullion is first melted with 2 per cent. of zinc, as in Parke's process for desilverising the metal, and then electrolysing only the zinc skimmings, which may contain about 20 per cent. of the lead originally used, with practically the whole of the precious metal; thus, the period of time necessary for electrolysis is considerably lessened.

The Electro-Extraction of Zinc.

Methods Available.—The principal ores of zinc are blende (sulphide) and calamine (carbonate). The methods proposed for dealing with them either require a preliminary roasting in the case of blende, to obtain the zinc in a soluble form, or they use the ore itself as anode.

Luckow adopted the latter method, and enclosed the ore, mixed with coke, in open chests, which were used as anodes in an electrolyte of a somewhat strong and acid solution of common salt or of zinc chloride (20 to 30 per cent. of zinc), with a similar case filled with coke, or else a zinc plate, as cathode. These were held in large rectangular troughs of wood or stoneware, while beneath the cathode case was a wooden vessel covered with webbing or basket-work, to retain the zinc which becomes detached during the deposition. The process must be well watched, so that no short-circuiting may result from the formation of arborescent growths upon the cathode, that the scum forming upon the surface may be frequently removed, and that the bath do not become acid, especially if the sulphate be employed instead of the chloride. A current of somewhat high electro-motive force is necessary.

Lambotte and Doucet chose the first alternative, and formed a neutral solution of zinc chloride by treating the roasted ore with crude hydrochloric acid, and freeing it from iron by the addition of chloride of lime and zinc oxide, which effected the precipitation of the latter as insoluble peroxide; were the iron, which is universally present in zinc ores, allowed to remain in the solution, it would be deposited with the zinc—a result to be carefully avoided. The electrolysis is conducted between insoluble (carbon) anodes and zinc cathodes.

Létrange has, however, devised probably the most practicable method; it is, of course, used to best advantage where power is cheap and fuel is not over expensive. Sulphides (blendes) are first roasted carefully at a low temperature—a very low red heat—so that as much of the sulphide as possible shall form soluble zinc sulphate, for this, at a high temperature, would be completely decomposed into insoluble zinc oxide, oxygen, and sulphur dioxide, the last two passing away in the gaseous form. This sulphate is extracted with water, and being neutral, forms the electrolytic solution. The residual zinc oxide, left after lixiviation, he proposes to place in a moist condition with other oxide, in towers through which pass the waste gases from the roasting furnace; thus, any sulphur dioxide formed during the calcination of a fresh batch of ore is absorbed by the zinc oxide, which it converts into zinc sulphite, and this salt, itself soluble, and, therefore, capable of electrolysis, is gradually converted into sulphate by absorption of oxygen from the air. These reactions are expressed in the following equations :—

1. Desired result of roasting the ore, ... $ZnS + O_4 = ZnSO_4.$

2. Effect of overheating zinc sulphate, ... $ZnSO_4 = ZnO + SO_2 + O.$

3. Effect of roasting ore at high temperature, . . . $ZnS + O_3 = ZnO + SO_2.$

4. Absorption of sulphur dioxide in towers, . . . $ZnO + SO_2 = ZnSO_3.$

5. Conversion of zinc sulphite into sulphate, . . . $ZnSO_3 + O = ZnSO_4.$

Thus, the zinc sulphide theoretically yields (with the oxygen in the calcining furnace) the acid which is necessary to dissolve the zinc oxide, even if the latter be not left in the soluble form as sulphate. This is an advantage, in that it allows the use of a higher roasting temperature, and hence increases the chance of converting iron into the insoluble condition of peroxide. The presence of galena in the blende is not injurious because lead remains insoluble in sulphuric acid or in a sulphate solution. The zinc solution is electrolysed between lead anodes, which are insoluble, and zinc cathodes. The solution is thus gradually deprived of zinc, and also becomes richer in acid, because, as the zinc sulphate is decomposed, the metal is deposited and the acid left free in the solution. But as zinc is not precipitable electrolytically from acid solutions, a point is soon reached at which action ceases; the solution partially spent, but still carrying much zinc, is, therefore, caused to flow over fresh zinc oxide (roasted blende or calamine), so that it may take up a further quantity of metal which, at the same time, neutralises the free acid; it is then ready to be passed once more through the electrolytic tanks. This cycle of operations is constantly maintained, so that the same acid is used again and again, until it has accumulated so large a proportion of foreign matter that the solutions are useless. The current to be employed must have a high electro-motive force, because, when the lead becomes peroxidised at the surface, which occurs almost immediately, the opposing electro-motive force set up between this surface covering and the zinc cathode is very considerable.

THE ELECTRO-REDUCTION OF ANTIMONY.

Borchers in 1887 appears to have obtained favourable results from the electrolysis of a solution of antimony sulphide in sodium sulphide. The sodium compound may contain polysulphides, but should be added in such a proportion that there is only one equivalent of sulphur present in any form, to each equivalent of sodium, as it is found that any increase in the amount of sodium produces a less conductive fluid, while on the other hand a reduction is attended by the precipitation of sulphur. The ore, or product containing antimony tersulphide, is treated with a solution of sodium sulphide in water, until the liquid has a density equal to 12° Baumé; it is then electrolysed in iron tanks, the walls of which serve as cathodes to receive the deposited metal; leaden anodes being insoluble are to be preferred. A current-pressure of 2 to $2\frac{1}{2}$ volts per vat is recommended.

The antimony is deposited in the pulverulent condition, or in the form of shining scales; any portion clinging to the iron tank-walls is readily removed by steel brushes, and the whole is washed free from admixed solution, first with water containing a small quantity of sodium sulphide, and ammonia or caustic soda, then several times with water, next with very dilute hydrochloric acid, and, finally, with water again, after which it is dried and fused together under glass of antimony.

This process, like most others of the kind, has to compete with a fairly economical extraction method by the dry way; and as a fusion is after all necessary to unite together the powdered deposit, it is a question whether it could ever compete with the older processes, unless a very cheap supply of electrical energy were available.

ELECTRO-SMELTING.

The electrolysis of fused metallic compounds instead of aqueous solutions has frequently been suggested, and many patent-specifications have been filed with this object in view. But the difficulties are considerable, and the advantages, generally speaking, not very clear, inasmuch as it is necessary to expend fuel to bring the charge to a state of fusion, so that one of the chief recommendations of electro-extraction methods is absent, namely, that there is no charge under the head of heating. Thus, even if adequate water-power were available for driving the dynamos, it is rarely that the process could be economically applied, when the relative costs of labour and supervision, and the interest on plant are taken into account.

There are a few metals which not only refuse to be deposited from aqueous solutions, but involve a somewhat costly extractive process, even by the old furnace methods. To these it would seem that electro-smelting might be advantageously applied, unless and until a sufficiently economical dry process is evolved. The two chief metals in this class are aluminium and magnesium, both of which have vast fields for their application, and for one at least—aluminium—there would probably be a very great demand, on account of its unusual and valuable properties, could it only be thrown upon the market at a cheap rate.

Magnesium Smelting.—But there are enormous difficulties in the way of dealing with these metals electrolytically on a large scale. The original proposition was to heat the chloride of the metal to its fusing point by placing it in a crucible in a suitably arranged fire; and then to pass a powerful current through the melted mass. In this manner Bunsen obtained magnesium by urging a current from 10 cells of his battery between gas-carbon electrodes, passed through the clay cover of a porcelain crucible, in which at the upper part they were separated by a porcelain partition, while at the lower they both dipped into the fused magnesium chloride. The anode carbon was plain, but the cathode was made with inverted steps upon the surface, so that it would retain the globules of melted (metallic) magnesium, which being specifically lighter than the liquid would otherwise float to the surface and become re-oxidised. Matthiessen, by melting three equivalents of potassium chloride and a little ammonium chloride with the magnesium salt, obtained a bath which was of lower specific gravity than the metal, so that the special arrangement of the cathode was rendered unnecessary.

Aluminium Smelting.—Aluminium was similarly produced but, having a high fusing point as compared with its chloride, required much care in production. The aluminium chloride is liquid, and volatilises at about 360° F., so that it is readily boiled away and lost. Advantage is taken of the higher fusing point of the double chloride of aluminium and sodium, but even with this it is safer to produce the metal in the state of powder (that is, at a low temperature—below its fusing point) and afterwards to melt it together under a flux such as cryolite, but this, nevertheless, is a troublesome operation.

It is clear that by such methods only small quantities could be dealt with, and the extraction was rather a laboratory experiment than a commercial process. Yet several inventors, taking this principle as a basis, have endeavoured to elaborate a method for producing the metal on a working scale; and many patents have been applied for to protect special means of effecting the removal of the chlorine evolved at the anode as fast as it is evolved, and so of preventing it from exerting a re-solvent action upon the deposited metal; as well as for filling the space above the liquid in the crucible with a reducing or neutral gas, to prevent oxidation of the aluminium, and so forth.

Siemens' Electric Furnace.—Another process, however, which is probably in no degree dependent upon electrolysis, but purely upon the reducing action of carbon at excessively high temperatures, has lately been introduced by the two brothers E. H. and A. H. Cowles. Before dealing with this, however, reference should be made to the Siemens' electrical furnace, with which a number of experiments were made (and recorded in 1882 in a paper before the British Association) by the inventor, conjointly with Professor Huntington, in the Metallurgical Department of King's College, London. This furnace of Sir William Siemens was the first embodiment of the principle subsequently adopted in modified form by Cowles.

The furnace depends for its power upon the intense

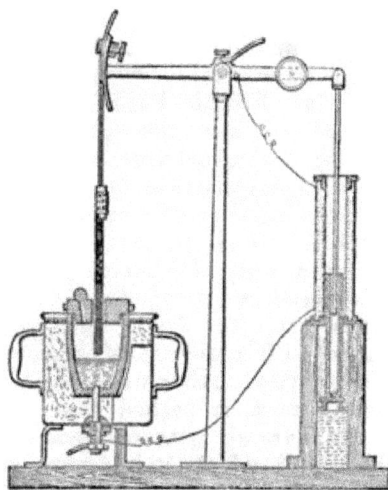

Fig. 98.—Siemens' electrical furnace.

heat of the electric arc. A crucible is bored at the bottom to receive a tightly-fitting stout carbon rod (fig. 98), such as is used in large electric arc-lighting installations; this forms the positive pole and is connected with a wire which passes from the dynamo as generator. The negative pole consists of a similar carbon rod, connected with the negative brush of the dynamo, and is suspended. from above, centrally within the crucible; an arrangement of a solenoid upon the same base plate controls the current, so that immediately the upper of the two carbon rods is sufficiently lowered to make contact with the other, the current flows, and, passing around the solenoid, automatically separates the poles and maintains them at a practically equal distance from one another during the whole time of fusion, in spite of the gradual wearing away of the carbon. Thus at the moment when the rods break contact, an intensely powerful electric arc is formed between them, and this is maintained evenly within the crucible so long as the current is continued; any substance packed around the lower carbon is in this manner rapidly heated to the highest temperature attainable. Comparatively large quantities of steel and of platinum (several pounds) were rapidly melted; and even tungsten, one of the most refractory metals known, could be fused in the arc, but only in small quantities and with the greatest difficulty, by placing it in a hollow scooped out of the lower carbon itself instead of in a crucible. Setting aside all questions of prime cost of plant and current working expenses, this furnace no doubt renders available the highest temperature that is to be obtained with the means now at command.

Cowles' Electric Furnace.—In the Cowles furnace, the two carbons are passed through opposite ends of a fire-brick chamber lined with charcoal dust, which is selected as a lining on account of its high resistance to fusion; but as the charcoal, itself but a poor conductor of electricity, rapidly becomes converted into graphite, which is vastly superior in this respect, and thus causes a leak of electricity unutilised between the poles, it was found necessary to soak the charcoal in milk of lime before use. On drying this product each grain of charcoal is coated with a deposit of lime, which effectually destroys the conductivity. All around the carbons, and within the space thus formed, is packed the mixture of the ore or compound to be reduced, with fragments of carbon, and with pieces of the metal which it is desired to alloy with the aluminium; the whole is then covered with fine charcoal, and finally with a luted iron lid, lined with fire-brick, and provided with an aperture for the escape of gases. The carbon rods are attached to stout copper cables leading to the poles of the dynamo, but placed in circuit with a current-measurer (a kind of ammeter for enormous currents), a switch by which the current may be broken, an automatic cut-out to prevent the passage of an over-powerful current, and a set of resistances which may be interposed if necessary at any time. Placing first the resistances in circuit the current is turned on, a short space only existing between the ends of the carbon electrodes, which are then withdrawn by degrees through the furnace walls as the operation proceeds, until at last the whole interior of the furnace is filled with a glowing mass.

The Cowles' Process.—In a paper read by Dagger before the British Association at Newcastle-on-Tyne, in 1889, a description is given of the most recent development of this process as it has been installed at Milton, in Staffordshire. The furnace, which is shown in vertical cross-section, as well as in part elevation, part longitudinal vertical section, in fig. 99, is constructed of fire-brick, as described, with a cast-iron cover, and measures internally 5 feet, by 3 feet, by 1 foot 8 inches; through each end-wall is built a cast-

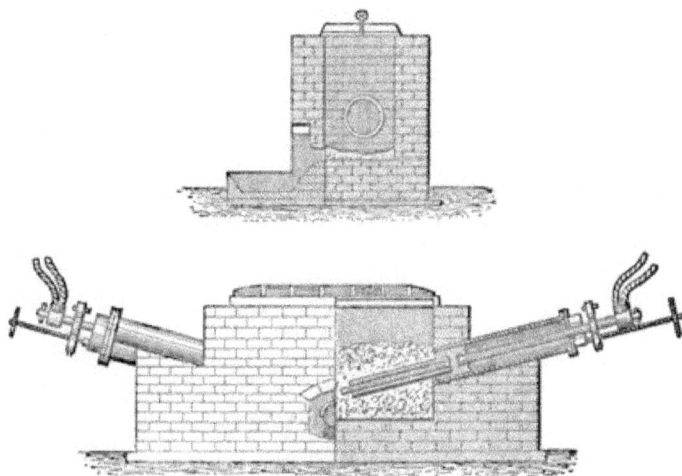

Fig. 99.—Cowles' electrical furnace.

iron tube of sufficient diameter to pass the bundles of electrodes. The latter consist of nine carbon rods, each $2\frac{1}{4}$ inches in diameter (or of five 3-inch rods), held together by a cast-copper or cast-iron head (the former if a copper alloy is being made, the latter for iron alloys), attached to a rod, actuated by gearing suitable for the withdrawal of the carbons from the furnace, and connected with one of the poles of the dynamo. The walls having been lined with finely-crushed oak charcoal, insulated by the lime process, the charge of aluminium compound, charcoal and alloying metal in the form of small turnings or granules, is introduced and covered with fine charcoal, broken fragments of carbon having been previously placed in position, in order to start the arc as soon as the current is switched on. The source of electrical energy is a large Crompton dynamo capable of furnishing a current of from 5,000 to 6,000 ampères at a pressure of from 50 to 60 volts. The current-volume at first is about 3,000 ampères, but is gradually increased to about 5,000 ampères in the space of half-an-hour, the whole process lasting about $1\frac{1}{2}$ hours. Gases, due to the reduction of the aluminium compound, and therefore consisting mainly of carbonic oxide, as explained by the annexed equation, pour from the aperture in the cast-iron cover, and burn with a long white flame.

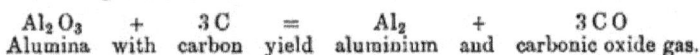

$$Al_2 O_3 \quad + \quad 3\,C \quad = \quad Al_2 \quad + \quad 3\,C\,O$$
Alumina with carbon yield aluminium and carbonic oxide gas.

When the operation is complete, the current is broken, and switched on to an adjoining furnace, while the contents of the first are either allowed to settle upon the hearth, or are tapped out into a receptacle in front, through a tap-hole, which is plugged during the passage of the current. The crude metal is then refined by metallurgical processes. The expenditure of power to produce a given weight of metal depends upon the nature of the alloy which is being produced, but is said to average eighteen horse-power, per pound of copper, per hour.

Whether the process will ever produce pure aluminium at a low cost is yet to be discovered; it is at present devoted to the production of aluminium bronze alloys, consisting of copper containing various proportions of aluminium, or ferro-aluminium, which is an alloy of the latter metal with iron. The presence of the alloying metal no doubt favours the reduction of the aluminium; just as in the case of heating together iron or copper with carbon and silica, even 'at temperatures obtainable with furnaces fed by solid or gaseous fuel, the silicon is reduced and combines with the assisting metal, although the heating of silica and carbon alone is barren of result; but the Cowles furnace is said to reduce alumina to the metallic state, even in the absence of alloying elements; such metal is not however quite pure, but is combined with carbon in the form of carbide. Although it is much to be desired that pure aluminium should be produced at a cheap rate, there is yet an enormous field for processes conducted upon the Cowles principle, even if alloys only could be produced satisfactorily, inasmuch as one of the principal uses of the metal would be found in its application to the production of valuable alloys, and to this end the initial production of an alloy of the right character will suffice, provided that its exact composition be known.

The use of this principle is capable of extension to many other metals and ores; and the economical conversion of heat into electricity, without incurring the great loss of energy involved in the intermediary use of the steam engine, would, doubtless, have the same effect in developing the electro-smelting industry that the discovery of the dynamo-electric machine had in the evolution of other branches of electro-metallurgy.

CHAPTER XVI.

THE RECOVERY OF CERTAIN METALS FROM THEIR SOLUTIONS OR FROM WASTE SUBSTANCES.

WHEN a solution has become spent, and is too highly charged with impurities to be any longer serviceable for electrolytic work, it is desirable to so treat the liquid that the valuable metal is recovered in a form suitable for re-application to the plating work. Such methods as are applicable to the more usual solutions are briefly sketched in this chapter; but it must be understood that they are merely general methods, and modifications of baths by the addition of fresh substances may render recovery by these means incomplete or perhaps impossible; and in many cases the metal may not be entirely regained, even under ordinary circumstances, the amount that is left depending upon the solubility of the precipitated compound in the solution from which it is to be separated.

COBALT.

The solution should be boiled down until fairly concentrated; if acid (reddens blue litmus-paper), sodium carbonate should be added until it is neutral, or nearly so, but it must still remain clear. The addition now of finely-divided barium carbonate, stirred up in water, will cause the separation of all the iron as peroxide; this is allowed to subside, and the clear liquid containing nearly the whole of the cobalt is poured off. This liquid must be neutral or slightly acidified with acetic acid, and to it must be added an excess of strong solution of potassium nitrite mixed with sufficient acetic acid to prevent it from turning red litmus-paper blue; the whole should be left in a warm place for one or two days. The precipitated double nitrite of cobalt and potassium is then separated from the supernatant liquid, and dissolved in hydrochloric acid; the addition of caustic alkali (potash or soda, not ammonia) to this solution brings down the cobalt practically pure as hydrated oxide, which may be filtered off and dissolved in any acid, the cobalt salt of which it is desired to use. From a simple solution of cobalt, free from iron and other metals, the pure cobalt may be at once precipitated as hydrated oxide by caustic alkali; but the solution must be boiled, if necessary, until all odour of ammonia has vanished, as the presence of this body interferes with complete precipitation.

COPPER.

From acid solutions, fragments of metallic iron suspended in the solution will precipitate all the copper in the metallic state, by simple exchange. Subsequently the copper may be dissolved in warm dilute sulphuric acid containing a small proportion of nitric acid, more of the latter being added by degrees as the action is observed to become less; and in this way it will become converted into copper sulphate, which may be afterwards separated in the form of pure crystals by slow evaporation. The precipitation by iron may be effected by single-cell deposition, as explained in an earlier chapter, by connecting the fragments of iron contained in sulphuric acid within a porous cell, with copper strips placed in the copper solution. This will cause the precipitated metal to deposit upon the copper instead of upon the iron, so that there is no risk of introducing impurities into the spongy copper.

GOLD.

From the cyanide solution the gold may be recovered as follows:—Hydrochloric acid is first added in excess, until no further precipitate of gold cyanide is produced; the precipitate is allowed to subside, and is washed twice by pouring water upon it, allowing it to settle, and then pouring away and renewing the water again; then, after filtering, it is dried and fused with an excess of dry sodium carbonate in a clay crucible. But this process is not to be recommended on account of the intensely poisonous nature of the hydrocyanic acid gas, so abundantly evolved during the first treatment with hydrochloric acid; but if it be adopted, this portion of the process must be conducted in a special draught-cupboard, or in the open air with the operator well to the windward of the vessel.

Böttger's method is preferable; he evaporates the whole solution to dryness in a porcelain or enamelled dish, placed over a saucepan containing boiling water. The residue is then to be crushed in a mortar and mixed with an equal weight of lead oxide (litharge) and a thirtieth part of its weight of charcoal powder introduced into a fire-clay crucible and heated to bright

redness in a small pot-furnace; a portion of the lead is reduced
to the metallic state and with it will be alloyed all the gold
contained in the charge. This alloy is then boiled with nitric
acid, in which the lead will dissolve together with any silver or
copper that may be present; the gold is left in a finely divided
condition, and after washing may be fused into a single homogeneous mass, or it may be re-dissolved at once to form a fresh
gold-bath.

A plate of zinc attached by wire to one of gold, and placed in
the original gold-bath, gradually deposits the precious metal
upon the gold strip, forming, in fact, a single-cell arrangement,
which may be used for recovering the gold directly from the
spent solution ; but the operation occupies a long time, and other
metals are liable to be precipitated subsequently. These latter,
however, may be partly separated by boiling with nitric acid
in which gold is insoluble (the separation can never be absolute,
because portions of them are locked up within the gold so as to
be out of the reach of the acid solvent).

LEAD.

From the solution of the oxide in potash, insoluble lead
carbonate is produced by bubbling carbonic acid gas through it ;
this gas is evolved by the action of hydrochloric (muriatic) acid
upon chalk or marble contained in a separate vessel, which must
be closed with a cork through which a tube is passed to conduct
the gas to the required spot. The addition of acetic acid to the
liquid until it is nearly neutral, followed by the introduction of
sodium bicarbonate, will produce the same effect. From the
lead acetate solution, sodium carbonate precipitates lead carbonate. This latter substance, however produced, may, after
washing, be dissolved in acetic acid, or in any other solvent
suited to the particular form of bath which is to be prepared.

MERCURY.

This metal is used largely in amalgamating the zinc plates of
the galvanic battery, from the residues of which it may subsequently be recovered by distillation. To obtain it pure, it
should be distilled *in vacuo*, or at least under reduced pressure;
but to recover it fairly clean, in a condition suitable for application once more to amalgamation, the fragments of zinc may be
introduced into an iron retort, A (fig. 100), with a tube passing
air-tight through one neck of a Woulffe's bottle, B, into water;

the other neck of the bottle is connected with a second tube, also dipping beneath water in a second vessel, *d*, as a safeguard against the escape of mercury.   On applying heat to the retort by a suitable fire the mercury slowly vaporises, and distilling over, should be entirely condensed in B, any escaping this being

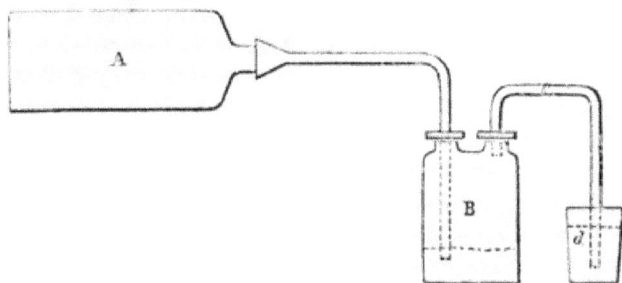

Fig. 100.—Mercury distillation apparatus.

collected in *d*. The residue in the retort still contains a little mercury, together with the bulk of the lead and copper originally present in the whole zinc plate, for being more electro-negative than the zinc, the latter dissolves in preference, and leaves the other metals to accumulate by the amalgamating effect of the mercury.

NICKEL.

From solutions of the double sulphate of nickel and ammonia the green double salt itself is precipitated in a practically pure condition in the form of minute crystals by the addition of successive quantities of a saturated solution of ammonium sulphate, until on standing, and after complete subsidence of the precipitate, the solution is quite decolorised.   Advantage is thus taken of the insolubility of the double salt in strong solutions of ammonium sulphate, a fact which was first observed by Unwin. The crystals have only to be filtered, drained free from the liquid clinging to them, and washed once or twice with a strong solution of ammonium sulphate, and they are ready to form a new bath.   The original liquid may with advantage be boiled down to half its bulk, or even less, before applying this treatment.

PLATINUM.

By adding to the bath an excess of a ferrous sulphate solution, and then a caustic alkali, the resulting dark-green coloured pre-

cipitate of hydrated ferrous oxide (Fe O) will reduce the platinum to the metallic state, and itself become converted to a proportionate extent into ferric oxide ($Fe_2 O_3$). This precipitation
should be effected in a covered or corked flask, which should then
be allowed to stand for some hours in a warm place, with occasional agitation. On the addition subsequently of an excess of
hydrochloric acid, the iron oxides will re-dissolve, but the
platinum powder precipitated with them will remain untouched.
This finely-divided platinum may then be filtered, washed, and
re-dissolved in aqua regia for further use.

SILVER.

As in the case of gold solutions, the addition of hydrochloric
acid precipitates silver cyanide; but the use of an excess of the
acid effects its further conversion into silver chloride. If this
precipitate be red in colour the presence of copper cyanide is indicated, which may, however, be effectually removed by boiling
it with moderately concentrated hydrochloric acid. The washed
silver chloride should then be dried, mixed with soda ash or
dried sodium carbonate, and fused in a clay crucible. The
acidifying process has the same objections on the score of danger
to health, and, therefore, requires the same precautions as the
similar method above described for the treatment of gold solutions.

A second method, allied to the dry process for recovering gold,
is also applicable to silver residues. The solution is evaporated
to dryness, and the residue is transferred to a crucible, but alone,
not mixed with litharge, and fused at a bright red heat. If
sufficiently heated, the silver compound, mixed with excess of
potassium cyanide and carbonate, will be broken up, and the
liquid silver will melt and sink through the fluid slag to the
bottom of the pot. When effervescence has ceased, the crucible
is removed from the fire, and its contents poured into a hemi-
spherical cast-iron ingot-mould, from which the solidified mass is
detached when cold, and broken by a light blow with a hammer,
to separate the metal from the slag. The crucible may be used
again and again, provided that an examination after each operation reveals no sign of a crack. The fluid contents of the pot
may, of course, be allowed to cool *in situ*, by setting the crucible
aside on a level surface; but to extract the silver button in this
case involves the loss of the pot by fracture. If the heat is
insufficient, the silver will remain as a grey spongy mass, from
which the slag may be removed by boiling with water containing
a small proportion of hydrochloric acid.

The silver should be practically "fine;" but to ensure the perfect purity of the product, it may be re-fused with carbonate of soda mixed with about one-tenth of its volume of nitre, which will attack the base metals (but not the silver), and cause them to pass into the slag; it will not affect gold or platinum, which are scarcely likely, however, to be present. If small quantities of these metals should by any chance become mixed with the silver, the button must be boiled with pure nitric acid (free from hydrochloric acid), which dissolves silver (and even platinum when present in small proportion with much silver), but leaves the gold as a dark-brown or black powder. But if more than 40 per cent. of gold be present, a certain proportion of silver (increasing with the percentage of gold) will be left with the latter. The silver solution is then filtered from the residue of gold, and is mixed with an excess of hydrochloric acid. In this way the insoluble silver chloride is formed, which must be filtered, washed, dried, mixed with an equal bulk of sodium carbonate and a little charcoal, and fused at a red heat. Pure silver will thus result, the gold having been left undissolved, while the platinum which had passed into the solution would not be precipitated by the hydrochloric acid, and, therefore, remains to be extracted from the liquid filtered from the silver chloride precipitate.

CHAPTER XVII.

THE DETERMINATION OF THE PROPORTION OF METAL IN CERTAIN DEPOSITING SOLUTIONS.

In this chapter it is proposed to give outlines of methods by which the depositing solutions of the more important metals in electro-metallurgical use may be assayed, to give at least an approximate idea of the quantities contained. For a full explanation of this subject, or for methods of dealing with complicated analytical problems, works on quantitative analysis must be consulted, as it is impossible and undesirable to treat of so wide a subject in detail within the limits of this book.

For this purpose, a fairly delicate balance is essential. It should be enclosed within a glass case, as in fig. 101, to protect

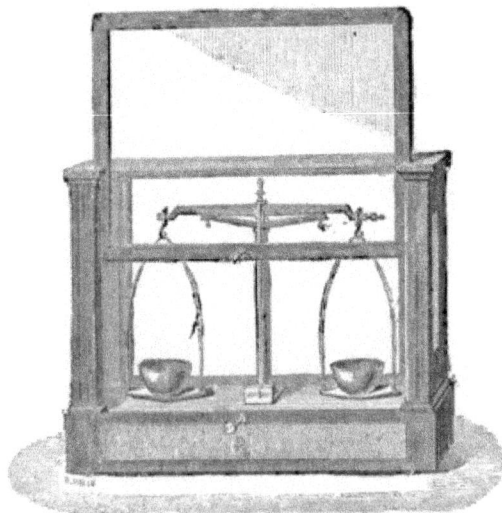

Fig. 101.—Balance.

it from dust and corrosive fumes, and should be capable of indicating a weight of $\frac{1}{100}$ of a grain. For electrolytic analysis, a platinum dish, about three-quarters of an inch to an inch in height, and about 3 inches in diameter, forms a convenient cathode, at once holding the solution and receiving the deposited metal. The anode consists of a circular plate of stout platinum foil about $2\frac{1}{2}$ inches in diameter, with several perforations to allow gas to escape from beneath it. The platinum sheet is fastened horizontally without solder to the end of a vertical platinum wire attached to the positive pole of the battery, the platinum dish making contact externally with a copper wire attached to the negative pole. Instead of this, a cylinder of platinum foil may be used as cathode, being suspended, with its main axis vertical, within a small beaker, the anode consisting of a coil of platinum wire placed within the cathode. The object of the electrolytic method is to continue the action of the current until every trace of the required metal is precipitated on the platinum cathode ; and, as the latter should have been weighed previously, the increase of weight shown after the deposition gives the number of grains of the metal in the quantity of solution taken. The platinum dish should be lightly covered with the two halves of a broken clock glass 4 inches in width—that no splashings be lost ; these glasses should be rinsed into the solution about a couple of hours before stopping the current. After deposition, the metal must be washed well with water, then rinsed with alcohol, and dried rapidly by heating to the temperature of boiling water.

ANTIMONY.

For the sulphide solution, measure out very accurately half a fluid ounce of the liquid, transfer it to the weighed platinum dish ; dilute it with from 1 to 2 ounces of water ; introduce the platinum anode, and pass a current from two Bunsen-cells through the liquid, until a drop of the solution, removed by a glass rod and placed upon a watch-glass, gives no yellow-coloured precipitate (but only white), when mixed with a few drops of hydrochloric acid. Electrolytic experiments of this nature may be conveniently started in the evening, and will be found to be finished on the following morning. When completed, the current is stopped, the anode is removed, the contents of the dish poured into a beaker ; the dish rinsed four or five times with pure water, then with alcohol, and finally with a few drops of ether ; it is, lastly, deposited in a warm place for a few minutes until the ether has quite evaporated, and is then re-weighed. The increase in weight shows the number of grains of metallic antimony in half-an-ounce of the original solution.

For the tartar emetic solutions, half-an-ounce should be taken and mixed with 1 ounce of yellow ammonium sulphide, and a like volume of water. It is then electrolysed as before.

COBALT.

This is a difficult assay for untrained hands—and, indeed, even the simplest methods of dealing with any metal can scarcely be expected to give more than approximate results, unless the operator has had some previous training. Probably the best method will be to measure half-an-ounce of the solution and proceed to obtain the cobalt as the cobalt-potassium nitrite, after the manner described in the last chapter (p. 306). The experiment must, of course, be conducted with great care; the solutions should be maintained fairly concentrated, and a considerable excess of the potassium nitrite must be used for the precipitation, because the precipitate is less soluble in a liquid containing this salt than in pure water, so that the time required for it to rest in a warm place is shortened—usually one night suffices. The precipitate may then be treated as described, and the second precipitate of hydrated cobalt oxide may be filtered through blotting-paper (see p. 54), washed well, dried on the paper, brushed off from the filter, by means of a camel's-hair brush, into a small weighed porcelain crucible (about an inch in height), heated to dull redness in a smokeless Bunsen-burner or spirit lamp, cooled and weighed in the crucible. The excess of weight over that of the clean crucible gives the weight of the precipitate which is an oxide of cobalt having the formula Co_3O_4. Multiplying this weight by 0·73, gives the weight of metallic cobalt in the half fluid-ounce of solution taken. If preferred, the nitrite precipitate may be dissolved in a little hydrochloric acid, the solution rendered neutral with ammonia, and a good excess of ammonium oxalate added to produce a clear solution of the double cobalt-ammonium oxalate, which may then be electrolysed in the platinum dish, and the deposited cobalt washed, dried, and weighed as above described for antimony. The solution should be kept warm during electrolysis by resting the dish upon a sauce-pan of water heated over a gentle flame. The electrolysis must be continued until a drop of the solution gives no black precipitate, or even brown colour, when mixed with a drop of ammonium sulphide in a watch-glass.

COPPER.

The simple acid copper solution lends itself so readily to the electrolytic assay, that this would certainly seem to be the most suitable. It is possible to separate every trace of copper from such a solution, so that the method may be made to give absolutely accurate results. The cyanide solution should be acidified with sulphuric acid and boiled until there is no longer any smell of prussic acid, resembling that of bitter almonds. Both these operations must be conducted cautiously and in a well-ventilated place, on account of the poisonous character of the evolved gas. The liquid is then ready for electrolysis. Half-an-ounce of either solution may be employed, and electrolysis is continued until the liquid is decolorised, and a drop removed from it strikes no blue colour with an excess of ammonia. The excess-weight of the platinum dish is that of metallic copper in the volume of solution taken.

GOLD.

One fluid-ounce of the solution is boiled with an excess of hydrochloric acid in a well-ventilated spot until there is no further smell of hydrocyanic acid; an excess of a clear solution of ferrous sulphate is then added, and the mixture is allowed to stand all night in a warm place. The precipitated gold powder is then filtered from the solution, which should now be quite free from precious metal. It is washed many times on the filter with pure water, and, after drying, is placed with the filter-paper in a small weighed porcelain crucible, and heated over a spirit-lamp or non-luminous gas-flame, until the paper is completely burnt to a white ash. The crucible is re-weighed with its contents, on cooling, and thus the weight of metallic gold in one fluid-ounce of the liquid is determined. The precipitate is pure gold, so that no further calculation is necessary.

LEAD.

Add to half-an-ounce of the solution 3 ounces of pure water, an excess of sulphuric acid (until a further addition no longer produces a white precipitate) and an equal volume of spirits of wine. Allow the mixture to stand for a few hours; filter off the white lead sulphate precipitate, and wash the precipitate on the filter several times with a mixture of equal volumes of spirits of wine and water. This washing must be very thorough, or the trace of residual acid left in the precipitate will char or weaken the paper when it is dried. A drop from the last washing but one should not redden blue litmus-paper. The paper and contents are dried, the heavy white powder is brushed into a weighed porcelain crucible, heated over a non-luminous flame, cooled and re-weighed. The increase of weight multiplied by 0·683 gives the weight of metallic lead in the sample taken.

NICKEL.

To half-an-ounce of the solution add a fair excess of ammonium sulphate, together with about 4 ounces of water, and then oxalic acid. If this should produce a precipitate, more ammonium sulphate must be added until the solution is clear. Excess of ammonia is now introduced, any precipitate is filtered off and washed with water containing ammonia, the washings being added to the filtrate. The filtered solution and washings are then mixed with a further small portion of ammonium oxalate, evaporated to convenient bulk in a porcelain dish, and electrolysed for metallic nickel as above described.

PLATINUM.

To half-an-ounce of the solution add an excess of a ferrous sulphate solution, and then excess of caustic potash. The mixture is kept in a warm place for an hour or two with occasional stirring; an excess of hydrochloric acid is now added, so that the liquid reddens blue litmus-paper, and the precipitated metallic platinum, which remains undissolved, is filtered and burned with the paper after the manner described under the heading of Gold in this chapter. The final precipitate should be pure metallic platinum.

SILVER.

The solutions may be treated electrolytically, using the current from one Bunsen-cell only, as a stronger current tends to deposit pulverulent or flaky metal which peels off during the process of deposition. The silver is, of course, deposited and weighed in the metallic state.

CHAPTER XVIII.

A GLOSSARY OF SUBSTANCES COMMONLY EMPLOYED IN ELECTRO-METALLURGY.

IN using the various chemical preparations, it is impossible to be too careful; as in all operations connected with electro-metallurgy, the most scrupulous cleanliness and thoughtful attention are necessary, especially on the part of one who is unaccustomed to handling chemical apparatus and reagents.

In opening fresh bottles of acid or of ammonia, especially in hot weather, the stopper must be first carefully cleansed externally, and must then be covered with a stout cloth before attempting to remove it, because it is frequently wetted with the liquid contained in the bottle, and if there be any pressure from within, drops of the fluid may be scattered in all directions when the stopper is loosened, and may fall upon the clothes, hands, or face, and are likely to cause blindness if they should happen to penetrate to the eye. In tropical climates the strong ammonia solution boils violently as soon as it is uncorked; and as this is evidence of a strong internal pressure, it is safer to surround the bottle in several folds of cloth before attempting to open it, lest the application of the slight force, sometimes required to loosen the stopper, may cause the bottle itself to burst should it be defective.

The following simple rules should be carefully observed :—

In opening any bottle, first dust the whole surface and then clear away from the neck all dirt or loose particles of sealing-wax, cork, or luting; see that the corkscrew does not break off particles of cork into the bottle.

In loosening the refractory stopper of a bottle containing any inflammable substance, on no account apply the heat of a flame; and in decanting any such liquid, or opening any such bottle, see that it is not in the vicinity of a lamp or flame of any kind, or even of any heated substance if the liquid be carbon bisulphide.

On removing the cork or stopper, never place it with the smaller end downwards upon the table, as it is very likely to pick up foreign matter from the surface and transfer it to the bottle as soon as it is re-inserted.

Never allow a bottle to remain open to the air longer than necessary, and always see that it is closed with its original stopper.

If the solid contents of a bottle require loosening, always apply a stout glass or glazed porcelain rod, carefully cleaned before insertion. On no account use a metal rod, unless it be made of platinum or other material which cannot be corroded by the substance with which it is brought in contact.

Never pour the contents of a bottle into a dirty vessel, or upon an unclean surface.

Never place chemical substances directly upon the metal pan of a balance, but always upon a tared plate of glass, though a sheet of clean glazed paper may suffice if the substance is dry and not deliquescent (that is, does not tend to become moist by exposure to damp air).

In returning an excess of any substance into store, see that it is placed in the right bottle or vessel.

Never mix a strong acid with a strong alkali, and even when both are diluted, let the mixture be effected gradually.

Never add water to strong sulphuric acid, but if it be desired to dilute the acid, let it be added little by little, with constant stirring, to the required bulk of water.

Never employ vessels of a kind which are used for domestic purposes to contain chemical reagents. Dangerous results may follow the leaving of acids or poisonous substances in ordinary tumblers or wine glasses. Let the vessels used for chemical work be restricted to this duty, and let no others be employed.

Note.—In the following glossary, the specific gravity of a substance refers to its weight as compared with that of an equal bulk of pure distilled water.

ACIDS.

Acetic Acid, $CH_3 . CO_2H$, or $C_2H_4O_2$.—A monobasic and somewhat feeble acid, the chief acid constituent of vinegar. It may be bought as *glacial acetic acid*, which is practically the pure substance, and is a crystalline solid at ordinary temperatures. It is more usually obtained somewhat diluted with water. The *acidum aceticum* of the British Pharmacopœia contains from 32 to 33 per cent. of the pure acid, and has a specific gravity of 1·044. With bases it forms *acetates*.

Benzoic Acid, $C_6H_5 . CO_2H$, or $C_7H_6O_2$.—A monobasic weak organic acid, obtained as colourless plates very slightly soluble in water, which should leave little or no residue when burnt upon a sheet of thin platinum held over a flame. Its salts are termed *benzoates*.

Boric Acid, $H_3 BO_3$.—Commonly known as *boracic acid;* a mineral body, and the acid basis of borax; it is a white crystalline solid, fairly soluble in water. Dissolved in warm spirits of wine and ignited, the liquid burns with a brilliant and characteristic green flame. It is tribasic, and its salts are called *borates*.

Citric Acid, $C_3 H_4 (OH) . (CO_2 H)_3$, or $C_6 H_8 O_7$.—A white translucently-crystalline, solid, tribasic, organic acid, readily soluble in water (100 parts in 75 of cold or 50 of hot water). It forms *citrates* with metallic oxides.

Hydrochloric Acid, HCl.—The pure body is a gas under ordinary conditions. It is intensely soluble in water, and its solution is sold under the above name or that of *muriatic acid*. The stronger solutions emit pungent white fumes in air, owing to a partial evaporation of the contained gas, which recondenses in the form of clouds in contact with atmospheric moisture. A saturated solution contains about 40 per cent. of the pure gas, and has a specific gravity of 1·2; the actual saturation percentage depends upon the temperature, as the acid is less soluble in hot water than in cold. The acid of commerce has usually a specific gravity of about 1·150, equal to 30 per cent. of pure HCl; the *acidum hydrochloricum* of the British Pharmacopœia contains about 32 per cent., with a specific gravity equal to 1·16. The liquid should be colourless, but the commercial muriatic acid is generally yellow, owing to the presence of iron and other impurities; only the pure acid, however, should be used for most electro-metallurgical processes, though the crude liquid may often be employed for cleansing purposes. The acid is monobasic, and its salts are termed *chlorides*.

Hydrocyanic Acid, HCN.—Commonly known as *prussic acid*. This also is bought as an aqueous solution of HCN, which is a liquid of low boiling point. It smells strongly of bitter almonds and is a deadly poison, so that its use in the arts is to be strongly deprecated whenever it can be avoided. The British Pharmacopœia solution contains 2 per cent. of the pure acid. When used, the greatest care must be exercised, as the vapour evolved by a strong solution is itself extremely poisonous, and, even when diluted considerably with air, produces giddiness and headache. Its salts are designated *cyanides;* the acid is monobasic.

Nitric Acid, HNO_3.—Commercially known as *aqua fortis*. It is a very powerful and corrosive monobasic acid, which must be handled with great care. It stains the skin and other animal substances a bright yellow, which becomes intensified by the application of an alkali or of soap; its oxidising power is so intense that the accidental fracture of a carboy of the acid has been known to set fire to straw which happened to surround it. The pure acid ($H N O_3$) has a specific gravity of 1·52. The acid commonly sold in commerce has a specific gravity of 1·45, equal to 77 per cent. of pure $H N O_3$, while a weaker acid, containing 70 per cent. (specific gravity = 1·42) is also to be had, and constitutes in fact the *acidum nitricum* of the British Pharmacopœia, while others even less concentrated are likewise to be procured. The stronger solutions fume in the air. When pure, the liquid should be colourless, but owing to partial decomposition into lower oxides of nitrogen, it frequently possesses a straw-coloured or yellow tint, becoming, in the commoner kinds, orange and, finally, green, when it is commercially termed *nitrous acid*. The salts of the acid are called *nitrates*.

When mixed with hydrochloric acid in the proportion of 1 to 3 ($H N O_3 : H Cl$) the liquid is known as *aqua regia*, and assumes an orange colour due to the presence of the gas nitrosyl chloride ($N O Cl$) formed by the union. This liquid is endowed with the highest oxidising powers, dissolving even the precious metals which resist the attack of either acid singly.

Oxalic **Acid**, ($C O_2 H)_2$ or $C_2 H_2 O_4$.—A dibasic organic acid, which forms a white crystalline solid containing 2 molecules of water ($C_2 H_2 O_4 . 2 H_2 O$). It is readily soluble in water or alcohol, and is a powerful poison. It forms *oxalates*.

Sulphuric Acid, $H_2 S O_4$.—Known as *oil of vitriol*. A dibasic and most powerful mineral acid. When perfectly pure, it is a colourless and odourless, oily liquid of specific gravity 1·842. It combines most energetically with water, and in doing so generates much heat, so that dilution even of the ordinary commercial acid with water must be effected with the greatest care. Large volumes must never be thoughtlessly mixed, nor should the water be added to the acid, lest a sudden generation of steam of explosive violence result, and the dangerously corrosive liquid be scattered in all directions. The acid is in all cases to be added to the water in a very gentle stream, and with constant stirring. The acid of commerce is diluted in various degrees, but is always concentrated. The salts formed from this acid are *sulphates*.

Tannic Acid, $C_{14} H_{10} O_9$.—A pale yellow solid and weak organic acid, readily soluble in water. It burns completely away when heated upon a metallic plate, and yields a blue-black colour when added to solutions containing iron. It forms *tannates*.

Tartaric Acid, $C_2 H_2 (O H)_2 . (C O_2 H)_2$ or $C_4 H_6 O_6$.—It is a colourless crystalline solid, which readily dissolves in water. It is a weak dibasic organic acid, forming salts which are known as *tartrates*.

Alcohol, $C_2 H_5 (O H)$ or $C_2 H_6 O$.—Pure or *absolute alcohol* has a strong affinity for water, so that the alcohol usually bought generally contains a small quantity of the diluent. It should be colourless and have a specific gravity of 0.7939. It is commonly sold under the name of *spirits of wine*, or *rectified spirit*, which contains about 16 per cent. of water, and has a specific gravity of 0.838. *Proof spirit* is a mixture carrying 49.24 per cent. of pure alcohol, and is the standard with which alcoholic liquids are compared in commerce. *Methylated spirit* contains a certain amount of methylic alcohol or wood spirit, and frequently has a quantity of resinous matter added to it to render it unfit for drinking, so that it may not be liable to excise duties, and shall yet be available for most of the purposes for which spirit is required in the arts. Such a product should on no account be used for electro-metallurgical work without previous thought as to the probable consequences of so doing. For example, the effect of washing an object in the sophisticated spirit and then plunging it into the electrolytic bath would be the introduction of undesirable organic impurities in the latter, and even the formation of a non-conductive coating upon the surface of the article itself; because water added to the impure alcohol throws down the resinous substances in the form of a white precipitate. Such spirit tends to burn with a smoky flame. Alcohol should, therefore, be tested by burning, when the flame should be almost non-luminous, and by the addition of water to a sample, which should produce no turbidity.

Aluminium, Al.—A white silvery element with an almost imperceptible bluish shade. It is extremely light (the specific gravity being only 2.58), is very malleable and ductile, takes a high polish, and is not liable to tarnish in air. It melts at about $1,300°$ F. At present somewhat costly, it is not often met with in commerce. Its principal common impurities are iron and silicon.

Aluminium Chloride, $Al_2 Cl_6$.—A white, sometimes yellowish, substance, which in the anhydrous condition absorbs water with great avidity; and, having once absorbed it, cannot be induced to part with it, the action of heat upon the hydrated crystalline

salt ($Al_2 Cl_6 . 12 H_2 O$) causing decomposition, with the formation
of alumina and hydrochloric acid. It must, therefore, be stored
in perfectly air-tight jars. In small quantities it volatilises at
356° to 365° F. without fusion, but in a large bulk it may be
induced to melt first.

Aluminium-Sodium Chloride, $Al_2 Cl_6 . 2 Na Cl$.—This salt is
made by heating the simple aluminium chloride with common
salt. It is more useful than the latter for electro-reduction by the
fusion method, because, melting at 365° F., it volatilises only at
a red heat; moreover, it is less readily attacked by aqueous
vapour.

Aluminium-Potassium Sulphate, $Al_2 (S O_4)_3 . K_2 S O_4 . 24 H_2 O$.—
Commonly known as *potash alum*. It is a crystalline substance,
with an astringent taste, and is readily soluble in water, 12 parts
of alum dissolving in 100 parts of water at the ordinary temperature, 357 parts at the boiling point. It is a double sulphate of
alumina and potash. *Ammonia alum* is exactly analogous, the
potassium sulphate being simply replaced by ammonium sulphate
($Al_2 (S O_4)_3 . (N H_4)_2 S O_4 . 24 H_2 O$), and is for most purposes
interchangeable with potash alum. Soda alum is similar, but is
more readily soluble in water.

Ammonium Hydrate, $N H_3 . H_2 O$.—Commonly termed *ammonia*
or *spirits of hartshorn*. It is simply water saturated with
ammonia gas ($N H_3$). It is a powerful alkali, and is, therefore,
useful to neutralise the acid properties of any substance. As it
has an overpoweringly pungent odour, care must be taken in
using the stronger solutions. It is obtainable in the market as
ammonia fortiss., with a specific gravity of 0·880, which is almost
saturated with the gas and contains about 36 per cent. of pure
$N H_3$. Bottles of the liquid must be opened cautiously in hot
weather, because the warmer water cannot dissolve so large a
volume except under pressure; and the excess is given off,
sometimes with considerable violence, directly the pressure is
released by the removal of the stopper. A weaker solution,
ammoniæ liquor fortior, containing 32·5 per cent. of $N H_3$ (specific
gravity = 0·891) is that recognised by the British Pharmacopœia,
and is safer for use in the heat of summer. By exposure,
ammonia gas is gradually evolved, so that it must be stored in
closely-stoppered bottles in order to preserve the strength of the
solution unimpaired. By regarding the formula as $N H_4 . O H$, it
becomes the hydrate of the hypothetical metal-like substance

ammonium ($N H_4$). This radical or group of elements, $N H_4$, behaves in its chemical relations exactly as a monovalent metal, combining with acids to form ammonium salts.

Ammonium Carbonate, $(N H_4)_2 C O_3$, and **Bicarbonate**, $(N H_4) H C O_3$, are white solid substances, soluble in water and smelling strongly of ammonia. The commercial salt is not a pure carbonate, but is in part carbamate ; it is, however, suitable for the purposes to which it is generally applied. As an alkaline carbonate it takes the place of ammonia itself in neutralising acids, carbonic acid gas being evolved while the ammonium salt of the stronger acid is formed.

Ammonium Chloride, $N H_4 Cl$.—A white substance occurring in commerce in the form of tough fibrous crystals, odourless, and soluble in water (100 parts of cold water dissolve 35 parts, and of boiling water 77 parts of the salt).

Ammonium Nitrate, $N H_4 N O_3$. — A colourless crystalline body, 200 parts of which dissolve in 100 parts of cold water. Heated gently it melts and is afterwards decomposed into nitrous oxide gas (laughing gas = $N_2 O$) and water, leaving no residue.

Ammonium Sulphide, $(N H_4)_2 S$, and **Hydrosulphide**, $N H_4 H S$, may be prepared by passing hydrogen sulphide gas into a solution of ammonia. It is at first colourless, but by gradual decomposition becomes yellow. It is generally bought as an amber-coloured, frequently almost orange, fluid. The many products of decomposition do not seriously interfere with its general utility except in so far as they weaken the solution of the sulphide itself.

Ammonium Sulphate, $(N H_4)_2 S O_4$.—A white crystalline solid of which 100 parts of cold water dissolve 50, of hot water 100 parts.

Ammonium Salts of the acids just described should leave no fixed residue when heated over a flame upon a piece of sheet metal.

Antimony, Sb.—A white, highly crystalline and extremely brittle metal. The commercial ingots have usually a very crystalline surface, which resembles the fern-like appearance of frost upon window-glass, from which the name *star antimony* is derived, and which is regarded by many as an infallible criterion of the purity of the metal, but which must not be relied upon too closely. This metal has a specific gravity of 6·8, and melts at 800° F. The principal objectionable impurities likely to occur in ordinary antimony are sulphur, arsenic, tin, lead, silver, bismuth, and iron. It is only procurable in the cast condition, because on account of its brittleness it cannot be rolled.

Antimony Chlorides.—There are two chlorides—the *trichloride* or *antimonious chloride*, $SbCl_3$, known as *butter of antimony* which is that more usually required for electro-metallurgical work ; and the *pentachloride* or *antimonic chloride*, $SbCl_5$. The former is a colourless crystalline substance melting at 164° F. It may be prepared in crude form, mixed with excess of acid, by dissolving antimony or the sulphide in hydrochloric acid to which a little nitric acid has been added to increase its oxidising action. The pentachloride is a colourless fuming liquid having a most unpleasant odour.

Antimony Sulphide, Sb_2S_3, occurs in nature as *stibnite* or *antimony glance*. It is usually bought as grey needle-shaped crystals sub-metallic in lustre. The pentasulphide, Sb_2S_5, requires no notice here.

Antimony-Potassium Tartrate, $KSbC_4H_4O_7$, commonly known as *tartar emetic*. A white crystalline substance, of which 100 parts of cold water dissolve 5 parts, while a like volume of hot water dissolves 50 parts.

Aqua Fortis.—See Acid, Nitric.

Aqua Regia.—See under Acid, Nitric.

Bees'-wax.—The substance of which the honey-comb is built up. It is usually of a yellow or brown colour, the specific gravity ranging from 0·958 to 0·960, and the melting point from 144° to 156° F. in different samples. It dissolves readily in oils and in ether, but not in water. Its chief adulterants are— mineral matter, starch or flour, and water, the presence of either of these being detected on melting the sample ; and resinous or fatty substances, vegetable waxes and paraffin which influence the specific gravity and the melting point. When melted it should give a clear liquid free from any cloudiness, and should not indicate the presence of water, by the formation of two layers of fluid.

Bismuth, Bi.—A highly crystalline and very brittle metal resembling antimony, but having a faint pinkish colour. Like antimony it cannot be rolled or drawn into a wire, but, on the contrary, may be crushed into a powder with an ordinary pestle and mortar. It melts at 515° F., and has a specific gravity of 9·759. It is one of the most useful constituents of fusible metal. The commonest objectionable impurities are—sulphur, iron, lead, copper, arsenic, and silver.

Bismuth Nitrate, $Bi(NO_3)_3 + 3H_2O$.—Made by dissolving the metal in dilute nitric acid; it is precipitated as a white basic bismuth sub-nitrate by the addition of much water.

Black Lead.—See Plumbago.

Brass is an alloy of copper and zinc, the percentage of the former varying from 70 to 60, but more usually from 70 to 73. Special alloys are made for certain purposes, many containing tin and lead. *Sterro metal* is a brass to which a small percentage of iron has been added, while other complex alloys are made containing iron and manganese in addition to other bodies, to give additional strength or stiffness to the metal. Ordinary brass, unless specially made from the purest virgin metal, generally contains notable proportions of iron and lead and sometimes of tin.

Bronze.—This term embraces the class of alloys of copper and tin, and includes bell-metal and gun-metal. The proportion of tin varies from 5 to 20, the average sample containing about 10 per cent. Some samples have a small percentage of zinc added, others manganese and iron; in fact the remarks made in regard to brass apply equally to bronze in this matter. Certain alloys are wrongly called by this term, for example, aluminium bronze, which contains 10 per cent. of aluminium, but no tin. Some forms of manganese bronze also contain only a nominal percentage of tin, and the alloy is then really a complex brass with a very small, but sufficient, percentage of manganese.

Cadmium, Cd.—A soft and very malleable bluish-white metal not unlike zinc, with which element it is commonly associated in nature. Its specific gravity should lie between 8·6 and 8·8, while its melting point is about 608° F. It is rarely used in the Arts in the metallic form, except in the manufacture of fusible alloys.

Cadmium Bromide, $CdBr_2$.—A white crystalline substance soluble in water.

Cadmium Chloride, $Cd Cl_2 . 2 H_2 O$.—A similar body, of which 140 parts are soluble in 100 of water. It is made by dissolving the metal in hydrochloric acid and evaporating.

Cadmium Sulphate, $3 Cd SO_4 . 8 H_2 O$.—A white crystalline substance, 59 parts dissolving in 100 parts of water.

Calcium Carbonate.—See Chalk.

Chalk is a natural rock composed of calcium oxide (lime) and carbonic acid, $Ca CO_3$. On strongly heating it the carbon dioxide (CO_2) is driven off, and the pure calcium oxide remains as *quick-lime* $(Ca O)$. Limestone has the same composition as chalk, and behaves similarly. A particular kind of lime, selected from that burnt in the neighbourhood of Sheffield, has found especial favour among metal-polishers. As the burnt lime, by contact with the air, rapidly absorbs first water, and thus becomes *slaked* $(Ca (OH)_2)$, and then carbon dioxide, and so becomes reconverted into calcium carbonate, the lime, which is to be stored for any length of time, must be preserved in air-tight cases until it is required for use. For polishing purposes it must be uniformly soft and free from gritty particles, which would give rise to scratches; treated with dilute hydrochloric acid, a sample of quicklime should dissolve with but slight effervescence, and leave no residue undissolved. Chalk or *whiting* should dissolve with brisk effervescence, but this also should leave no appreciable residue.

Cobalt, Co.—A metal similar to, and generally occurring in nature with, nickel. It has a specific gravity of from 8·5 to 8·7, and a high melting point, approximating that of iron. It is readily soluble in sulphuric (dilute), hydrochloric, and nitric acids, forming cobalt sulphate, chloride, and nitrate respectively.

Cobalt Chloride, $Co Cl_2 . 6 H_2 O$.—A dark red crystalline body readily soluble in water; it may be prepared by dissolving the metal, its oxide or carbonate, in just sufficient hydrochloric acid. It is well to use a deficiency of the latter in order to ensure the neutrality of the solution.

Cobalt Nitrate, $Co (N O_3)_2 . 6 H_2 O$.—A pink crystalline soluble substance prepared like the chloride but with nitric acid. It is readily procurable in the market.

Cobalt Sulphate, $CoSO_4 . 7H_2O$.—It resembles the two last-named salts in the manner of preparation (but using sulphuric acid); 100 parts of cold water dissolve 35 parts of the salt.

Cobalt-Ammonium Sulphate, $CoSO_4 . (NH_4)_2 SO_4 . 6H_2O$.— A pink salt which may be made by adding the right proportion of ammonium sulphate to cobalt sulphate (47 : 100), and then dissolving them and evaporating them together until a good crop of crystals has separated.

Copper, Cu.—A red metal with a fusing point of about 1996° F., and a specific gravity of 8·9 to 8·95 according to its condition—whether it is simply cast or has been afterwards rolled or hammered. By reason of its extreme malleability and ductility it may be readily obtained in the form of rolled plate or foil, and of rod or wire. It is most readily attacked by nitric acid, but is slowly dissolved when immersed in heated hydrochloric or sulphuric acids. The metal is frequently found *native* (that is in the metallic state in nature), but is most usually smelted from ores in which it is combined with sulphur as sulphide, or with oxygen and perhaps other acid substances as oxide, or oxidised compounds. Such metal often contains small percentages of sulphur, lead, bismuth, arsenic, antimony, and iron, with sometimes traces of silver and gold, and, more rarely, of nickel, cobalt, and tin. There are two classes of copper salts—one, *cupric*, containing more oxygen or other electro-negative element, formed from the oxide CuO, and yielding generally blue- or green-coloured salts and solutions; the other, *cuprous*, prepared from the sub-oxide, Cu_2O, and giving nearly colourless solutions. The former, which are more usual, alone need be referred to here.

Copper Acetate, $Cu(C_2H_3O_2)_2 . H_2O$.—Dark green crystals, moderately soluble in water, formable by dissolving cupric oxide or carbonate in acetic acid.

Copper Carbonate, $CuCO_3$.—Occurs in nature as malachite and allied minerals. The artificial carbonate is a green substance, insoluble in water, but readily decomposed by mineral acids (as well as by many of organic origin) yielding the copper compound of the added acid, and carbon dioxide gas, the evolution of which gives rise to great effervescence. The so-called carbonate is usually mixed with hydrated oxide and has the formula—$CuCO_3 . Cu(OH)_2$.

Copper Chloride, $Cu Cl_2 . 2 H_2 O$.—Blue-green crystals, readily soluble in water. May be prepared by treating an excess of oxide or carbonate with hydrochloric acid, filtering, and evaporating the resulting solution.

Copper Cyanides.—The *cupric cyanide*, $Cu (C N)_2$, precipitated by potassium cyanide from copper sulphate solutions, is very unstable, rapidly changing by exposure into *cupro-cupric cyanide*, $Cu (C N)_2 . Cu_2 (C N)_2$, and cyanogen gas; the former when solid form green crystals, which are again decomposed at the temperature of boiling water into cuprous cyanide, $Cu_2 (C N)_2$, and cyanogen; and this latter forms several double cyanides with potassium—for example, $Cu_2 (C N)_2 . K C N . H_2 O$, $Cu_2 (C N)_2 . 2 K C N$, and others, which are for the most part very soluble in water.

Copper Nitrate, $Cu (N O_3)_2 . 3 H_2 O$.—Blue crystals, very soluble in water, and rapidly absorbing moisture from the air. Excess of metal, oxide or carbonate treated with nitric acid yields the salt.

Copper Sulphate, $Cu S O_4 . 5 H_2 O$.—Commercially known as *blue vitriol;* it is the commonest compound of copper. It forms blue crystals, of which 100 parts of cold water dissolve about 40, and of hot water about 200 parts. It is so largely used in the Arts that it may be procured everywhere; it may be made the starting point for making other compounds of the metal. By adding to an aqueous solution of the salt a quantity of sodium carbonate, dissolved in water, a green solid precipitate of copper carbonate is produced; this may be allowed to subside, filtered, washed well, and dissolved in any acid which will produce the required salt.

Glue.—The ordinary best glue in the market, made from bones, should be used for moulding. It should be quite transparent, although perhaps dark in colour, hard, and brittle when sharply struck, and must be free from particles of foreign matter. When soaked in water it should swell and absorb about five or six times its weight of the water. *Gelatine* is only a specially pure and clean form of glue, and *isinglass* is similar in composition.

Glue, Marine.—There are several descriptions of this useful cementing material. A commonly employed glue is made by dissolving a little india-rubber very carefully and with the aid of heat in twelve times its weight of coal-tar naphtha, adding to it twenty times its weight of shellac, and finally pouring it upon a flat cold surface to solidify and harden. It is only necessary to warm the glue and to apply it to the gently-heated surfaces that are to be united. It also makes a good waterproof and non-conductive lining when painted thickly upon the interior surfaces of tanks for cold solutions.

Gold, Au.—A yellow metal of high specific gravity (19·26) and fusing point (2015° F.). It is the most malleable and ductile of metals, and combines with these properties that of a very fair electric conductivity. In nature it occurs in the metallic state, almost invariably associated with silver, and often with copper and iron. In commerce it is met with as fine gold, and in various alloys of which the principal has the standard value of the British sovereign gold—91·67 of gold to 8·33 of copper. These alloys are described as being so many carats fine; thus, if an alloy contain 22 parts of pure gold in 24, it is said to be 22-carat gold, if it contain $\frac{18}{24}$ of its weight of the pure metal it is 18-carat gold, and so forth; the remaining metal may be copper or silver, separately or together, according to the colour which the metal is required to have; the sovereign is made of 22-carat gold $\left(\frac{22}{24} = \frac{91·6}{100}\right)$.

Gold is insoluble in nitric, hydrochloric, or sulphuric acid alone, but readily dissolves in a mixture of the two former (aqua regia), and in a very finely-divided condition may be made to dissolve in a mixture of the first and third. To prepare pure gold from the alloy on a small scale is a simple matter. If the alloy contain less than 40 (or more safely 30) per cent. of the precious metal, mere prolonged boiling in nitric acid will dissolve the copper and silver, but leave the gold untouched in the form of a black powder, which is very heavy, and requires only to be washed several times, by stirring it up with repeated additions of cold water, allowing it to settle and pouring off one batch of water each time before adding fresh, and then to be dried and fused to yield the metal practically in a state of purity. The solution must be effected in a glass or glazed earthen- or stone-ware vessel, which will not be attacked by the strong acid; and should be carried on in the open air or in a well-ventilated fume-cupboard. But if the alloy contain a larger percentage of gold, the other metals are not completely removed by the acid, and the original mixture must be treated with aqua regia. The gold will now be in solution; any undissolved white residue is probably silver chloride, which is formed by the agency of the hydrochloric acid, but is insoluble in the liquid. It should be allowed to subside, washed once or twice by decantation and filtered. The solution should now be transferred to an evaporating dish or other vessel, in which it may be evaporated to the consistency of a thick syrup, by placing it over a saucepan of boiling water. By this time the bulk of the nitric acid will have been boiled away, and the residue will be a strong, but more or less impure, solution of gold chloride containing hydrochloric acid. The liquid is now diluted, and a quantity of a solution of ferrous sulphate is added, and the mixture is allowed to stand for a day or two in a warm place. The ferrous salt becomes converted into a ferric compound at the expense of the gold oxide, and the gold should thus be liberated completely as pure precipitated metal, of dark brown or black colour, which may be washed, dried, and fused as before. The fusion may be made in a small clay crucible under a cover of a few grains of borax by way of flux for residual impurities. Oxalic acid is sometimes substituted for ferrous sulphate as a precipitant.

Gold Chloride, $Au\,Cl_3 . 2\,H_2\,O$.—A most soluble and deliquescent yellow crystalline substance, which may be prepared as described in the latter portion of the last paragraph, but using pure gold instead of alloyed metal as the basis.

Gold Cyanides.—On adding a neutral gold chloride solution to one of potassium cyanide, there is produced *potassium auricyanide*, $2\,K\,Au(CN)_4 . 3\,H_2\,O$, which in the solid condition forms colourless tabular crystals, that are very soluble in hot water, but decompose at about 400° into *potassium aurous cyanide*, $Au\,CN\,.\,K\,CN$, and cyanogen gas. This latter body, potassium aurous cyanide, is formed by dissolving aurous oxide, $Au_2\,O$, or even finely-divided gold in potassium cyanide solution; it is a colourless crystalline and very soluble salt. Simple *aurous cyanide*, $Au\,CN$, is an insoluble lemon-yellow substance.

Gold, Fulminating, $Au_2O_3 . (N\,H_3)_4$.—A brown or green powder, obtainable by adding ammonia or ammonium carbonate to a solution of gold chloride. It should never be allowed to become dry, for in this condition it is liable to explode with great violence. So long as it is moist, there is no danger attending its use.

Gold Sulphide, Au_2S_3.—Obtained by passing sulphuretted hydrogen gas into a solution of the chloride; it then appears as a black precipitate, soluble in alkaline sulphides.

Graphite.—See Plumbago.

Gutta-percha.—A gum prepared from the exudation of certain trees in the Malay Peninsula and Islands. It is usually procurable in sheets. As a moulding material it is valuable, because at the temperature of boiling water it is extremely plastic and may be worked into any required shape, which it will retain on cooling, when it again becomes hard yet somewhat elastic. It is a non-conductor of electricity.

Iron, Fe.—The fusing point of the pure metal is very high. The iron of commerce is never pure. In the condition in which it is smelted from the ore as *pig-iron* or *cast-iron* it is more readily fusible, but is highly charged with impurities, derived from the ore, fuel, and flux, and contains varying proportions of carbon, sulphur, phosphorus, silicon, and manganese, frequently accompanied by other elements also, the foreign matter in the aggregate amounting to from 4 to 7 or more per cent. In this condition it is hard, brittle, and unworkable. By refining away the greater proportion of these impurities, the melting point is greatly raised, and at the same time the metal becomes soft, ductile, and malleable, and is known as *malleable-* or *wrought-iron*, which may be rolled into sheet of any degree of thinness. Wrought-iron is the purest form of marketable iron, but even this is not pure, containing perhaps 0·5 per cent. of foreign substances. Between wrought- and cast-iron is another form—*steel*—the characteristics of which are chiefly governed by the percentage of carbon, which may range from 1½ per cent. in the harder varieties of tool steel to practically nothing in *mild-steel*.

The latter of these alone, from its greater purity, enters into competition with wrought-iron as a rival in the manufacture of anodes.

The metal is soluble in either of the three common mineral acids, and forms two classes of salts, one (*ferric*) with more oxygen, of which the peroxide, Fe_2O_3, is typical, the other (*ferrous*) of which the protoxide, FeO, is the basis. The latter are readily converted into the former by the addition of oxygen, even by absorption from the air; but unless there be an excess of acid in the bath the effect of the peroxidisation will be the precipitation of basic salt (see p. 244). It is for this reason that neutral ferrous solutions rapidly become turbid with a yellowish slimy deposit.

Iron Chlorides.—*Ferrous chloride*, $Fe\,Cl_2 . 4\,H_2\,O$, and *ferric chloride*, $Fe_2\,Cl_6 . 12\,H_2\,O$, are both very soluble crystalline bodies —the former bluish, the latter yellow in colour.

Iron Sulphate.—*Iron protosulphate, ferrous sulphate*, or *green vitriol*, $Fe\,S\,O_4 . 7\,H_2\,O$, is a green crystalline substance, often yellowish on the exterior, owing to the formation of ferric compounds with the aid of atmospheric oxygen. On account of this tendency to peroxidation, this and other ferrous compounds should not be exposed more than necessary to the air. 100 parts of cold water dissolve about 70 parts of the salt, of hot water 330 parts. The *ferric* sulphate, $Fe_2\,(S\,O_4)_3$, demands only casual mention in this place.

Iron-Ammonium Sulphate, $Fe\,SO_4 . (N\,H_4)_2\,S\,O_4 . 6\,H_2\,O$, is a body similar to the last, but with a bluer shade of colour, and is much less liable to alteration by exposure to air, and is, therefore, preferable to the former for many reasons. 100 parts of cold water dissolve 16 parts of this salt.

Lard.—The pure white lard is used ; as it is frequently adulterated with water and solid substances, it should be melted and allowed to stand for some time in this condition. The bulk of the water and heavier matter will sink to the bottom, and the purified fat, which should now be quite transparent, may be drawn off from above, or removed by ladles into vessels wherein it may be allowed to solidify. It should melt at a temperature of about 110°·5 F.

Lead, Pb.—One of the softest of metals, it is very malleable, but, having a low tenacity, is deficient in ductility ; it may be rolled into sheet, but not drawn into wire. Its fusing point is 633° F., and its specific gravity 11·25 in the cast state, or 11·39 when it has been condensed by rolling. On account of its ready fusibility it may be cast into slabs, or it may be rolled into sheet for use as anodes. Commercial lead, frequently very nearly pure, is never absolutely so. It always contains at least a trace of silver, often with varying proportions of antimony, tin, copper, iron, and sulphur. It is readily dissolved in nitric acid, and slowly in boiling hydrochloric acid. Sulphuric acid, except of the most concentrated description, is almost without action on the metal. Both chloride and sulphate of lead are practically insoluble in their respective acids, so that very soon the metallic surface becomes coated with a deposit which prevents further action. The salts of lead are formed from the basis of the monoxide (litharge = PbO), in which the metal is divalent, although two other oxides, *red lead*, Pb_3O_4, and *peroxide*, PbO_2, are known.

Lead Acetate, $Pb(C_2H_3O_2)_2$.—A readily soluble white crystalline substance, easily formed by dissolving lead oxide or carbonate in acetic acid. It is very commonly known as *sugar of lead*.

Lead Nitrate, $Pb(NO_3)_2$, forms soluble white crystals (100 parts of water dissolve about 54 parts in the cold, or 135 parts when heated).

Lime.—See Chalk.

Magnesium, Mg.—A white divalent metal, readily becoming dull in moist air. When ignited at a slightly elevated temperature, it continues to burn with a most brilliant white light, until it is completely converted into oxide. It has a very low specific gravity (1·75), and fuses at 850° F. It forms one oxide, magnesia, MgO, which is the basis of the various salts of the metal.

Magnesium Chloride, $MgCl_2 . 6H_2O$.—A most soluble and deliquescent crystalline salt (100 parts of water dissolve 280 parts in the cold, or 782 parts of the body when heated). It must be stored in a closely-stoppered bottle.

Magnesium Sulphate, $MgSO_4 . 7H_2O$.—Commonly known as *Epsom salts*. It forms white crystals, easily procurable, of which 100 parts of cold water dissolve about 70.

Mercury, Hg.—Frequently called *quicksilver*. It is the only known metal which is liquid at ordinary temperatures; it solidifies at $-38°·9$ F., and boils at 680° F. Its specific gravity at the normal temperature is 13·59. Mercury has a great tendency to dissolve other metals, and so to form *amalgams*; it must not, therefore, be stored in, or allowed to come into contact with, clean surfaces of any metal commonly in use except iron or platinum, with which it does not combine. Gold and silver are especially liable to be dissolved, and as articles of jewellery are thus readily spoiled by mercury, the greatest care must be taken in using it. Gold becomes white and dulled by it, and requires the application of strong heat to effect its removal; and the surface is then left "dead," so that it must be re-polished. Mercury, therefore, should not unnecessarily be introduced into the work-room containing electro-plated goods awaiting treatment; but if used for any purpose it should be carefully preserved from contact with any article liable to be spoiled by it. In consequence of its proneness to combine with other metals, mercury is rarely quite pure. If it be required clean, it may be spread in a shallow dish and covered with dilute nitric acid, with which it should be stirred from time to time. The base metals, such as zinc, copper, and lead, being more electro-positive than mercury, tend to dissolve first; but a certain amount of the mercury itself dissolves also, and forms mercurous nitrate. This sub-nitrate assists in the removal of the other metals by simple exchange; gold and silver, which are more negative than mercury, are, of course, unaffected by the treatment. The solution which has been used for cleaning mercury may be used again and again to treat fresh samples, so long as it contains either an excess of acid, or an appreciable quantity of mercury in solution. This may be ascertained by the blue litmus-paper test in the former case, or in the latter, by adding a drop of hydrochloric acid to a little of the solution placed in a test tube, when a dense white precipitate of mercurous chloride (calomel) is at once produced if mercury be present. Mercurous nitrate solution may be substituted for nitric acid at the outset if preferred. The most satisfactory method of purification, however, is to distil the mercury from a glass retort, and, preferably, under diminished atmospheric pressure, effected by adopting a system of hermetically-joined retort and condenser connected to an air pump. In this way the boiling point of the mercury may be greatly lowered, and the probability of simultaneous distillation of small quantities of zinc and lead is diminished. A rough test, commonly applied to indicate the presence of any considerable percentage of base metal, is conducted by placing a drop of the mercury upon an inclined surface of smooth glass or glazed porcelain; if pure, it should retain its spherical shape and roll over the surface, leaving no trace behind; but if impure, it assumes an elongated form and tends to leave a grey trail behind it, or, in other words, it is said to *tail*. Mercury may be monovalent or divalent, and thus forms two oxides, *mercurous* ($Hg_2 O$) and *mercuric* ($Hg O$), with their corresponding salts. As these salts deposit mercury on base metal by simple immersion, and the reduced mercury then amalgamates with the remainder of the other metal, their solutions must be used in the operating room with as much circumspection as quicksilver itself.

Mercury Chlorides—*Mercurous chloride* or *calomel*, $Hg_2 Cl_2$, and *mercuric chloride* or *corrosive sublimate*, $Hg Cl_2$.—The former is a heavy white powder insoluble in water; the latter an extremely poisonous white, crystalline body, of which about 7 parts dissolve in 100 parts of cold, 53 in a like weight of hot water.

Mercurous Nitrate, $Hg_2 (NO_3)_2 . 2 H_2 O$.—A white, crystalline, very poisonous, substance, which may deposit basic salt when treated with water, but is readily soluble in water containing a little nitric acid. It is best made by treating an excess of mercury with cold dilute nitric acid. The hot concentrated acid tends to produce mercuric nitrate, $Hg (NO_3)_2$.

Nickel, Ni.—A white metal of specific gravity 8·9, and very high fusing point. Formerly it could be obtained only in the cast condition, but by improved methods of treatment it is now readily procurable rolled into sheet of any required size. The chief impurities affecting its use are iron, copper, cobalt, and arsenic. It forms two oxides, but the chief salts belong to the monoxide group (NiO), in which it is divalent. Nickel is slowly dissolved by sulphuric or hydrochloric acids, rapidly by nitric acid, the attack being always favoured by heating.

Nickel Carbonate, $Ni C O_3$.—An insoluble pale apple-green powder. An impure carbonate containing an excess of oxide is produced by adding potassium or sodium carbonate to the solution of a nickel salt.

Nickel Chloride, $Ni Cl_2 . 6 H_2 O$.—Green soluble crystals, resulting from the solution of oxide, metal, or carbonate in hydrochloric acid.

Nickel Citrate, $Ni (C_6 H_5 O_7)_2 . 14 H_2 O$.—A soluble green body, formed by dissolving nickel oxide or carbonate in citric acid.

Nickel Nitrate, $Ni (NO_3)_2 . 6 H_2 O$.—Green crystals, of which 50 parts are soluble in 100 of cold water. May be formed like the chloride, substituting nitric for hydrochloric acid.

Nickel Sulphate, $Ni SO_4 . 7 H_2 O$.—The most generally known and used salt of nickel. It is full green in colour, and is soluble in water to the extent of 37 parts in 100.

Nickel-Ammonium Sulphate, Ni SO_4 . $(N H_4)_2 SO_4$. 6 $H_2 O$.— Resembles the last, but 100 parts of water dissolve only 5·5 parts of the salt. It may be made by dissolving together nickel sulphate and ammonium sulphate, and evaporating the solution until crystals are obtained.

Phosphorus, P.—A non-metallic elementary substance procurable in two modifications—vitreous and amorphous. The *vitreous phosphorus* is sold in colourless or yellowish translucent sticks which gradually become slightly opaque, especially upon the surface. It is poisonous, and is insoluble in water, but dissolves very readily in certain liquids, of which carbon bisulphide is a type. It is a most oxidisable body, and takes fire spontaneously when exposed to the air ; it must, therefore, be preserved under water, and should only be removed from it when required for use, and then all operations must be conducted rapidly and carefully. If it is required to cut the blocks into smaller fragments, they should be placed singly in a shallow dish containing sufficient water to cover them completely ; they may then be cut with a penknife, but on no account should they be so cut except under water, as the friction of the knife may suffice to inflame the phosphorus when in contact with air. Fragments must be prevented from clinging under the finger nail, as should they inflame subsequently, very troublesome sores may be produced. The pieces should be rapidly dried between pieces of blotting-paper and used without delay, being handled as little as possible ; it is safer for those unaccustomed to work with chemical substances to hold them with light brass tongs.

Phosphorus is soluble in oils and in carbon bisulphide ; its solution in the latter substance is used occasionally to assist in the metallisation of electrotype moulds (see pp. 153, 157), but it is a most dangerous liquid to work with, owing to the readiness with which the solvent evaporates and leaves upon any object a thin film of phosphorus which often takes fire spontaneously. This solution and its destructive properties have long been known under the name of *Greek fire*. This, and indeed all operations involving the use of stick-phosphorus, should be undertaken only by experienced persons, and should, if possible, be excluded from common workshop use.

The *amorphous* (or red) phosphorus, which is prepared by heating the vitreous variety to 464° F. with suitable precautions for the exclusion of air, is not spontaneously inflammable at ordinary temperatures, and is not poisonous; but as it is insoluble in carbon bisulphide, it is useless for the purposes to which this element is usually applied in electro-metallurgy.

Plaster of Paris, from the mineral *gypsum*.—This is a more or less pure *calcium sulphate*, $CaSO_4$. Its use as a plaster depends upon the property possessed by the anhydrous substance (all trace of water having been expelled by the application of heat) of taking to itself a quantity of water in chemical combination, to form the hydrated salt, $CaSO_4 . 2H_2O$. In doing this a considerable amount of heat is evolved, expansion ensues, and the cream formed by the admixture of water and the powdered material sets into a substance, which rapidly hardens as the combination becomes complete and the excess of liquid is absorbed. Since the value of the plaster is dependent upon its power of absorbing water, it must never be allowed to remain exposed to the moisture of the air, from which it would slowly extract its full measure of water of hydration, but must be preserved in well-closed vessels. Gypsum, which has been overburnt, refuses to absorb water, and is, therefore, useless. A sample of the plaster when made into a cream with water should become warm, and in the course of half-an-hour set into a firm, solid, but porous mass. For moulding purposes the plaster must be free from foreign matter, especially from gritty particles.

Platinum, Pt.—A heavy, brilliantly-white metal, unalterable in air, very ductile and fusible, and of extremely high fusing point. Its specific gravity is 21·5. It dissolves only in aqua regia, being unaffected by either hydrochloric or nitric acid alone, and forms two sets of salts, corresponding to the oxides PtO and PtO_2 respectively.

Platinic Chloride, $PtCl_4 . 5H_2O$.—Red crystals soluble in water; but the substance usually known by this name is *hydroplatinic chloride*, $PtH_2Cl_6 . 6H_2O$. It results from evaporating a solution of the metal in aqua regia, together with a good excess of hydrochloric acid, and thus forms red brown, very soluble—and, indeed, deliquescent—crystals.

Plumbago, sometimes known as *graphite* or *black-lead*.—It is an impure natural variety of carbon; and is found very abundantly in Cumberland and in Ceylon. Being a conductor of electricity, it is largely used for facing non-conductive surfaces,

22

which are to receive an electro-deposit of any metal. It should be very finely crushed, even to an impalpable powder. As some varieties are very inferior conductors, samples should be tested for efficiency in this respect before final selection for use.

Potassium Acetate, $K C_2 H_3 O_2$.—White soluble crystals; 100 parts of cold water dissolving about 230 of the salt.

Potassium Carbonate, $2 K_2 C O_3 . 3 H_2 O$.—White crystals very soluble in water; often used in the anhydrous state, when 100 parts of water dissolve about 105 parts of the solid. It is decomposed, with effervescence, by the addition of acid.

Potassium Bicarbonate, $K H C O_3$.—A much less soluble salt (25 in 100) which may be formed by passing carbon dioxide (carbonic acid gas) through a strong solution of the normal carbonate.

Potassium Citrate.—White soluble crystals formed by just neutralising citric acid with potassium carbonate.

Potassium Cyanide, $K C N$.—A white opaque solid, generally bought in irregular lumps or in sticks. It is very soluble in water; and owing to its becoming decomposed by even the weakest acids, carbonic acid among the number, it gradually alters by exposure to the air, especially in large towns where the atmosphere is laden with carbon dioxide, slowly evolving hydrocyanic acid, which imparts to it the peculiar and characteristic faint smell of bitter almonds. It is a deadly poison, and must be used with the utmost caution. Taken internally in minute quantities it may cause instant death, while the solution passing into the blood through cuts in the hand give rise to painful sores and blood-poisoning; even the fumes, in a badly-ventilated room cause headache and depression. The commercial cyanide is rarely pure, that known as *gold cyanide* is the best, the *silver cyanide* is inferior. It should always be tested before use, as it frequently contains less than half its weight of the pure salt.

Potassium Ferro-Cyanide, $K_4 Fe C_6 N_6 . 3 H_2 O$. *Yellow prussiate of potash.*—Yellow crystals, of which 25 parts dissolve in 100 of water. A very commonly procurable substance. It gives a deep blue precipitate of Prussian blue when mixed with a solution of ferric chloride.

Potassium Hydrate, K H O—*Caustic Potash.*—A most powerful caustic alkali bought, like the cyanide, in sticks or cakes. It is soluble in water with evolution of much heat, and substances, moistened with the solution, give rise to a peculiar slimy sensation of the skin when touched. It should never be allowed to enter the mouth, as even dilute solutions almost immediately remove the lining of tender skin. Should such an event happen, the mouth should be at once rinsed several times with water and then with very dilute acetic acid. This body, whether in the solid state or in solution, must be carefully stored in well-closed vessels as it rapidly becomes converted into carbonate by absorption of carbonic acid from the air, and thus loses its caustic properties.

Potassium Iodide, K I.—An intensely soluble, white crystalline substance, 150 parts of which dissolve in 100 of cold water. It is decomposed by nitric acid with separation of iodine, which colours the solution yellow if dilute, or produces a dark almost black precipitate if it be concentrated. Strong sulphuric acid has a similar effect.

Potassium Binoxalate, $K C_2 H O_4$—*Salt of Sorrel.*—White crystals, not largely soluble in cold water, but imparting to it an acid reaction, turning blue litmus-paper red.

Potassium Sulphocyanide, K C N S.—A very soluble white salt, absorbing much heat (or, as it is more commonly said, producing great cold) when dissolved in water. Its solution gives a blood-red colour when mixed with ferric chloride.

Potassium Bitartrate, $K C_4 H_5 O_6$—*Cream of Tartar.*—A somewhat insoluble acid salt, 100 parts of water dissolving only 0·5 parts in the cold or 7 at the boiling temperature. It is colourless when pure, but the commercial crude *tartar* or *argol*, which is a bye-product in the wine industry, is usually stained purple. The pure salt may be made from this by dissolving it in water, filtering it and allowing it to crystallise on cooling.

Potassium-Sodium Tartrate, $K Na C_4 H_4 O_6 . 4 H_2 O$—*Rochelle-* or *Seignette-salt.*—A very soluble white crystalline substance, which may be made by adding 4 parts of potassium bitartrate and 3 of crystallised sodium bicarbonate, little by little, to 12 parts of boiling water, and then cooling in order to allow crystals to deposit.

Rosin, or *Colophony*.—One of a large series of bodies termed resins, exuded by certain trees. Common rosin is deep amber to brown in colour, and should be translucent and brittle. It becomes slightly but distinctly softened at a temperature of 120° F., and as the temperature rises increases in softness until it becomes viscous, and, finally, liquid at about the temperature of boiling water.

Silver, Ag.—A very white and unalterable metal with a specific gravity of 10·4 to 10·5 and a fusing point of 1873° F. It is extremely malleable and ductile, and is at the same time the best known conductor of heat and electricity. It combines readily with sulphur, and is thus rapidly covered with a black tarnish of silver sulphide in the atmosphere of towns. Alloyed with copper it is used for silver coinage, the amount of alloy varying in different countries, the English standard being 92·5 of silver to 7·5 of copper. To prepare *fine silver* (*i.e.*, pure silver) from such an alloy, the metal should be dissolved in nitric acid in a glass or glazed earthenware vessel; any black residue is gold, of which there is frequently a small quantity present, and must be filtered off. To the solution (which is blue, owing to the presence of copper) a common salt solution, or better, dilute hydrochloric acid, is slowly added so long as it continues to produce a white curdy precipitate. The liquid is stirred well to promote the subsidence of the latter, and then allowed to settle; a little more of the salt or acid is now added, which should produce no further precipitate (if it should do so, more must be added, until the whole of the silver has thus been thrown down). Any addition of common salt beyond that necessary for complete precipitation only tends to re-dissolve the silver chloride formed, which is fairly soluble in brine; thus, hydrochloric acid is to be preferred as a precipitant, because a moderate excess is without action on the silver salt. The blue copper solution is now poured away from the heavy silver chloride, which is then stirred up with fresh water, and allowed to subside. This washing *by decantation* is repeated several times, the wash waters being disregarded. The chloride is then collected on a filter, dried, and mixed with an equal bulk of dried sodium carbonate, transferred to a fire-clay crucible, and heated to a bright-red heat in a clear charcoal- or coke-fire. As soon as fusion commences effervescence will be observed, due to the mutual decomposition which occurs between the silver chloride and sodium carbonate, whereby sodium chloride and silver carbonate are produced, the latter body being dissociated at the temperature of the operation into carbon dioxide, oxygen, and metallic silver, the two former escaping in the gaseous state, the latter sinking through the slag by virtue of its higher specific gravity, and collecting into a fused mass at the bottom of the pot. When the contents of the crucible are tranquil they are poured, with the aid of a pair of large bent iron tongs, into an iron ingot-mould of cup-shape, from which, when cold, the silver and slag (that is, the fused salt) are readily removed and separated one from the other. Silver forms one set of salts derived from the oxide, Ag_2O, in which the metal is monovalent. Moist silver salts should not be allowed to come into contact with clean surfaces of base metals, which will decompose them by simple exchange; nor should they be exposed unnecessarily to white light, by which many of them are gradually decomposed and darkened in colour.

Silver Carbonate, Ag_2CO_3.—A pale yellow, insoluble substance, formed by adding a carbonate of soda solution to one of a soluble silver salt such as the nitrate.

Silver Chloride, $AgCl$—*Horn Silver.*—A white substance gradually passing through a gradation of shade from violet to black by exposure to white light. It is practically insoluble in water, but dissolves to some extent in solutions of sodium chloride, and readily in ammonia, and in sodium hyposulphite or potassium cyanide solutions. It is formed, as described above under the head of fine silver, by adding hydrochloric acid or common salt to a solution of silver nitrate.

Silver Cyanide, $AgCN$.—A white insoluble salt, best formed by gradually adding a potassium cyanide solution to one of silver nitrate, carefully watching the formation of the precipitate, and allowing it to subside after each addition of cyanide, so that, immediately another drop of the potash salt fails to produce a further precipitate or cloudiness in the liquid, all further addition is stopped, otherwise the silver cyanide will begin to re-dissolve in the excess of the precipitant. The liquid is then washed several times by decantation, as in the case of the chloride. To obtain the pure cyanide, only distilled water must be used; ordinary spring- or river-water, or even rain-water, contain chlorides, which cause the contamination of the cyanide by silver chloride. The essentials for success are pure substances, and precisely the right proportion between the silver nitrate and potassium cyanide. The silver cyanide dissolves readily in ammonia and sodium hyposulphite as well as in potassium cyanide.

Silver Iodide, Ag I.—Has a pale yellow colour; it is readily formed by adding potassium iodide to silver nitrate solutions. It is insoluble in water, and practically even in strong ammonia; strong potassium iodide liquor, however, dissolves a fair proportion of the salt.

Silver Nitrate, Ag N O$_3$—*Lunar Caustic.*—A white crystalline body, obtainable readily in crystals, but sometimes fused into sticks. It dissolves readily in water. In making solutions of this or of any other silver salt, only distilled water should be employed; all other waters, owing to the presence of chlorine, produce a cloudiness or even a distinct precipitate of silver chloride.

Silver Oxide, Ag$_2$ O.—A deep brown, or almost black, insoluble powder, obtained by adding caustic soda or potash to silver nitrate solution.

Silver Sulphate, Ag$_2$ S O$_4$.—Brilliant white crystals, only slightly soluble in cold water, but more so in boiling water: they are also soluble in strong sulphuric acid, from which they are partly reprecipitated by the addition of water.

Sulphur, S, formerly better known as *brimstone.*—Obtainable as *flowers of sulphur* and as stick or *roll* sulphur, both of which are pale yellow in colour, but the latter variety is crystalline and soluble in carbon bisulphide, while the former is amorphous and insoluble. It melts at 238° F., and will yield sharp impressions of objects, so that it is occasionally useful in obtaining casts. Mixed with iron sulphide it forms the basis of *Spence's metal*, of which medallions and plâques have been sometimes made, and may thus come into the hands of the electrotyper. Being very fragile, objects made of sulphur or Spence's composition, must not be subjected to pressure in taking casts from them with other moulding materials. Sulphur is, of course, very inflammable, and, therefore, requires care in melting.

Sodium Carbonate, Na$_2$ C O$_3$. 10 H$_2$ O—*Washing Soda* or *Soda Crystals.*—Very soluble, colourless, alkaline crystals. It behaves chemically like potassium carbonate. An impure kind, containing, *inter alia,* caustic soda and various foreign salts, is sold as a non-crystalline powder under the name of *soda ash,* which is suitable for fluxing in obtaining fine silver or gold, but should not be employed in making up electrolytic baths. A similar variety, commonly known as *refined alkali*, is purer, but still not always safe.

Sodium Bicarbonate, $Na\,HCO_3$.—A white soluble powder, whose relation to the carbonate is analogous to that between the corresponding potassium salts.

Sodium Chloride, $Na\,Cl$—*Common Salt; Table Salt; Rock Salt;* or *Bay Salt,* the latter are not always pure.—The pure salt should form white cubical crystals, of which 100 parts of cold water dissolve 36 parts, hot water taking up slightly more. The natural varieties, or rock salt, frequently contain a considerable percentage of iron, which imparts a brown or purple tint to the body; while salt obtained from the sea water is often found to contain magnesium compounds and other bodies.

Sodium Citrate.—Soluble colourless crystals formed by neutralising citric acid with sodium carbonate.

Sodium Hydrate, $Na\,O\,H$—*Caustic Soda.*—White soluble lumps of a highly caustic character resembling potassium hydrate (which see) in properties and effects.

Sodium Phosphate.—There are three principal phosphates— the *orthophosphate,* $Na_3\,PO_4\,.\,12\,H_2\,O$; the *pyrophosphate,* $Na_4\,P_2\,O_7\,.\,10\,H_2\,O$; and the *metaphosphate,* $Na\,PO_3$. All of them are white bodies soluble in water; but the orthophosphate, or rather the *disodium-orthophosphate,* $Na_2\,HPO_4\,.\,12\,H_2\,O$, in which one atom of hydrogen takes the place of one of sodium, is that more commonly met with. The pyrophosphate is sometimes used in making up baths.

Sodium Sulphite, $Na_2\,SO_3\,.\,7\,H_2\,O$.—White soluble crystals, with an alkaline reaction.

Sodium Bisulphite, $Na\,HSO_3$, is a similar body, but with an acid reaction. Both are compounds of soda with sulphurous acid.

Sodium Hyposulphite, $Na_2\,S_2\,O_3\,.\,5\,H_2\,O$.—Colourless soluble crystals, which have the property of dissolving silver salts by forming a soluble double hyposulphite of silver and soda.

Sodium Tartrate, $Na_2\,C_4\,H_4\,O_6\,.\,2\,H_2\,O$.—White and soluble crystals, much more soluble in hot than in cold water.

Tin, Sn.—A white, soft, very fusible and malleable metal, too
weak to possess any great ductility, with specific gravity 7·29,
and fusing point 442° F. It is not readily tarnishable, and,
therefore, retains its brilliancy for a long time when exposed to
the air. Easily soluble in hydrochloric acid or aqua regia, but
converted into an insoluble white oxide by nitric acid. It forms
two series of salts, corresponding to the oxides *stannic*, SnO_2, in
which it is tetravalent, and *stannous*, SnO, where it is divalent.
Commercially, tin frequently contains traces of lead, tungsten,
iron, copper, antimony, or arsenic.

Tin Tetrachloride, $SnCl_4$.—*Stannic Chloride.*—A colourless,
fuming liquid, boiling at 248° F. It forms several solid hydrates,
mostly crystalline, by the addition of water; such are the *butter
of tin* and *oxymuriate of tin.*

Tin Dichloride, $SnCl_2$. $2H_2O$—*Stannous Chloride* or *Tin Salt.*
—A white soluble crystalline substance, formed by dissolving
tin in hydrochloric acid.

Sodium Stannate, Na_2SnO_3.—A white substance soluble in
water, formed by fusing either stannic oxide (the dressed ore,
tin stone, or *cassiterite* may be used) with caustic soda; or the
metal itself with caustic soda to which sodium nitrate has been
added. The aqueous solution, when evaporated, yields crystals
containing water of crystallisation.

Varnish—*Lacquer Varnish.*—The formulæ for this varnish,
which is used for protecting metallic surfaces from tarnish, are
almost innumerable. Perhaps the best are those in which seed
lac is dissolved in from 8 to 10 parts of the strongest spirits of
wine, freed as far as possible from water, and coloured by the
addition of dragon's blood or gamboge (say from $\frac{1}{8}$ to $\frac{1}{2}$ of a part),
or by mixtures of these, according to the particular tint that it is
desired to obtain.

Stopping-off Varnish.—Since the object of this class of varnish
is to prevent the formation of an electrolytic deposit upon any
desired portion of an article, the requirements are evidently—
a non-conductive material, easily applied and with facility removable, which shall not be attacked by the solution in which it is
to be immersed, and which should of preference be coloured in
such a way that the portions of the surface protected by it may
be seen at a glance, for convenience of application. To resist
ordinary bath-solutions used cold, almost any varnish is applic-

able, and common copal varnish, mixed with a colouring medium, will be found suitable; or asphalt varnish may be used, as for the back of electrotype plates. But for hot cyanide solutions other materials are required, and for this work, a mixture of 44 per cent. rosin with 26 of bees'-wax, 17 of sealing-wax, and 13 of jeweller's rouge, may be applied with advantage, provided that the best materials alone are employed in preparing it.

Zinc, Zn.—A bluish-white divalent metal, fusible at 773° F., easily distilled at a higher temperature. Specific gravity, 7·1. It is brittle in the cold, and above 250° F., but near the temperature of boiling water, it is sufficiently malleable to admit of rolling into thin sheets. It forms one class of salts only, from the monoxide, ZnO. The chief impurities in the commercial metal are lead, iron, cadmium, and arsenic.

Zinc Acetate, $Zn(C_2H_3O_2)_2 . 3H_2O$.—Very soluble white crystals.

Zinc Carbonate, $ZnCO_3$.—A white insoluble substance, prepared by adding sodium carbonate in excess to a solution of any zinc salt. It is decomposed by acids with effervescence.

Zinc Chloride, $ZnCl_2$. — A very soluble and deliquescent white, opaque, soft mass; may be prepared by dissolving zinc or its oxide or carbonate in hydrochloric acid.

Zinc Cyanide, $Zn(CN)_2$.—A white substance insoluble in water, but readily so in potassium cyanide solutions. Prepared by precipitating a solution of zinc sulphate with one of potassium cyanide; of course, carefully avoiding excess of the latter.

Zinc Oxide, ZnO.—A white substance, becoming yellowish when strongly heated, but returning to its original colour on cooling. The *hydroxide*, or hydrated oxide, $Zn(OH)_2$, is precipitated when an exact equivalent of caustic alkali is added to a solution of a zinc salt; an excess of the alkali tends to redissolve the precipitate.

Zinc Sulphate, $ZnSO_4 . 7H_2O$— *White Vitriol.*—Soluble white crystals; the commonest salt of zinc. It may be prepared like the chloride, of course substituting sulphuric for hydrochloric acid. 100 parts of water dissolve about 50 of the salt in the cold, and nearly 100 at the boiling point.

ADDENDA.

TABLE XXVIII.—Giving Data for Calculating the Weight and Thickness of Deposit Produced by a known Current-Volume in a Given Time for certain of the Commoner Metals.

METAL.	Atomic Weight.	Equivalent Weight.	Specific Gravity.	Electro-Chemical Equivalent.	Weight deposited per hour by current of 1 Ampère.		Thickness of Deposit produced in 1 hour by current of :—	
					Grammes.	Grains.	1 Ampère per Square Decimetre.	1 Ampère per Square Inch.
							Mm.	Inches.
Aluminium. . . .	27	9·0	2·6	0·00009317	0·3354	5·175	0·0129	0·007875
Antimony	122	40·6	6·8	·00042025	1·5130	23·350	·0222	·013584
Arsenic	75	25	5·8	·00025880	0·9317	14·378	·0161	·009806
Bismuth	210	70	9·8	·00072464	2·6087	40·258	·0266	·016251
Cadmium	112	56	8·6	·00057971	2·0870	32·207	·0243	·014811
Chromium	52·5	17·5	7·0	·00018116	0·6522	10·052	·0093	·005687
Cobalt	59	29·5	8·7	·00030538	1·0994	16·966	·0126	·007715
Copper (Monovalent)	63·5	63·5	8·9	·00065735	2·3665	36·520	·0265	·016208
,, (Divalent) .	63·5	31·7	8·9	·00032867	1·1832	18·260	·0133	·008104
Gold	196·6	65·5	19·3	·00067806	2·4410	37·670	·0127	·007721
Iridium	197	49·2	21·1	·00050932	1·8335	28·296	·0087	·005291
Iron (Divalent) . .	56	28	8·1	·00028986	1·0435	16·103	·0128	·007826
Lead	207	103·5	11·4	·00107140	3·8571	59·525	·0338	·021134
Magnesium. . . .	24·3	12·1	1·7	·00012526	0·4509	6·959	·0265	·016190
Manganese. . . .	55	27·5	8·0	·00028468	1·0248	15·816	·0128	·007821
Nickel	59	29·5	8·5	·00030538	1·0994	16·966	·0129	·007894
Palladium	106·5	26·6	11·4	·00027536	0·9913	15·298	·0087	·005291
Platinum	197	44·3	21·2	·00045859	1·6509	25·478	·0078	·004743
Silver	108	108	10·6	·0011180	4·0249	62·113	·0380	·023142
Tin (Divalent). . .	118	59	7·3	·00061077	2·1988	33·932	·0302	·018414
Zinc	65	32·5	7·1	·00033644	1·2112	18·691	·0171	·010415

NOTE.—These figures are based on Lord Rayleigh's number for the electro-chemical equivalent of hydrogen = 0·000010352.

TABLE XXIX.—Showing the Value of Equal Current Volumes as Expressed in Ampères per Square Decimetre, per Square Foot, and per Square Inch of Electrode Surface.

Ampères per Square Decimetre.	= Ampères per Square Foot.	= Ampères per Square Inch.	Ampères per Square Decimetre.	= Ampères per Square Foot.	= Ampères per Square Inch.	Ampères per Square Decimetre.	= Ampères per Square Foot.	= Ampères per Square Inch.
0·05	0·46	0·0032	0·8	7·43	0·0516	6·20	57·6	0·4
0·054	0·5	0·0035	0·86	8	0·0555	6·46	60	0·4167
0·077	0·72	0·005	0·9	8·36	0·0581	7	65·0	0·4516
0·1	0·93	0·0064	0·93	8·64	0·06	7·53	70	0·4861
0·11	1	0·0069	0·97	9	0·0625	7·75	72·0	0·5
0·15	1·44	0·01	1	9·29	0·0645	8	74·3	0·5161
0·2	1·86	0·0129	1·08	10	0·0694	8·61	80	0·5555
0·22	2	0·0139	1·09	10·08	0·07	9	83·6	0·5806
0·3	2·79	0·0193	1·24	11·52	0·08	9·30	86·4	0·6
0·31	2·88	0·02	1·39	12·96	0·09	9·69	90	0·6250
0·32	3	0·0208	1·55	14·4	0·1	10	92·9	0·6452
0·4	3·71	0·0258	2	18·6	0·1290	10·76	100	0·6944
0·43	4	0·0278	2·15	20	0·1389	10·85	100·8	0·7
0·46	4·32	0·03	3	27·9	0·1935	12·40	115·2	0·8
0·5	4·64	0·0323	3·10	28·8	0·2	13·95	129·6	0·9
0·54	5	0·0348	3·23	30	0·2083	15·50	144·0	1
0·6	5·57	0·0387	4	37·1	0·2581	20	185·8	1·2903
0·62	5·76	0·04	4·30	40	0·2778	21·53	200	1·3889
0·65	6	0·0417	4·60	43·2	0·3	30	278·7	1·9355
0·7	6·50	0·0452	5	46·4	0·3226	31·0	288	2
0·75	7	0·0486	5·38	50	0·3478	32·3	300	2·0833
0·77	7·20	0·05	6	55·7	0·3871	46·5	432·0	3

By this table the current density may be expressed in ampères per square decimetre, square foot, or square inch, any one of them be ng given. Thus a current of 1 ampère per square decimetre has the same electrolytic value as one of 9·29 ampères per square foot, or 0·0645 per square inch. To find the value of intermediate numbers not shown above, add together the various numbers representing the hundreds, tens, units, and decimals of the given quantity. Thus 27·5 ampères per square decimetre (= 20 + 7 + 0·5) is equivalent to 185·8 + 65 + 4·64 = 255·44 ampères per square foot, or 1·2903 + 0·4516 + 0·0323 = 1·7742 ampères per square inch.

TABLE XXX.—Showing the Specific Gravities of Sulphuric, Nitric, and Hydrochloric Acids corresponding to varying percentages of Pure H_2SO_4, HNO_3, and HCl in the Liquids respectively.

%	H_2SO_4	HNO_3	HCl.	%	H_2SO_4	HNO_3	HCl.	%	H_2SO_4	HNO_3	HCl.
100	1·8426	1·500	...	66	1·568	1·3783	...	33	1·247	1·1895	1·1640
99	1·8420	1·498	...	65	1·557	1·3732	...	32	1·239	1·1833	1·1584
98	1·8406	1·496	...	64	1·545	1·3681	...	31	1·231	1·1770	1·1536
97	1·8400	1·494	...	63	1·534	1·3630	...	30	1·223	1·1709	1·1484
96	1·8384	1·491	...	62	1·523	1·3579	...	29	1·215	1·1648	1·1433
95	1·8376	1·488	...	61	1·512	1·3529	...	28	1·2066	1·1587	1·1382
94	1·8356	1·485	...	60	1·501	1·3477	...	27	1·1980	1·1515	1·1333
93	1·8340	1·482	...	59	1·490	1·3427	...	26	1·190	1·1467	1·1282
92	1·8310	1·479	...	58	1·480	1·3376	...	25	1·182	1·1403	1·1232
91	1·8270	1·476	...	57	1·469	1·3323	...	24	1·174	1·1345	1·1182
90	1·8220	1·473	...	56	1·4586	1·3270	...	23	1·167	1·1286	1·1131
89	1·8160	1·470	...	55	1·448	1·3216	...	22	1·159	1·1227	1·1081
88	1·8090	1·467	...	54	1·438	1·3163	...	21	1·1516	1·1168	1·1031
87	1·802	1·464	...	53	1·428	1·3110	...	20	1·144	1·1109	1·0981
86	1·794	1·460	...	52	1·418	1·3056	...	19	1·136	1·1051	1·0931
85	1·786	1·457	...	51	1·408	1·3001	...	18	1·129	1·0993	1·0882
84	1·777	1·453	...	50	1·398	1·2947	...	17	1·121	1·0935	1·0832
83	1·767	1·450	...	49	1·3886	1·2887	...	16	1·1136	1·0878	1·0783
82	1·756	1·446	...	48	1·379	1·2826	...	15	1·106	1·0821	1·0734
81	1·745	1·4424	...	47	1·370	1·2765	...	14	1·098	1·0764	1·0684
80	1·734	1·4385	...	46	1·361	1·2705	...	13	1·091	1·0708	1·0635
79	1·722	1·4346	...	45	1·351	1·2644	...	12	1·083	1·0651	1·0586
78	1·710	1·4306	...	44	1·342	1·2583	...	11	1·0756	1·0595	1·0537
77	1·698	1·4269	...	43	1·333	1·2523	...	10	1·068	1·0540	1·0487
76	1·686	1·4228	...	42	1·324	1·2464	...	9	1·061	1·0485	1·0438
75	1·675	1·4189	...	41	1·315	1·2402	...	8	1·0536	1·0430	1·0389
74	1·663	1·4147	...	40	1·306	1·2341	1·1966	7	1·0464	1·0375	1·0340
73	1·651	1·4107	...	39	1·297	1·2277	1·1922	6	1·039	1·0320	1·0292
72	1·639	1·4065	...	38	1·289	1·2212	1·1878	5	1·032	1·0267	1·0244
71	1·627	1·4023	...	37	1·281	1·2148	1·1840	4	1·0256	1·0212	1·0196
70	1·615	1·3978	...	36	1·272	1·2084	1·1786	3	1·019	1·0159	1·0148
69	1·604	1·3945	...	35	1·264	2·2019	1·1738	2	1·013	1·0106	1·0098
68	1·592	1·3882	...	34	1·256	1·1958	1·1689	1	1·0064	1·0053	1·0049
67	1·580	1·3833	...								

Note.—The sulphuric acid numbers are quoted from Otto, those for nitric acid from Ure; while the hydrochloric acid figures are compiled by interpolation from Ure. Liquid hydrochloric acid is practically saturated with 40 per cent. of H Cl gas.

TABLE XXXI.—Showing the Specific Gravities of Solutions
corresponding to the Degrees of the Baumé Hydrometer.

Degree Baumé.	= Specific Gravity.	Degree Baumé.	= Specific Gravity.	Degree Baumé.	= Specific Gravity.	Degree Baumé.	= Specific Gravity.
0	1·000	19	1·147	37	1·337	55	1·596
1	1·007	20	1·157	38	1·349	56	1·615
2	1·014	21	1·166	39	1·361	57	1·634
3	1·020	22	1·176	40	1·375	58	1·653
4	1·028	23	1·185	41	1·388	59	1·671
5	1·034	24	1·195	42	1·401	60	1·690
6	1·041	25	1·205	43	1·414	61	1·709
7	1·049	26	1·215	44	1·428	62	1·729
8	1·057	27	1·225	45	1·442	63	1·750
9	1·064	28	1·235	46	1·456	64	1·771
10	1·072	29	1·245	47	1·470	65	1·793
11	1·080	30	1·256	48	1·485	66	1·815
12	1·088	31	1·267	49	1·500	67	1·839
13	1·096	32	1·278	50	1·515	68	1·864
14	1·104	33	1·289	51	1·531	69	1·885
15	1·113	34	1·300	52	1·546	70	1·909
16	1·121	35	1·312	53	1·562	71	1·935
17	1·130	36	1·324	54	1·578	72	1·960
18	1·138						

Note.—The specific gravity of a solution is rapidly ascertained by floating in it a weighted glass tube closed at both ends, with a bulb in the centre and a long thin glass tube above, which carries a graduated scale upon it. This instrument sinks deeper in solutions of low density than in those of high gravity; and the actual specific gravity is found by the level at which the liquid stands on the graduated portion when the apparatus is floating freely in it. *Hydrometers* of this kind are sometimes graduated so that the specific gravity is read off direct from the scale, others are graduated by Baumé's method; and the reading may then be converted into the number representing the true density, by reference to the above table.

TABLE XXXII.—Densities of Solutions of Crystallised Copper
and Zinc Sulphates.

Copper Sulphate.				Zinc Sulphate.			
Percentage by Weight of $CuSO_4 . 5 H_2O$.	Density.	Percentage by Weight of $CuSO_4 . 5 H_2O$.	Density.	Percentage by Weight of $ZnSO_4 . 7 H_2O$.	Density.	Percentage by Weight of $ZnSO_4 . 7 H_2O$.	Density.
2	1·0126	14	1·0923	5	1·0289	35	1·2285
4	1·0254	16	1·1063	10	1·0588	40	1·2674
6	1·0384	18	1·1208	15	1·0899	45	1·3083
8	1·0516	20	1·1354	20	1·1222	50	1·3511
10	1·0649	22	1·1501	25	1·1560	55	1·3964
12	1·0785	24	1·1659	30	1·1914	60	1·4451

Note.—The pure salt is supposed to be dissolved in pure distilled water.

TABLE XXXIII.—Showing the Specific Electrical Resistances* of
Different Sulphuric Acid Solutions at various Temperatures
(*Fleeming Jenkin*).

Specific Gravity of Acid.	TEMPERATURES (FAHRENHEIT).							
	32°	39°·2	46°·4	53°·6	60°·8	68°	75°·2	82°·4
1·10	1·37	1·17	1·04	0·92	0·84	0·79	0·74	0·71
1·20	1·33	1·11	0·93	0·79	0·67	0·57	0·49	0·41
1·25	1·31	1·09	0·90	0·74	0·62	0·51	0·43	0·36
1·30	1·36	1·13	0·94	0·79	0·66	0·56	0·47	0·39
1·40	1·69	1·47	1·30	1·16	1·05	0·96	0·89	0·84
1·50	2·74	2·41	2·13	1·89	1·72	1·61	1·32	1·43
1·60	4·32	4·16	3·62	3·11	2·75	2·46	2·21	2·02
1·70	9·41	7·67	6·25	5·12	4·23	3·57	3·07	2·71

TABLE XXXIV.—Showing the Specific Electrical Resistances* of
different Copper Sulphate Solutions at various Temperatures
(*Fleeming Jenkin*).

No. of parts of Copper Sulphate dissolved in 100 parts of water.	TEMPERATURES (FAHRENHEIT).						
	57°·2	60°·8	64°·4	68°	75°·2	82°·4	86°
8	45·7	43·7	41·9	40·2	37·1	34·2	32·9
12	36·3	34·9	33·5	32·2	29·9	27·9	27·0
16	31·2	30·0	28·9	27·9	26·1	24·6	24·0
20	28·5	27·5	26·5	25·6	24·1	22·7	22·2
24	26·9	25·9	24·8	23·9	22·2	20·7	20·0
28	24·7	23·4	22·1	21·0	18·8	16·9	16·0

*Note.—By the term *Specific Resistance* in the above tables is meant the absolute resistance in ohms of a column of the liquid 1 square centimetre in cross-section, and 1 centimetre long; in other words, it is the resistance of a cubic centimetre of the liquid. The diminution of resistance accompanying a rise of temperature should be especially marked.

TABLE XXXV.—THE ELECTRICAL RESISTANCE OF PURE COPPER WIRES OF VARIOUS DIAMETERS.

No. of Wire B.W.G.	Resistance of 1 foot in ohms.	Number of feet required to give Resistance of 1 ohm.	No. of Wire B.W.G.	Resistance of 1 foot in ohms.	Number of feet required to give Resistance of 1 ohm.	No. of Wire B.W.G.	Resistance of 1 foot in ohms.	Number of feet required to give Resistance of 1 ohm.	No. of Wire B.W.G.	Resistance of 1 foot in ohms.	Number of feet required to give Resistance of 1 ohm.
0000	·0000516	19358	7	0·000329	3043·4	17	0·00316	316·1	27	0·04159	24·0
000	·0000589	16964	8	·000391	2557·1	18	·00443	225·5	28	·05432	18·4
00	·0000737	13562	9	·000486	2057·7	19	·00603	165·7	29	·06300	15·9
0	·0000922	10857	10	·000593	1686·5	20	·00869	115·1	30	·07393	13·5
1	·000118	8452·6	11	·000739	1352·5	21	·01040	96·2	31	·10646	9·4
2	·000132	7575·1	12	·000896	1116·0	22	·01358	73·6	32	·13144	7·6
3	·000159	6300·1	13	·001180	847·7	23	·01703	58·7	33	·16634	6·0
4	·000188	5319·9	14	·001546	647·0	24	·02200	45·5	34	·21727	4·6
5	·000220	4545·9	15	·002053	487·0	25	·02661	37·6	35	·42583	2·4
6	·000258	3870·3	16	·002520	396·8	26	·03286	30·4	36	·66537	1·5

TABLE XXXVI.—ACTUAL DIAMETERS CORRESPONDING TO THE NUMBERS OF THE BIRMINGHAM WIRE GAUGE.

Gauge Number.	Diameter.		Gauge Number.	Diameter.		Gauge Number.	Diameter.	
	Inches.	Milli-metres.		Inches.	Milli-metres.		Inches.	Milli-metres.
0000	0·454	11·53	11	0·120	3·05	24	0·022	0·56
000	·425	10·79	12	·109	2·77	25	·020	·51
00	·380	9·65	13	·095	2·41	26	·018	·46
0	·340	8·63	14	·083	2·11	27	·016	·41
1	·300	7·62	15	·072	1·83	28	·014	·36
2	·284	7·21	16	·065	1·65	29	·013	·33
3	·259	6·58	17	·058	1·47	30	·012	·305
4	·238	6·04	18	·049	1·24	31	·010	·254
5	·220	5·59	19	·042	1·07	32	·009	·229
6	·203	5·16	20	·035	0·89	33	·008	·203
7	·180	4·57	21	·032	·81	34	·007	·178
8	·165	4·19	22	·028	·71	35	·005	·127
9	·148	3·76	23	·025	·63	36	·004	·102
10	·134	3·40						

TABLE XXXVII.—ACTUAL DIAMETERS CORRESPONDING TO THE NUMBERS OF THE NEW IMPERIAL WIRE GAUGE.

Gauge Number.	Diameter.		Gauge Number.	Diameter		Gauge Number.	Diameter.	
	Inches.	Milli-metres.		Inches.	Milli-metres.		Inches.	Milli-metres.
7/0	0·500	12·70	13	0·092	2·34	32	0·0108	0·27
6/0	·464	11·79	14	·080	2·03	33	·0100	·25
5/0	·432	10·97	15	·072	1·83	34	·0092	·23
4/0	·400	10·16	16	·064	1·63	35	·0084	·21
3/0	·372	9·45	17	·056	1·42	36	·0076	·19
2/0	·348	8·84	18	·048	1·22	37	·0068	·17
0	·324	8·20	19	·040	1·02	38	·0060	·152
1	·300	7·62	20	·036	0·91	39	·0052	·132
2	·276	7·01	21	·032	·81	40	·0048	·122
3	·252	6·40	22	·028	·71	41	·0044	·112
4	·232	5·89	23	·024	·61	42	·0040	·102
5	·212	5·38	24	·022	·56	43	·0036	·091
6	·192	4·88	25	·020	·51	44	·0032	·081
7	·176	4·47	26	·018	·46	45	·0028	·071
8	·160	4·06	27	·0164	·42	46	·0024	·061
9	·144	3·66	28	·0148	·38	47	·0020	·051
10	·128	3·25	29	·0136	·35	48	·0016	·041
11	·116	2·95	30	·0124	·32	49	·0012	·030
12	·104	2·64	31	·0116	·29	50	·0010	·025

TABLE XXXVIII.—Comparison of Centigrade and Fahrenheit Thermometers.

Fah.	Cent.	Fah.	Cent.	Fah.	Cent.	Fah.	Cent.	Fah.	Cent.	Fah.	Cent.
Deg.	Deg.	Deg.	Deg.	Deg.	Deg.	Deg.	Deg.	Deg.	Deg.	Deg.	Deg.
32	0	66	18·9	100	37·8	133	56·1	166	74·4	199·4	93
33	0·5	66·2	19	100·4	38	134	56·7	167	75	200	93·3
33·8	1	67	19·4	101	38·3	134·6	57	168	75·5	201	93·9
34	1·1	68	20	102	38·9	135	57·2	168·8	76	201·2	94
35	1·7	69	20·5	102·2	39	136	57·8	169	76·1	202	94·4
35·6	2	69·8	21	103	39·4	136·4	58	170	76·7	203	95
36	2·2	70	21·1	104	40	137	58·3	170·6	77	204	95·5
37	2·8	71	21·7	105	40·5	138	58·9	171	77·2	204·8	96
37·4	3	71·6	22	105·8	41	138·2	59	172	77·8	205	96·1
38	3·3	72	22·2	106	41·1	139	59·4	172·4	78	206	96·7
39	3·9	73	22·8	107	41·7	140	60	173	78·3	206·6	97
39·2	4	73·4	23	107·6	42	141	60·5	174	78·9	207	97·2
40	4·4	74	23·3	108	42·2	141·8	61	174·2	79	208	97·8
41	5	75	23·9	109	42·8	142	61·1	175	79·4	208·4	98
42	5·5	75·2	24	109·4	43	143	61·7	176	80	209	98·3
42·8	6	76	24·4	110	43·3	143·6	62	177	80·5	210	98·9
43	6·1	77	25	111	43·9	144	62·2	177·8	81	210·2	99
44	6·7	78	25·5	111·2	44	145	62·8	178	81·1	211	99·4
44·6	7	78.8	26	112	44·4	145·4	63	179	81·7	212	100
45	7·2	79	26·1	113	45	146	63·3	179·6	82	213	100·5
46	7·8	80	26·7	114	45·5	147	63·9	180	82·2	213·8	101
46·4	8	80·6	27	114·8	46	147·2	64	181	82·8	214	101·1
47	8·3	81	27·2	115	46·1	148	64·4	181·4	83	215	101·7
48	8·9	82	27·8	116	46·7	149	65	182	83·3	215·6	102
48·2	9	82·4	28	116·6	47	150	65·5	183	83·9	216	102·2
49	9·4	83	28·3	117	47·2	150·8	66	183·2	84	217	102·8
50	10	84	28·9	118	47·8	151	66·1	184	84·4	217·4	103
51	10·5	84·2	29	118·4	48	152	66·7	185	85	218	103·3
51·8	11	85	29·4	119	48·3	152·6	67	186	85·5	219	103·9
52	11·1	86	30	120	48·9	153	67·2	186·8	86	219·2	104
53	11·7	87	30·5	120·2	49	154	67·8	187	86·1	220	104·4
53·6	12	87·8	31	121	49·4	154·4	68	188	86·7	221	105
54	12·2	88	31·1	122	50	155	68·3	188·6	87	250	121
55	12·8	89	31·7	123	50·5	156	68·9	189	87·2	302	150
55·4	13	89·6	32	123·8	51	156·2	69	190	87·8	400	204
56	13·3	90	32·2	124	51·1	157	69·4	190·4	88	482	250
57	13·9	91	32·8	125	51·7	158	70	191	88·3	572	300
57·2	14	91·4	33	125·6	52	159	70·5	192	88·9	752	400
58	14·4	92	33·3	126	52·2	159·8	71	192·2	89	932	500
59	15	93	33·9	127	52·8	160	71·1	193	89·4	1112	600
60	15·5	93·2	34	127·4	53	161	71·7	194	90	1292	700
60·8	16	94	34·4	128	53·3	161·6	72	195	90·5	1472	800
61	16·1	95	35	129	53·9	162	72·2	195·8	91	1652	900
62	16·7	96	35·5	129·2	54	163	72·8	196	91·1	1832	1000
62·6	17	96·8	36	130	54·4	163·4	73	197	91·7	2282	1250
63	17·2	97	36·1	131	55	164	73·3	197·6	92	2732	1500
64	17·8	98	36·7	132	55·5	165	73·9	198	92·2	3182	1750
64·4	18	98·6	37	132·8	56	165·2	74	199	92·8	3632	2000
65	18·3	99	37·2								

TABLE XXXIX.—Avoirdupois Weight.

	= Ounces.	= Drachms.	= Grains.	= Grammes.
1 Pound, ...	16	256	7,000	453·25
1 Ounce, ...	1	16	437·5	28·33
1 Drachm, . . .	0·062	1	27·34	1·77

TABLE XL.—Troy Weight.

	= Ounces.	= Pennyweight.	= Grains.	= Grammes.
1 Pound, ...	12	240	5,760	372·96
1 Ounce, ...	1	20	480	31·08
1 Pennyweight, . .	0·05	1	24	1·55

TABLE XLI.—Apothecaries' Weight.

	= Ounces.	= Drachms.	= Scruples.	= Grains.	= Grammes.
1 Pound, . . .	12	96	288	5,760	372·96
1 Ounce, . . .	1	8	24	480	31·08
1 Drachm, . . .	0·125	1	3	60	3·88
1 Scruple, . . .	0·042	0·33	1	20	1·29

TABLE XLII.—Imperial Fluid Measure.

	= Quart.	= Pints.	= Fluid Ounces.	= Fluid Drachms.	= Minims.	= Weight in Grains.	= Cubic Inches.	= Litres.	= Cubic Centimetres.
1 Gallon, . .	4	8	160	1,280	76,800	70,000	277·276	4·541	4,541
1 Quart, . . .	1	2	40	320	19,200	17,500	69·319	1·135	1,135·2
1 Pint, . . .	0·5	1	20	160	9,600	8,750	34·659	0·567	567·6
1 Fluid Ounce,.	0·025	0·05	1	8	480	437·5	1·733	0·0284	283·8
1 Fluid Drachm,	0·0031	0·0062	0·125	1	60	54·7	0·217	0·0035	35·5
1 Minim, . .	0·00005	0·0001	0·0021	0·0167	1	0·91	0·0036	0·00006	0·59

TABLE XLIII.—For the Inter-Conversion of Certain Standard Weights and Measures.

Value to be Converted.	LENGTH		AREA				VOLUME				WEIGHT	
	Of Inches equals Millimetres.	Of Millimetres equals Inches.	Of Square Inches equals Square Decimetres.	Of Square Decimetres equals Square Inches.	Of Cubic Inches equals Cubic Centimetres.	Of Cubic Centimetres equals Cubic Inches.	Of Gallons equals Litres.	Of Litres equals Gallons.	Of Cubic Inches equals Pints.	Of Pints equals Cubic Inches.	Of Grains equals Grammes.	Of Grammes equals Grains.
0·05	1·2	0·002	·003	0·77	0·82	·003	0·227	·011	·0015	1·73	·003	0·772
·1	2·5	·004	·006	1·55	1·64	·006	·454	·022	·003	3·47	·006	1·543
·2	5·1	·008	·013	3·10	3·28	·012	·909	·044	·006	6·93	·013	3·086
·3	7·6	·012	·019	4·65	4·92	·018	1·363	·066	·009	10·40	·019	4·630
·4	10·2	·016	·026	6·20	6·55	·024	1·817	·083	·011	13·86	·026	6·173
·5	12·7	·020	·032	7·75	8·19	·031	2·272	·110	·014	17·33	·032	7·716
·6	15·2	·024	·039	9·30	9·83	·037	2·726	·132	·017	20·79	·039	9·259
·7	17·8	·028	·045	10·85	11·47	·043	3·180	·154	·020	24·26	·045	10·803
·8	20·3	·031	·052	12·40	13·11	·049	3·635	·176	·023	27·73	·052	12·346
·9	22·9	·035	·058	13·95	14·75	·055	4·089	·198	·026	31·19	·058	13·889
1·0	25·4	·039	·064	15·50	16·39	·061	4·543	·220	·029	34·66	·065	15·432
2	50·8	·079	·129	31·00	32·77	·122	9·087	·440	·058	69·32	·130	30·865
3	76·2	·118	·193	46·50	49·16	·183	13·630	·660	·087	103·98	·194	46·297
4	101·6	·157	·258	62·00	65·54	·244	18·174	·880	·115	138·64	·259	61·729
5	127·0	·197	·323	77·50	81·93	·305	22·717	1·100	·144	173·29	·324	77·162
6	152·4	·236	·387	93·00	98·31	·366	27·261	1·321	·173	207·95	·389	92·594
7	177·8	·275	·452	108·50	114·70	·427	31·804	1·541	·202	242·61	·454	108·025
8	203·2	·315	·516	124·00	131·09	·488	36·348	1·761	·231	277·27	·518	123·459
9	228·6	·354	·581	139·50	147·47	·549	40·891	1·981	·260	311·93	·583	138·891
10	254·0	·394	·645	155·01	163·86	·610	45·434	2·201	·289	346·59	·648	154·323
20	508·0	·787	1·290	310·01	327·72	1·221	90·869	4·402	·577	693·18	1·296	308·647
30	762·0	1·181	1·935	465·02	491·58	1·831	136·304	6·603	·866	1039·77	1·944	462·970
40	1016·0	1·575	2·581	620·02	655·44	2·441	181·738	8·804	1·154	1386·36	2·592	617·294
50	1270·0	1·969	3·226	775·03	819·30	3·051	227·173	11·005	1·443	1732·95	3·240	771·617
60	1523·9	2·362	3·871	930·04	983·15	3·662	272·607	13·206	1·731	2079·5	3·888	925·941
70	1777·9	2·756	4·516	1085·04	1147·01	4·272	318·042	15·467	2·020	2426·1	4·536	1080·264
80	2031·9	3·150	5·161	1240·05	1310·87	4·882	363·477	17·608	2·308	2772·7	5·184	1234·588
90	2285·9	3·543	5·806	1395·05	1474·73	5·493	408·911	19·809	2·596	3119·3	5·832	1388·911
100	2539·9	3·937	6·451	1550·06	1638·59	6·103	454·346	22·010	2·885	3465·9	6·480	1543·235
200	5079·8	7·874	12·903	3100·12	3277·18	12·206	908·692	44·019	5·770	6931·8	12·959	3086·470
300	7619·8	11·811	19·354	4650·18	4915·78	18·308	1363·037	66·029	8·666	10397·7	19·440	4629·704
400	10159·7	15·748	25·805	6200·24	6554·37	24·411	1817·383	88·039	11·541	13863·6	25·920	6172·939
500	12699·7	19·685	32·257	7750·30	8192·96	30·514	2271·729	110·048	14·438	17329·5	32·399	7716·174
1000	25399·4	39·371	64·514	15500·6	16385·92	61·028	4543·458	220·097	28·862	34659	64·799	15432·348

From this table any ordinary conversions up to 2000 units may be readily made. For example:—It is required to find the number of cubic centimetres equal to 1728 cubic inches.

$$1728 = 1000 + 700 + 20 + 8 \text{ cubic inches.}$$

But a reference to the sixth column shows that

1000 cubic inches	=	16,385·92 cubic centimetres	
*700 ,, ,,	=	11,470·10 ,,	,,
20 ,, ,,	=	327·72 ,,	,,
8 ,, ,,	=	131·09 ,,	,,
Add together and 1728 ,, ,,	=	28,314·83	

The Bronzing of Copper and Brass Surfaces.

It is often desired to give newly-deposited copper the appearance of age and to destroy the brilliant metallic lustre which it possesses at first. The methods of accomplishing this end are numerous. In all cases it is desirable to start with a clean metallic surface, freed from grease by immersion in potash, or by any other suitable cleansing process. To obtain a red bronze tone, the metal is brushed over with finely-powdered crocus, or a mixture of crocus and black-lead, made up into a paste with a little water, and is then heated on a metal plate above a clear fire until the powder has become dark. After cooling, the whole surface is thoroughly brushed; if necessary the process may be repeated to produce a darker colour. The bronzing is due to the oxidation of the copper superficially by the heated crocus; a better lustre is obtained by finally rubbing persistently with a brush which is from time to time passed over the surface of a cake of bees'-wax. Slightly wetting the clean surface of copper articles with very dilute nitric acid, or with a solution of ferric chloride and nitrate, or with a solution of copper nitrate, followed by drying and heating, also effects the required oxidation and produces a brown bronze.

Dark brown or black bronzing has sometimes been effected by merely brushing the surface with plumbago or vegetable black, conveyed in a suitable medium followed by a varnish or lacquer. The formation of the black copper sulphide on the surface of the metal, by painting with dilute alkaline sulphide solution (such as ammonium sulphide), gives the same appearance. The superficial precipitation of other metals, such as platinum, gold, or arsenic, is often adopted also; a very weak solution of platinic chloride, or of gold chloride, or a solution of 1 ounce of arsenious acid (*white arsenic*), and 1 ounce of ferrous sulphate in 12 ounces of water answering the purpose well. After applying any of these solutions, the object must be well washed and dried, and finally lacquered. A mere bronze-coloured varnish, recommended by Hutton for bronzing brass work, is made by dissolving 5 parts of aniline purple and 10 parts of fuchsine in 100 parts of methylated spirit, and then adding 5 parts of benzoic acid, and boiling until the liquid has attained the desired colour.

* Note that the equivalent of 700 cubic inches is found by multiplying the figure for 70 by 10.

Green bronzes are made by converting the surface of the article into the green basic acetate or carbonate of copper, and may be produced by exposing the article for a time to the vapour of acetic acid. Any acid vapour in moderation will afford the same result. One method, recommended by Napier, consists in enclosing the object immediately above a little dry bleaching-powder contained in a closed vessel, until the required effect is produced.

ANTIDOTES TO POISONS.

Most of the metallic-plating and the cleansing solutions are extremely poisonous, and stress has already been laid upon the danger both of using domestic drinking utensils for any purposes connected with the work of electro-plating, and of dipping the bare arm or hand into any of the depositing liquids. Unforeseen accidents, however, may occur and may demand the application of speedy remedial measures. Amateur doctoring is to be strongly deprecated, and medical aid should be sought at once; but upon a sudden emergency it may be necessary to administer relief, pending the arrival of the physician. In any case of poisoning by swallowing, simple emetics should at once be given—for example, luke-warm water, mustard and water, ipecacuanha, or even zinc sulphate (of the latter, from 10 to 30 grains are often given), the two first-named are better for domestic application; while these are preparing, the patient may often induce vomiting by thrusting the fore-finger as far as possible down the entrance to the throat. The nature of the subsequent remedies will depend upon the character of the poison, thus :—

Acids.—Mineral acids, such as sulphuric, nitric, hydrochloric, or glacial acetic acids, require an alkali to neutralise them; magnesia, chalk, whiting, lime water, or carbonate of soda may be administered stirred up with water. Failing these, the acid must be diluted by copious draughts of water; olive oil, milk, or white of egg may then be given.

Alkalies.—Caustic alkalies demand neutralisation with a mild acid, as vinegar, or the juice of an acid fruit, such as the lemon or lime, or by extremely dilute acetic, citric, or tartaric acids. Then oil or white of egg may be taken.

Antimony.—For the chloride solution magnesia or sodium carbonate are used; for tartar emetic, a vegetable astringent is to be applied; very strong tea may answer the purpose; then barley water or the like; small doses of stimulants being given from time to time.

Arsenic.—Freshly made hydrated ferric oxide with magnesia are often employed.

Copper.—White of egg mixed with water, plenty of milk, water, or barley-water or the like should be taken. Some have used calcined magnesia stirred with water.

Cyanides.—Freshly precipitated peroxide of iron with an alkaline carbonate, such as potassium carbonate; plenty of fresh air should be available; the coldest possible water should be poured over the head and down the spine; and the atmosphere around the patient may with advantage contain a little chlorine; for example, a little dilute acid may be poured upon bleaching powder in a saucer placed at some distance to windward of the patient.

Lead.—A very dilute solution of sulphuric acid, or a solution of magnesium or sodium sulphate may be administered; some use sodium phosphate; the object in each case being the formation of an insoluble salt of lead. This should be followed by an active purgative. Milk or white of egg may be plentifully taken.

Mercury.—Albuminous fluids (white of egg) should be given in sufficient quantity, mixed preferably with milk ; a large excess of the albumen is not advisable, the quantity generally recommended being the white of one egg to each 4 grains (about) of mercuric chloride taken. Then barley-water or its equivalent is allowable.

Oxalic Acid and Oxalates.—Lime-water or chalk may be used; but alkaline carbonates should not be applied, because they form intensely poisonous oxalates.

Silver Nitrate.—Common salt in solution forms insoluble silver chloride.

Zinc Salts.—Warm demulcent drinks, such as barley-water, to be given.

In all the above cases the application of the special remedy must be preceded by the use of strong emetics, except perhaps in the case of strong acids, when water should be taken to effect dilution before inducing the vomiting.

Acids which have been spilled upon the hands or upon the floor of the room should be neutralised with chalk after dilution. The vapour of acid in the atmosphere of a room may be neutralised by the vapour of ammonia.

INDEX.

A.

Ampère, value of the, 36.
Ampèreage, best, for electro-deposition, 88.
,, surface-, interconversion of units, 348.
Ampèremeter, 95.
Animal forms, reproduced in copper, 181.
Anion, meaning of, 27.
Annealing, effect of, on hammered metals, 118.
,, ,, on iron deposit, 248.
,, ,, on nickel ,, 230.
Anode, meaning of, 27.
,, plate, form of, 108.
,, slime (copper), 140.
Anodes, alteration of position at first, 206.
,, antimony, 266.
,, arrangement of, in artwork, 178.
,, brassing, 274.
,, choice of, 99.
,, coating on, by lead, 264
,, cobalt, 242.
,, copper, 140.
,, ,, behaviour in refining, 283.
,, ,, refinery, 287.
,, ,, -regulus, use of, 293.
,, ,, size for, 140.
,, gold, 217.
,, incrustation on, in brassing, 276.
,, insoluble, use of, 28, 29, 97.
,, iron, 247.
,, lead, effect of, in copperbath, 179.
,, nickel, 235.
,, ,, arrangement of, 238.
,, silver, 197.
,, ,, appearance during electrolysis, 192.
,, ,, arrangement of, 203.
,, size of, 100.
,, ,, effect of, 139.
,, soluble, effect of, 30.
,, ,, use of, 29, 97.

Anodes, supplementary, use of, 181.
,, suspension of, 107, 108.
,, tin, 263.
,, unlike, effect of, 30.
,, various, effect of, 30.
,, wire-skeleton, for statuary, 177.
,, zinc, 257.
Antidotes to poisons, 358.
Antimony and its compounds, 323.
,, anodes, 266.
,, behaviour in copper refining, 284.
,, deposited, nature of, 266.
,, deposition of, 264.
,, explosive, 266.
,, extraction of, 299.
,, solutions, assay of, 313.
Antique silver, 209.
Apothecaries' weight, 355.
Aqua fortis, 122, 319.
,, regia, 320.
Argol, 339.
Armature, dynamo, 74.
,, direction of current in, 72.
,, varieties of, 76.
Arrangement of baths, copper-refining, 287.
,, ,, electrotyping, 161.
,, ,, plating, 96.
,, of rooms, 102.
Arsenic, behaviour in copper-refining, 284.
,, effect of, in brassing-bath, 271.
Art-electrotyping, 175.
Articles, cleansing of, 117.
Assay of depositing solutions, 312.
,, electrolytic, 312.
Astatic galvanometer, 94.
Atomic-weight, definition of, 14.
,, weights of elements, 18.
Atoms, meaning of term, 13.
Autogenous soldering of lead, 105.
Avoirdupois weight, 355.

B.

Backing of copper electrotypes, 173.
,, -metal for electrotypes, 174.
Balance, plating-, 110.

Balance, plating-, correction in use of, 113.
,, sensitive, 312.
Barometer dials, dead-gilding of, 223.
,, ,, silvering of, 188.
Base-bullion, refining of, 296.
Basis-metal, influence of colour on gilding, 222.
,, use of term, 125.
Baths (see Solutions).
,, arrangement of, in copper-refining, 287.
,, cyanide, spontaneous alteration of, 193.
,, electrotype, arrangement of, 161.
,, old, recovery of metal from, 312.
,, plating, arrangement of, 96.
Battery, bichromate, 51.
,, Bunsen's, 49.
,, costliness of, 38.
,, Cruickshank's, 4.
,, Daniell's, 44.
,, ,, Breguet's, 46.
,, ,, gravity, 47.
,, ,, Kuhlo's, 47.
,, ,, Meidinger's, 47.
,, ,, post-office, 47.
,, depolarisation of, 42.
,, direction of current in, 23.
,, economical arrangement of cells, 56.
,, effect of size of plates, 55.
,, for brassing, 270.
,, ,, cadmium-plating, 259.
,, ,, cobalt-plating, 242.
,, ,, copper-depositing, 136.
,, ,, electrotype, 163.
,, ,, gilding, 213.
,, ,, iron-depositing, 249.
,, ,, nickel-plating, 231.
,, ,, silvering, 189.
,, ,, tinning, 261.
,, ,, zinc-depositing, 255.
,, Grove's, 48.
,, injurious fumes from, 51, 102.
,, invention of, 3.
,, Leclanché's, 52.
,, local action on zinc, 39.
,, maximum efficiency of, 59.
,, parts of, 41.

Battery, polarisation of, 41.
,, position of, in plant, 103.
,, practical hints on, 53.
,, principle of, 23, 39.
,, screws, 59.
,, secondary, 85.
,, single and two-fluid, 43.
,, size of, effect on current, 55.
,, Smee's, 43.
,, switch-board for, 60.
,, thermo-electric, 62.
,, ,, Clamond's, 68.
,, ,, direction of current in, 63.
,, ,, Noé's, 69.
,, ,, reversal of current in, 65.
,, ,, wastefulness of, 70.
,, weakening of, 41.
,, Wollaston's, 40.
,, -zincs, amalgamation of, 39.
,, ,, need for purity, 39.
Baumé's hydrometer, value of degrees, 350.
Bay salt, 343.
Beardslee's cobalting solution, 241.
Becquerel's cobalting solution, 241.
,, electro-chromy, 269.
,, ,, -gilding, 214.
,, electrolytic works, 5.
,, ore-treatment, 294.
Bees'-wax, 324.
,, cracking of, on cooling, 152.
,, use of, in moulding, 152.
Benzene, use of, in cleansing, 117.
Benzoic acid, 318.
Benzoline, use of, in cleansing, 117.
Bertrand's bismuth solution, 268.
,, cadmium solution, 258.
Bessemer's copper-plating, 5.
Bichromate battery, 51.
Binding-screws, 59.
Birmingham wire-gauge, value of numbers, 353.
Bisulphide of carbon for bright-plating, 9, 196.
Bismuth, 324.
,, behaviour of, in copper-refining, 284.
,, deposition of, 268.
,, use of, in fusible alloys, 155.

382

THE END.

BELL AND BAIN, PRINTERS, GLASGOW.